Watson's Advanced
Textile Design

Watson's Advanced Textile Design

Compound Woven Structures

Z. J. GROSICKI, MSc, FTI

Department of Fibre Science, University of Strathclyde

WOODHEAD PUBLISHING LIMITED

Cambridge   New Delhi   Philadelphia

Published by Woodhead Publishing Limited
80 High Street, Sawston, Cambridge CB22 3HJ, UK
www.woodheadpublishing.com

Formerly published by Butterworth & Co. (Publishers) Ltd
First published 1913
Second edition 1925
Third edition 1947
Fourth edition 1977
Reprinted by Woodhead Publishing Limited, 2004

British Library Cataloguing in Publication Data
A catalogue record for this book is available from the British Library.

ISBN 978-1-85573-996-3

Printed and bound by Replika Press Pvt Ltd, India.

Preface

This book is a companion volume to *Watson's Textile Design and Colour* published in its new version last year. It represents a continuation of the subject and deals with compound woven structures.

As in the case of the first volume this book has not been revised so much as re-written and about two thirds of the content is either entirely new or differently presented. The changes were not wrought in a fit of iconolastic frenzy but were dictated by modern advances in the technique of manufacture and by the need to organise the subject on a more logical basis. In the original work by W. Watson published in 1913 there was a tendency to group quite disparate constructions together because they were used for similar purposes and to provide highly detailed information of the methods of manufacture as then known. In this book the basic criterion of the classification is the structural principle and, in consequence, common headings group together such structures as are 'genetically' related to one another irrespective of their end uses. With reference to the methods of manufacture, the functional purpose of the various special mechanisms involved is outlined in general but only a few detailed descriptions are given. The reason for this is two-fold; firstly, in many instances several different solutions are available to achieve a given operation and it is felt that as long as the advantages and limitations of each method are provided there is no need to burden a designer with the actual minutiae of mechanical action; secondly, frequent improvements in methods rapidly invalidate a given mode of operation without, in fact, changing its purpose and it is again felt that it is the reason for pursuing a given course of action which is the important factor rather than the manner of achieving it.

Apart from some constructions discussed in Appendix II all those considered in the body of the book are still being produced together with some variants too numerous to mention in detail. Some structures are produced in a large volume, some only in a small way and it must be appreciated that the size of a chapter does not necessarily bear a relation to the amount of fabric made. It is undesirable to tailor the subject matter in accordance with the yardage of goods produced at present as, apart from a loss of substance, the balance may be rapidly upset due to the fickleness of fashion and taste in the textile field which is such that what is almost extinct at the moment may acquire unprecedented vogue in a few years' time.

End uses of woven fabrics are in some cases inextricably connected with the nature of the construction but in other instances a given type of cloth may admirably serve many different purposes. For this reason the areas of employment for some products are not particularly strongly stressed in the full knowledge that, with the availability of new materials and with novel settings, many structures can serve a function quite different from the one envisaged originally.

The gathering of information for this book has been aided by constant interest and involvement in the problems of the weaving industry and by many firms in Britain and in several Western and Eastern European countries who have willingly revealed their methods. The author is particularly indebted to E. G. Taylor, Esq., for his invaluable assistance in proof reading the script and for the many friendly arguments which have helped to crystallise several important issues. He also wishes to acknowledge the advice and a selection of samples used in a number of chapters and obtained from Messrs. J. & J. Crombie, Ltd. of Aberdeen, Brocklehurst & Whiston of Macclesfield, Lister & Co. Ltd. of Bradford and J. H. Fenner & Co. Ltd. of Hull. Several illustrations from publications by Messrs. Grob & Co. Ltd. of Horgen in the chapter on Leno Weaving; by D. Crabtree & Son, Ltd. of Bradford and by Platt Saco Lowell Ltd. of Oldham in the chapters concerned with carpet weaving were used and grateful thanks are offered for their kind permission to do so. The permission of the Textile Institute to reproduce some illustrations and script from a paper previously published by the author in the J. T. I. on Lappet and Swivel Weaving is also acknowledged. Finally, thanks are due to my wife who with good grace has permitted me to neglect the house, the garden and even the dog over the past two years of sustained work on this volume.

Z. G.

Contents

1 **Designing and Card-cutting Systems**   1

2 **Figuring with Extra Threads**   11

Principles of figuring with 'extra' materials. Extra warp figuring. Extra weft figuring. Figuring with extra warp and extra weft. Imitation extra thread effects.

3 **Backed Cloths**   42

Weft-backed cloths. Warp-backed cloths. Interchanging figured backed fabrics. Backed cloths with wadding threads. Imitation backed cloths.

4 **Figured Pique Fabrics**   65

5 **Stitched Figuring Weft Constructions**   73

Figured book muslin fabrics. Patent satin structure.

6 **Damasks and Compound Brocades**   83

Damasks. Figured warp rib brocades. Multi-weft brocades. Multi-warp brocades.

7 **Stitched Double Cloths**   103

Self-stitched double cloths. Wadded double cloths. Centre-stitched double cloths.

8 **Interchanging Double Cloths**   136

Interchanging double plain cloths. Interchanging double twill and sateen stripe designs. Cut effects in interchanging double cloths.

9 **Multi-layer Fabrics**   158

Treble cloths. Multi-ply belting structures.

10 **Figured Double and Treble Cloths**   173

Figured interchanging double cloths. Figured interchanging treble cloths.

11 **Tapestry Structures**   190

Simple weft face tapestries. Repp-stitched weft face tapestry structures. Combined warp and weft tapestry structures.

12 Gauze and Leno Structures 207

The principle of Leno structure. Leno weaving with flat steel doups with an eye. Leno weaving with flat steel slotted doups. Equalisation of yarn tension in open and crossed sheds. Jacquard Lenos. Madras muslin structures. Leno structures in a slider frame and needle device.

13 Weft Pile Fabrics 257

All-over or plain velveteens. Weft plushes. Corded velveteens. Figured weft pile fabrics.

14 Terry Pile Structures 274

Terry ornamentation

15 Warp Pile Fabrics Produced with the Aid of Wires 287

All-over or continuous pile structures. Figuring with one series of pile threads. Figuring with several series of pile threads.

16 Warp Pile Fabrics Produced on the Face-to-Face Principle 320

All-over or continuous pile structures. Figured pile structures.

17 Spool and Gripper Axminster Carpets 349

The spool Axminster system. The gripper Axminster system. The spool-gripper system.

Appendix I Traditional Loom Mountings and Special Jacquards 371

Heald and harness mountings. The bannister harness. Sectional jacquard and harness arrangements. Inverted hook jacquards. Working comber-boards. String doup mountings for leno weaving

Appendix II Uncommon Woven Structures 393

Lappet weaving. Swivel weaving. Ondule fabrics. Wilton pile hook loom. Chenille Axminster pile. Woven pile fabrics produced by thermal shrinkage. Tuck fabrics.

Index 431

1

Designing and Card-cutting Systems

Until comparatively recently, most of the jacquard fabric weaving was done on coarse pitch machines many of which were using harness arrangements deliberately designed to weave efficiently only one class of construction and no other. There were two main factors which led to the development of the specialised types of jacquard harness.

The first factor was the comparatively small figuring capacity of the coarse pitch machine. This meant that, in the production of certain traditional goods such as table cloths, bed covers etc., where single-repeat design was often desired, several jacquards had to be used in tandem above the loom to provide the necessary figuring scope. This method uses vast amounts of cards, increases the cost of card cutting and introduces faults due to the difficulty of synchronising perfectly the several shedding motions and card presentation systems. The packs of cards obstruct light and make access to the machines more difficult. To obviate the disadvantages, selection systems were devised to make one needle control several ends simultaneously (see Appendix I), and thus expand the size of the repeat in width. Also, special motions were invented to make one card serve for several picks in succession thus increasing the length of the repeat obtainable from a given number of cards.

The second factor was connected with the difficulty of preparing compound structure designs for an ordinary jacquard machine. An ordinary jacquard is a most versatile machine, capable of producing any structure but in order to do that, the operation of each end must be worked out in full over the complete length of the repeat. Considering that a repeat size of, say, 1000 ends × 1000 picks is not uncommon, working a design out fully and painting every weave mark on design paper is a laborious and tedious task. To reduce the tedium tappet-controlled healds were added to take care of such ends in the structure which performed a regular interlacing repeating over a short length. Thus, in a structure in which alternate ends operate in a simple order, say, plain weave, and only the other alternate ends perform intricate figuring lifts, repeating over hundreds of picks, the jacquard is made to operate only the intricately working ends. In this way the figuring capacity of the machine is increased because a 400s jacquard takes control of half the total number of the ends and designs repeating on 800 ends can be produced as the other 400 ends are controlled by the healds. Also, the design painting

1

becomes very much simplified; the alternate plain weaving ends need not be indicated as their operation has nothing to do with the jacquard mechanism and the design can be painted solid, the marks indicating the lifts of the intricately figuring ends only. Card cutting of such solid painted designs is also easier and faster. The difference in the work involved in the preparation of designs for the ordinary jacquards and for the heald and harness systems will be readily appreciated by reference to A and B in *Figure 1.1* which respectively show a fully worked-out design, and one in which only the intricately

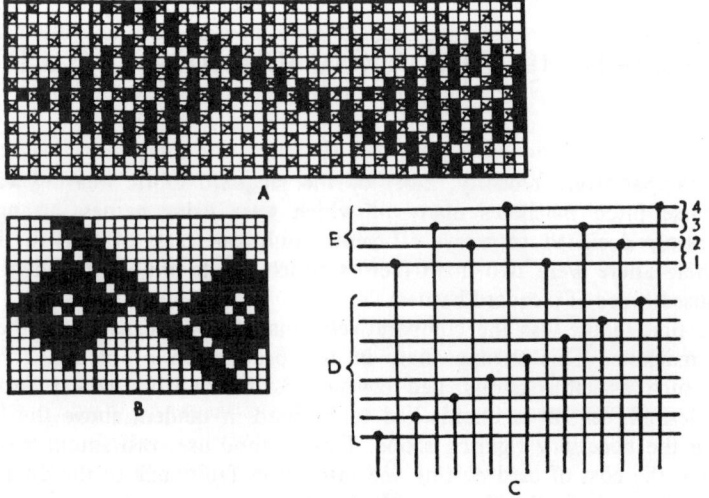

Figure 1.1

figuring ends are jacquard-controlled. An example of a heald and harness system which makes the reduction of work achieved at B possible is given at C in the form of a comber-board draft diagram in which D represents one row of jacquard harness cords and E a set of four healds used to operate the plain weaving ends. The healds are tappet-controlled and, therefore, the plain weaving ends, not being operated from the jacquard, need not be considered when the design is painted as already stated above.

The simple heald and harness mounting was followed by other modifications such as the working comber-boards, lifting rods, inverted hook operation etc. (see Appendix I), each adding a degree of sophistication devised to simplify the task of designing and card cutting until highly specialised types of machines were evolved, each of which was fitted specifically to produce one particular structure. At present very few of the special harness mountings are still in operation and these are dealt with under appropriate chapter headings. The majority of them, and this includes mountings which are only very rarely encountered as well as those no longer in use, are grouped together mainly for historical interest in Appendix I.

The decline of the special mountings was due to several causes. One of them was the general reduction in demand for some classes of jacquard work, arising out of improvements in the cloth printing and finishing techniques and in other methods of fabrication whereby imitations could be produced more cheaply. Due to greater concentration there was less room for specialisation. Another reason was the more frequent changes in the fashion or taste, so that some structures enjoying a vogue over a period of time were quickly replaced by others and a manufacturer who could not easily adapt to the required changes would have idle machinery. Therefore, versatile, rather than specialised machinery became necessary. Yet another cause of the decline of the special harness mounting was connected with the low speed of operation of most of these systems. Although they were admirably suited to reduce the labour of design painting and to increase the figuring scope they were generally cumbersome and many could only operate satisfactorily using the single-lift principle of action which made them uncompetitive compared with other systems of weaving where speeds of operation were generally rising in a spectacular manner.

At present most jacquard cloth manufacturers use modern ordinary harness machines providing complete versatility in respect of the type of structure produced and built in a fine pitch to give a large figuring capacity. (See *Watson's Textile Design and Colour* published by Newnes-Butterworths). These machines may be constructed to provide double-lift open-shed action and, selecting from a continuous paper roll, can operate satisfactorily at speeds of up to 300 picks per min and thus become competitive in the respect of speed with other types of shedding motions. However, as they are normally built with ordinary repeating ties without auxiliary shedding devices to take care of repetitive operation of ends outwith the jacquard, it would seem that the preparation of designs for such machines would need to revert to the laborious system of painting-in the lifting sequence of each end in the repeat in full. If this were so then the advantage of versatility which these machines possess would be partially destroyed, as laboriously prepared designs would require long runs to spread the high cost of designing over as great a length of cloth as possible. Frequent changes in design styles and short runs demanded by the modern market would tend to overburden the cloth with high designing charges to the point of non-competitiveness. Fortunately, developments in card-cutting machinery have kept pace with the developments in the jacquards, and specialised semi-automatic systems exist which can interpret a condensed, solid painted design into a fully worked-out structure. Indeed, recently a system has been devised in which a designer's sketch can be correctly developed into a fully worked-out design with the aid of a computer which when linked with a card-cutting machine makes the whole process automatic after the sketch has been provided.

Simplified and condensed designing

Almost every figured fabric, whether a single-cloth brocade or a compound structure, is built up of well-defined areas of design in which certain basic weaves are employed. Thus, in a floral brocade the ground may be woven in the plain weave, the petals in a weft sateen, the leaves in a warp twill, the stem in a weft twill, etc., the parts, all distinct and different, forming a harmonious whole. If a designer, instead of entering every weave mark in each different area, which is known as producing a fully worked-out design, adopted a colour code to denote each weave his task would be very much simplified as each area could then be painted solid in a different colour without tediously marking-in every weave interlacing. The burden of correct interpretation of the design in ordinary card-cutting systems would rest upon the card cutter who would have a very difficult task indeed. Considering the following colour code with reference to the floral brocade given as an example above, paper (unpainted area of design paper) = plain weave; red = weft sateen; blue = warp-faced twill; black = weft-faced twill, the card cutter, on reading the design across and encountering a given colour area would need to change the pattern of selection from one weave to another remembering at which point of its own repeat each weave existed in any given horizontal row of the design. If the weaves as mentioned above repeated on two, eight, six and four picks respectively then when cutting for example the seventh horizontal row of the

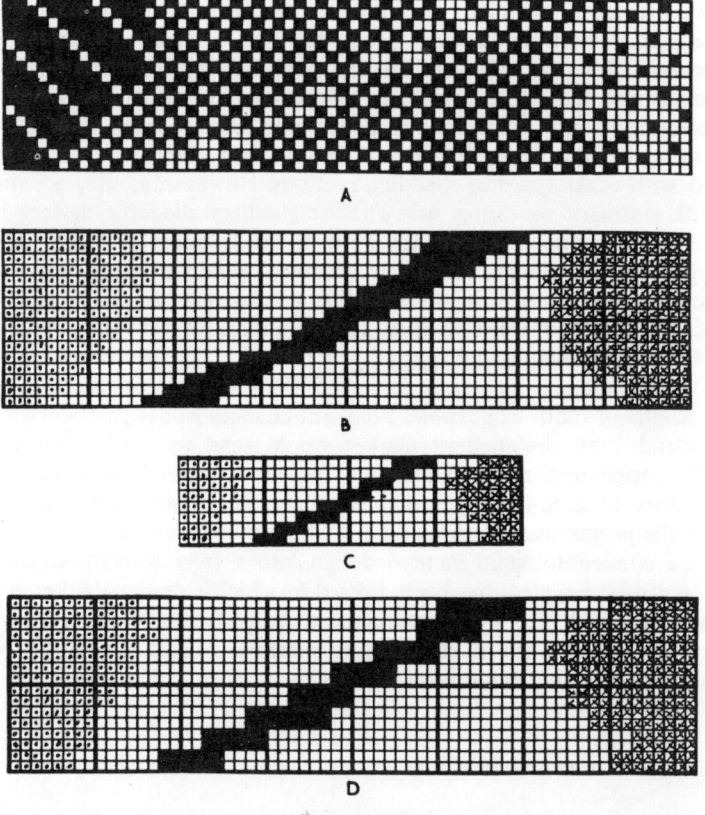

Figure 1.2

simplified design the card cutter would need to remember that, paper = first pick of the plain, red = seventh pick of the sateen, blue = first pick of the warp-faced twill, and black = third pick of the weft-faced twill. This is not an easy task, especially if not four, but eight, ten or more colours are used and if the constituent weaves are more complex than the four simple structures given by way of an example. This argument can be followed more clearly by reference to *Figure 1.2,* where at A, a small portion of a fully worked-out design, and at B, an equal portion of the identical design in simplified form, are shown. The code adopted at B is as follows: paper = plain weave, crosses = weft sateen, dots = warp-faced twill, solid marks = weft-faced twill.

The point of the argument outlined above is that merely to transfer the burden of fully working-out a design from the designer to the card cutter is not likely to be a profitable proposition. However, if the simplified design can be interpreted by mechanical means then the designer's task is made easier without transferring the problem to someone else and this, in fact, is the way in which the modern card-cutting systems operate.

Simplification of a design by colour coding the different weave areas reduces the labour of design preparation considerably but it does not reduce the scale over which the design has to be painted. Reduction of the size of a design can be obtained, however, by a process of condensation in which two or more ends are represented by one vertical row, and two or more picks, by one horizontal row of the design. The actual size of the repeat in the cloth is not diminished, as the design itself is only scaled down temporarily and is reconstituted in full again during card cutting by mechanical means which ensure that the multiplying factor in both directions used in the reconstitution is equal to the one used originally for the condensation. The condensation results in a further saving of labour during design painting and even a very small degree of condensation of, say, two in each direction means that a design formerly requiring 1320 vertical and horizontal rows of design paper can now be produced on only 660 vertical and horizontal rows, or, one quarter of the previous area. This can be observed by reference to *Figure 1.2* where at C, a design condensed according to the above formula, can be compared with the uncondensed original at B.

In compound fabrics the degree of condensation used is frequently dictated by the nature of the construction and, thus, in a cloth in which 3 figuring ends and 1 stitching end, and 2 figuring picks and 1 stitching pick form one structural unit the condensation by 4 warp-wise and by 3 weft-wise is natural. In simple cloth structures, however, such as brocades and damasks the degree of condensation permissible depends primarily on the density of the settings. This is due to the fact that although the design is reconstituted in the full size after card cutting, it will have lost in the process of condensation the fine definition of the outlines and become coarser in appearance. The reason for the coarser appearance is that in condensing by, say, two in each direction, the smallest step available is that of two instead of the step of one available in designing to the full size. This is shown by comparing B with D in *Figure 1.2* in which B is the original full size simplified design and D is reconstituted in simplified form from the condensed version shown at C. Although the general shape of the figures in both designs is the same, B is finer than D. In very finely set fabrics a considerable degree of condensation is possible without detriment to the appearance. For example if a brocade is set with 100 ends per cm and 60 picks per cm then it could be condensed by 3 warp-wise and by 2 weft-wise and still retain a sufficiently fine appearance because the minimum size of step under the above circumstances would be approximately 0.3 mm in both directions which is small enough to permit accurate definition of figure outline. Usually, fabrics composed of staple fibre yarns can stand somewhat coarser treatment than filament yarn fabrics where any inaccuracy in figure development is more easily discernible due to the general clarity of effect.

Card-cutting systems for continuous paper roll jacquards

There are several systems of card cutting such as the Lisage, the Dactyliseuse and the Uhlig, specially adapted to permit speedy punching of large designs for fine pitch jacquards. They range from comparatively simple manually controlled machines to quite complex devices in which operational control is obtained from punched card information rolls. The speed with which they can produce fully cut cards ranges from about 200 to 600 per shift depending on the type of machine used and on the complexity of the design. Faster operation is obtained from the more sophisticated and, therefore, more expensive systems. The high cost of the machines and their high production capacity has meant that the smaller jacquard fabric manufacturers would have to suffer the considerable

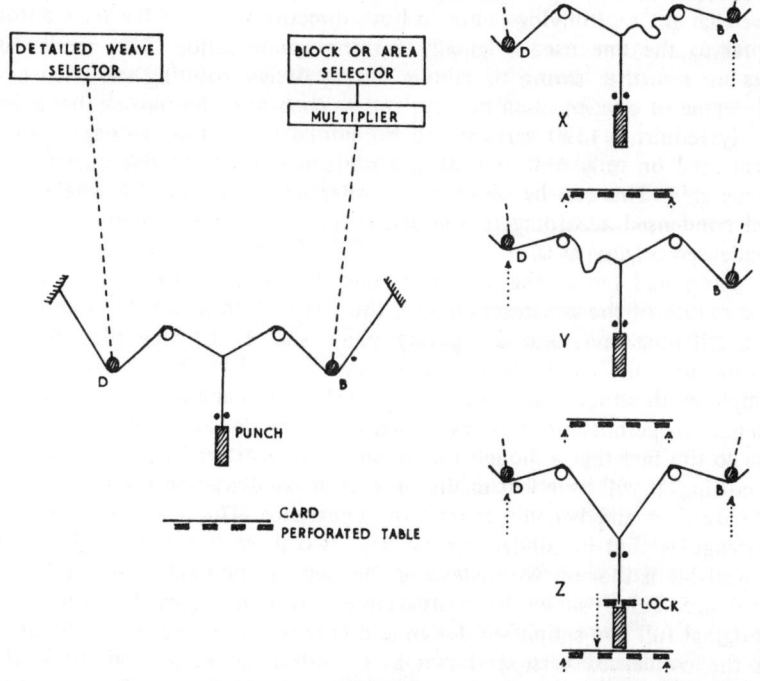

Figure 1.3

expense of installing such a machine without being able to utilise it fully. This has led to the establishment of specialist card-cutting firms who are able to install and utilise in full, the most efficient systems by providing a service to a number of manufacturers.

Each of the many systems available uses its own specific version of punch control and there is no intention here to describe the intricate mechanisms involved. As, however, they all use a similar basic principle of operation the explanation of it will provide sufficient information to appreciate the procedures involved.

The purely schematic diagram in *Figure 1.3* shows that the punch is controlled from two sources simultaneously. If only one of the two punch-connection supports is raised (D or B) the punch cannot descend into an operative position against the card because it is still suspended by the other support. The two alternative situations of the above order are shown at X and Y. It is only when both the supports are raised that the punch can descend to a position in which it is locked fast. *Figure 1.3Z*, and the card on pressing against it is perforated. It will be appreciated that although reference is made to one punch in each case their number is equal to the figuring capacity of the machine for which the cards are being cut so that for a 1200s jacquard, 1200 punches will be in operation.

The two sources of punch control are the area or block selector and the weave selector. The area selector selects punches which correspond to a given colour of paint in the simplified design in a horizontal row so that, for example, punches 400 to 820 are selected in a given design. Simultaneously the weave selector carries a card which indicates what weave is to be worked in this area, the weave selection being given for all the punches in the repeat so that if the weave for the area being cut is the plain weave then every odd punch of the 1200 punches will be positively activated. However, as only those punches which receive a positive activation from both control sources at once can cut, the card will be perforated only by the odd punches in the area of 400 to 820 because only these punches are free to descend into an operative position. This principle of operation is illustrated in *Figure 1.4*, where an example based on the simplified design given at B, in *Figure 1.2* is used.

Figure 1.4

In this example it is assumed that the block selector has been programmed to select the unpainted (paper) area for cutting; therefore, supports B *(Figure 1.3)* are raised and correspond with the unpainted area of the first horizontal row; this is shown by the vertical dashes. The full horizontal row is now overlaid with the weave selection which is the first pick of the plain weave and this is indicated by the horizontal dashes and represents the lifts of supports D in *Figure 1.3*. The actual cutting can, therefore, take place only in respect of ends indicated by crosses, as in *Figure 1.4*, which result from the overlap of the vertical with the horizontal dashes because it is only at these points that a dual 'affirmative', necessary to drop a punch, is given. Hence the first card is cut with the plain weave inserted in the required area. The second card is now presented, the area selector activates punches in accordance with the second row of the design, the weave selector activates all the even punches for the second pick of the plain weave and the second card is punched providing plain weave working for the required area, and so on until all the cards are perforated. By the end of the repeat all the areas requiring plain weave working will have been cut correctly. The cards are now brought back to card 1 position, the area selector is now programmed to select the second weave area, say, the warp-faced twill; on the weave selector the plain weave indicator cards are replaced by the 5 up, 1 down twill cards and the jacquard cards run through the machine again. Now the areas in which the warp-faced twill working is required will be cut, and so on; the number of times the cards pass through the machine being equal to the number of different weave areas in the design.

The example described above represents a simple brocade structure but it will be readily appreciated that the system is equally applicable to all structures, including compound cloths, and the only difference in operation will be the substitution of a compound weave on the detailed weave selector head for a simple weave.

If the simplified design, which serves as a programming guide for the area selector has been condensed in the warp direction then the multiplier, indicated in *Figure 1.3*, is brought into use to reconstitute the design in full size. In uncondensed designs the multiplier is not used as each unit of control in the area selector controls its own punch directly. If a design has been condensed by 2 warp-wise then a repeat of 600 vertical rows of design represents 1200 ends which require 1200 punches to ensure that each end works the detailed weave correctly. The area selector, however, has been programmed for 600 units of design and if the previous 'one to one' connections were retained then only 600 punches could be activated. To obviate this the multiplier is brought into operation which ensures, when suitable connections are made, that one control unit in the area selector activates two adjacent punches; if the warp-wise condensation was by three then the multiplier connections are trebled, and so on.

When the design has been condensed weft-wise then to obtain the correct reconstitution of the length of the repeat the same area selection is retained for a number of cards equal to the degree of condensation.

As the detailed weave selector controls every punch individually and presents the correct weave lifts for every end and every card it will operate correctly irrespective of the manipulations in the area selector associated with the design condensation and reconstitution. The change over from one weave to another on the detailed weave selector occupies very little time in the operation because after a comparatively short period all the standard weave combinations are accumulated in a form of 'library' from which they are drawn as required. Any new, unusual weave not held must be punched out manually before the commencement of the card-cutting operation so that it can be inserted upon the detailed weave control and applied to the appropriate area selected by the area selector.

Programming of the area selector in some systems takes longer and is one of the factors which reduces the speed of card preparation. The greater the number of different weave or constructional areas in the cloth the more complex the programming, although when the programming is carried out by means of punched paper strip the difference in time required to programme the area selector for, say, six weave areas as opposed to ten, is only marginal. The programming in most systems is carried out away from the actual cutting machine so that one design is being cut whilst another is being programmed. The main factor which determines the total time taken to cut a set of cards is the number of passages through the machine which each set has to undergo and that is dependent upon the number of weave areas, as already explained, because in one passage only one weave area is cut fully. In the more sophisticated systems facilities exist for programming as many as 17 different weave areas into the area selector which is usually more than satisfactory. In the manually-programmed systems there is no upper limit to the number of different structural areas which may be incorporated but as in such systems the programming operation is rather slow it usually becomes uneconomic as compared with the ordinary piano card-cutting machine to programme-in for more than four or five different weave areas.

Card repeating

The operations described in the above paragraphs are only required to produce an original design. If it is required to furnish several jacquard weaving machines with the same design then the original set is used for the purpose of obtaining replica sets by repeating, without the need to go through the time consuming operations as before. The replica sets can be produced on special repeating machines which are similar to the jacquards except that the harness cords are connected to the punches instead of the mail eyes in the loom and complete replica sets are obtained in a single run through.

Some of the special card-cutting systems can also be used as repeater machines if desired. This is achieved by activating all the punches at once in the area selector and by running the originally obtained set through the detailed weave selector whereupon all the punches corresponding to the holes in the original set will act, thus cutting each card correctly in a single passage.

Computerised card cutting

The latest development in card cutting consists of adapting a computer to process a design from the designer's sketch on squared paper onwards. As the computer can be made to shift a pattern unit, or reflect it about a horizontal or vertical axis, or displace it angularly the artist needs to provide only the basic minimum unit in all symmetrically constructed or repetitive patterns. This unit can be presented in solid painted simplified form and it can be condensed.

Computers and associated systems for this purpose are offered by a number of manufacturers and although the equipment from the various sources differs in some detailed aspects the basic method of operation is similar in each case. A design is painted solid in up to 12 colours to indicate different colour or constructional areas in the cloth and is placed upon a reading table. A photo-electronic scanner reading the design is capable of absorbing the information from one complete horizontal row of design in, at most, two seconds. This is transmitted to a control storage unit in an associated computer where it can be modified according to requirements as suggested in the foregoing paragraph. If the original design was condensed, e.g. by four warp-wise and by two weft-wise, it will be suitably expanded. Fully worked-out detailed weaves each associated with a different block colour area are transmitted from a previously assembled 'stock' of weaves stored on a 'floppy' disc or on magnetic tape to the central control where they are correctly superimposed. Clearly, after a few weeks of operation the stock of detailed weaves will become quite extensive and this can be drawn upon at any time; if, however, new structures are required they are digitised and simply added to the library as necessary. Full superimposition of all the detailed weaves on all the block colour areas takes place simultaneously for each horizontal row so that a complete card is punched in a single operation. If one row of the condensed design represents two, three, or four picks then the same row is held by the lock selection control whilst two, three, or four successive variants of the detailed weave are superimposed on it, each for a different successive card.

High-speed card-punching machines with, normally, 1344 punches are coupled to the computer. The punches in these machines are activated electromagnetically and the rate of production achieved varies between 30 and 90 fully punched cards per min. Most makers also offer additional equipment for checking the design and for correcting the mistakes.

In essence this system is similar to the mechanical systems described earlier in as much as the dual activation of punches is concerned, and apart from the improved speed of operation it offers the very considerable advantage of high-speed conversion of the read-out into punch selection which cannot be achieved with the mechanical devices. It also simplifies the procedure by producing fully cut patterns in a single run through the machine. It is envisaged that due to the high cost and high production capacity of this type of equipment an even greater concentration of the specialised card punching services will occur than has taken place upon the introduction of the mechanically operated systems.

The ultimate in design computerisation consists of direct selection of jacquard needles from a simplified and condensed design via a minicomputer. This form of control has already been introduced in prototype versions. As one minicomputer per jacquard is likely to prove too costly studies are in progress to find out the number of jacquards that one computer can control by suitable programming and it is possible that an economically viable solution will emerge shortly in which case instant pattern changes will become a possibility.

2

Figuring with Extra Threads

PRINCIPLES OF FIGURING WITH 'EXTRA' MATERIALS

A distinguishing feature of fabrics in which extra materials are employed is that the withdrawal of the extra threads from the cloth leaves a complete ground structure under the figure. This is illustrated in *Figure 2.1,* where the lower portion of the extra warp figured stripe, lettered A, is shown with the extra ends removed, leaving a perfect plain ground texture. The principle of this type of structure is also shown diagrammatically at C where an extra thread figuring float is made on the face of a plain ground weave. The figuring ends in stripe B

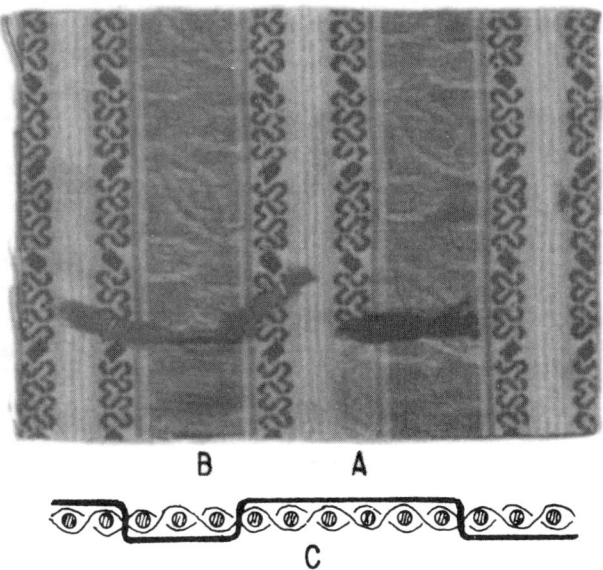

Figure 2.1

are not extra, but are simply crammed in the reed, and, as shown in the lower portion of the stripe, their withdrawal completely destroys the cloth structure since only the weft picks remain. The formation of a figure by means of extra

11

threads thus does not detract from the strength or wearing quality of a cloth, except so far as the extra threads are liable to fray out, whereas in ordinary fabrics, in which the figure is formed by floating the weft or warp threads loosely, the strength of the cloth is reduced somewhat in proportion to the ratio of figure and ground.

One of the advantages of figuring with extra materials is that bright colours—in sharp contrast with the ground—may be brought to the surface of the cloth in any desired proportion. Pleasing colour combinations may thus be conveniently obtained, since the extent of surface allotted to the figuring colour may be readily proportioned in accordance with the degree of its contrast with the ground shade, without the latter being affected.

Methods of introducing extra figuring threads

The extra threads may be introduced either as weft or warp, or the two methods may be employed in combination. When the extra material is introduced as warp then a separate beam is required for each warp on account of the different take-up rates between the extra and the ground ends. For extra weft figuring the weaving machine must have the capacity to insert more than one colour or kind of weft. The form of the design may render it necessary for the extra threads to be inserted in continuous order with the ground threads, or in intermittent order, while where they are introduced the arrangement of the figuring and ground threads may be 1-and-1, 1-and-2, 1-and-3, etc., according to the structure of the cloth and solidity of figure required. In extra weft figures, for looms with changing boxes at one side only, similar results to the 1-and-1 order may be produced by wefting 2-and-2; while the 2-and-4 order may be substituted for the 1-and-2, with, however, less satisfactory results as regards the solidity of the figure.

Methods of disposing of the surplus extra threads

The disposal of the extra warp or weft threads, in the portions of the cloth where they are not required to form figure, is of great importance, and one or other of the following methods may be employed:

(1) The extra yarn is allowed to float loosely on the back in the ground of the cloth. This method is suitable when the space between the figures is not excessive, when the ground is dense and when the fabric is used in situations that do not render the long floats on the back objectionable. It is not applicable to cloths in which the ground is so light and transparent that the positions of the extra threads on the back can be perceived from the face side.

(2) The extra yarn is allowed to float loosely on the back, and is afterwards cut away. This method is eminently suitable for light ground textures, but if the extra threads float somewhat loosely on the surface in forming the ornament, it is necessary for them to be bound in at the edges of the figure, or the loose figuring floats will readily fray out from the surface. The firm interweaving of the extra yarns at the edges, however, makes the

outline of the figure less distinct, and is rather objectionable unless employed in such a manner as to assist in forming the figure.

(3) In compact fabrics the extra threads are bound in on the underside of the cloth, either between corresponding floats in the ground texture, or by means of special stitching threads.

(4) The extra threads are interwoven on the face of the cloth in the form of small auxiliary figures or floats thus adding to the fullness of the texture.

Comparison of extra warp with extra weft figuring

In extra warp figuring there are two or more series of warp threads to one series of weft threads, and the method has the following advantages and disadvantages, as compared with the extra weft principle:

Advantages: (1) The productivity of a loom is greater because only one series of picks is inserted, and a faster running loom can be used. (2) No special picking, box, and uptake motions are required. (3) There is theoretically no limit to the number of colours that can be introduced. (4) In an intermittent arrangement of the extra ends either spotted or stripe patterns can be formed, whereas a similar arrangement in the weft can only be used to form spots (except in special cases) because of the objectionable appearance of horizontal lines.

Disadvantages: (1) Two or more warp beams may be required instead of one. (2) If an ordinary jacquard and harness are employed a smaller width of repeat is produced by a given size of machine, because the sett of the harness requires to be increased in proportion to the number of extra ends that are introduced in a design. (3) In dobby weaving the drafts are usually more complicated. (4) Stronger yarn is required for the figure, and the threads are not so soft, full, and lustrous; extra ends are subjected to greater tension during weaving than extra picks, and, as a rule, there is less contraction in length than in width, and the result is that extra warp effects usually show less prominently than extra weft figures. (5) If the extra threads have to be removed from the underside of the cloth, it is more difficult and costly to cut away extra ends than extra picks.

EXTRA WARP FIGURING

The chief advantage of the extra warp method is in productivity but at present it is mostly utilised for continuous styles arranged one end of ground, one of extra, except dobby effects which are still produced in a considerable variety of intermittent figuring arrangements. The reason for the decline of the fancier jacquard stripe styles lies in the fact that each different design frequently requires the harness to be re-tied or otherwise modified which is costly in itself and which often leads to further costs by increasing the length of the weaving machine downtime. Additional costs are incurred by the need to draw-in new warps into the newly re-tied harness which is more expensive than knotting-in, a procedure used when standard harness setts remain unchanged between one warp and another. Many elaborate styles similar in appearance to extra warp can be produced by means of extra weft figuring using standard harness setts and achieving the necessary variety by weft pattern changes. The higher actual weaving cost due to the lower rate of cloth production may be partially or entirely offset by the use of weaker and, therefore, often cheaper materials for the figuring yarns.

The above argument does not apply quite so strongly in dobby styles where draft changes only necessitate the change in the number of slider wires per heald frame although the economics of the drawing-in of new warps as opposed to knotting-in have to be taken into account despite the fact that both can be in dobbies performed by machines.

Continuous figuring in one extra warp

Figure 2.2 represents an extra warp figured fabric, in which the ends are arranged continuously in the order of 1 extra, 1 ground. The example is a style

Figure 2.2

in which the extra ends are floated on the back during weaving, but are cut away in the finishing processes, and the figure is therefore stitched at the edges. The stitches, however, are so arranged that they soften the outline of the figure, and do not detract from its appearance. A in *Figure 2.3* shows a portion of the extra warp figure, and B the weave of the ground ends, while C illustrates the method of constructing a squared paper design of the figure and ground in full. The solid marks indicate the lifts of the extra ends which are drawn on the odd harness mails, while the lifts of the ground ends are represented by the dots, a crepe ground weave being formed. The blank circles in A show a sateen binding weave which is inserted on the figure to stop the long warp floats. In the cloth the ground ends and picks per unit space are equal, so that, including the extra ends, there are twice as many ends as picks per unit space, and 8 × 4 design paper is therefore suitable in constructing the design in full, as shown at C in *Figure 2.3*.

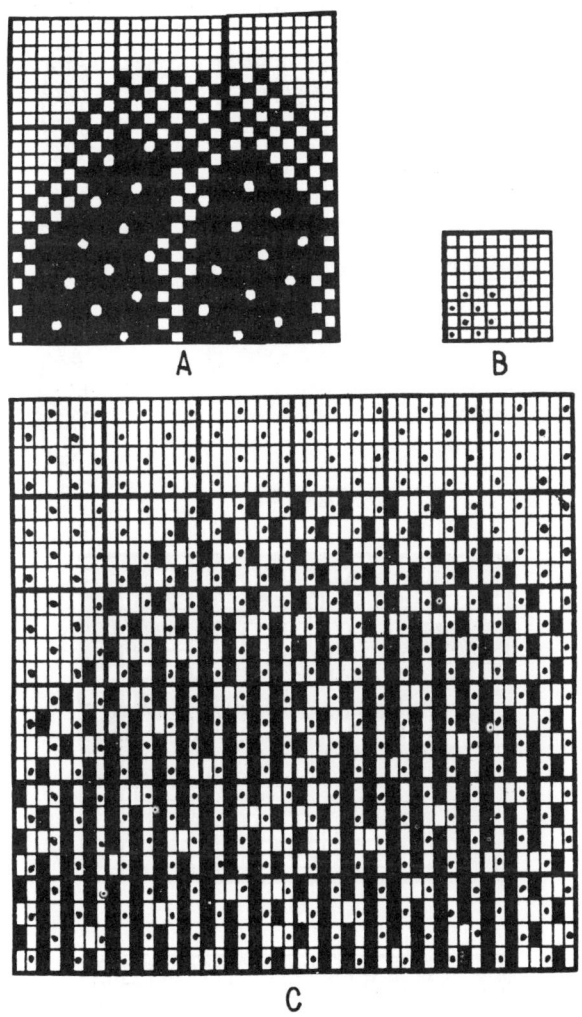

Figure 2.3

A typical continuous 1 extra, 1 ground style is represented by the Alhambra quilt structure used mainly for bed covers and consisting of plain ground weave worked between fine ends and coarse, soft spun picks. The extra figuring ends are thicker than the ground ends, or they may be woven two or three per mail in decked mail eyes (B, *Figure 2.4*) to produce distinct floats as illustrated at A, in *Figure 2.4*. A wide variety of qualities is produced in this structure and the following are suitable particulars for a medium quality cloth: Ground warp—30 tex, 12 ends per cm, 20 to 25 per cent crimp; extra warp—38 tex, 2 ends per mail, 12 double ends per cm, 6 to 8 per cent crimp; weft—190 tex, 12 picks per cm.

Cotton yarns are normally used throughout although good effects have been obtained with two-fold viscose rayon staple yarns for the figuring ends. This is a

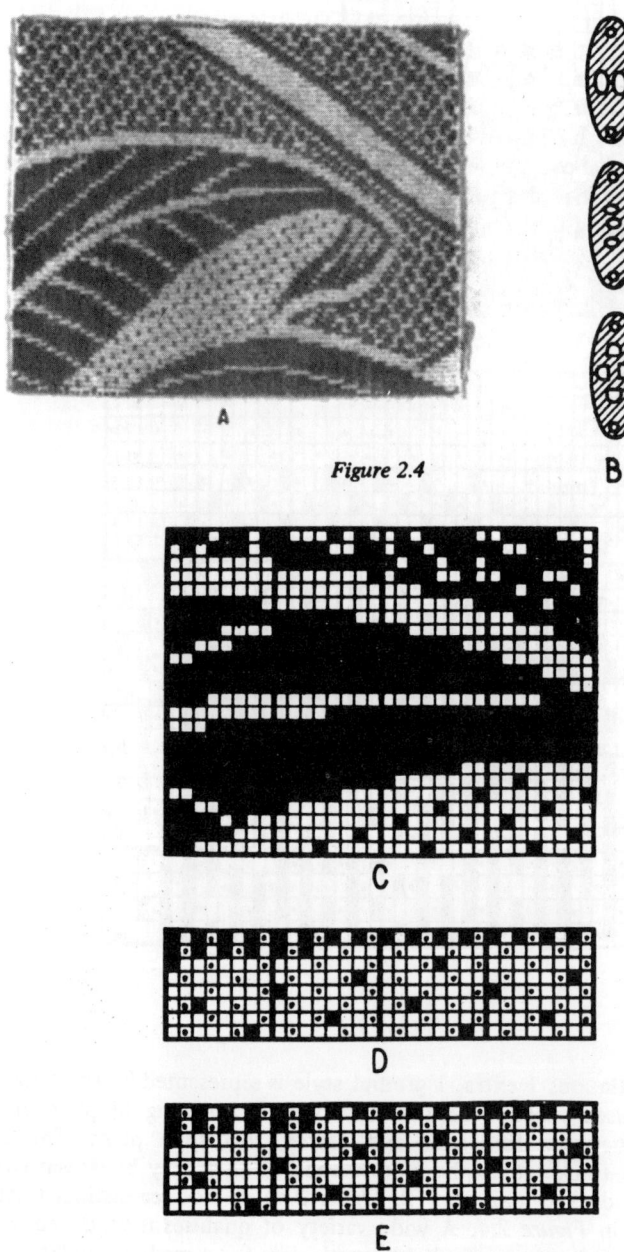

Figure 2.4

Figure 2.5

structure which at one time was extensively woven on a heald and harness mounting to increase the figuring scope of the jacquard and to simplify the design painting and card cutting (see Appendix I).

A portion of the fabric illustrated in *Figure 2.4* is shown in *Figure 2.5* where at C a simplified design showing only the lifts of the extra figuring ends is given. This shows the bold floats of the extra warp and contrasting ground areas. Where the ground area is very large the extra warp is stitched-in in sateen order to reduce the length of the extra warp float on the back. Weaves other than sateen can also be employed for this purpose and a broken weave is used in the ground area above the leaf in *Figure 2.4A*. At D, in *Figure 2.5*, a small portion of the simplified design is worked out in full, showing the plain weaving ground ends and using the technique already fully described in connection with *Figure 2.3*.

Binding in extra ends between face floats

In the ground of ordinary extra warp figured fabrics, it is usually necessary for the extra threads to be invisible from the face side, and they can be floated loosely on the back, or if the ground weave is suitable, be bound-in between corresponding warp floats. Thus, assuming that 2-and-2 twill ground is formed, the structure represented at E in *Figure 2.5* shows how the binding lifts of the extra warp can be concealed on the surface by placing them between two long floats of the ground structure.

Intermittent figuring in one extra warp

The cloths represented in *Figures 2.6, 2.7, 2.9* and *2.10* illustrate the intro-duction of one series of extra ends intermittently, and show various ways of forming either stripes or detached figures. In *Figure 2.6* the stripe figure is due

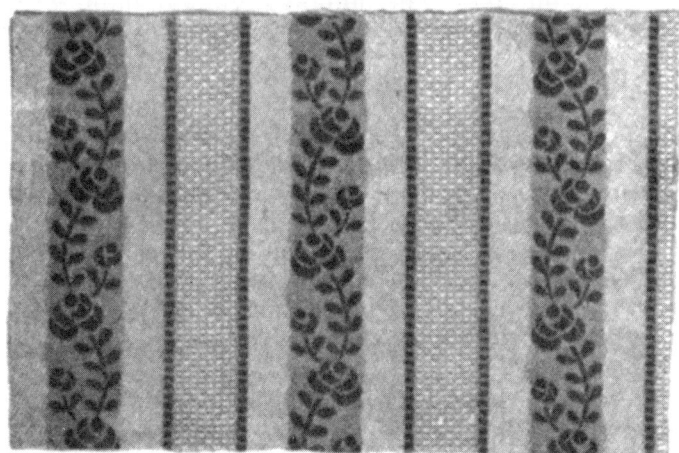

Figure 2.6

to the continuous manner in which the extra ends are floated. In *Figure 2.7* the extra figure is not continuous, but the parts are so near together that the figure has a striped appearance which is enhanced by the stripiness of the other parts of the design.

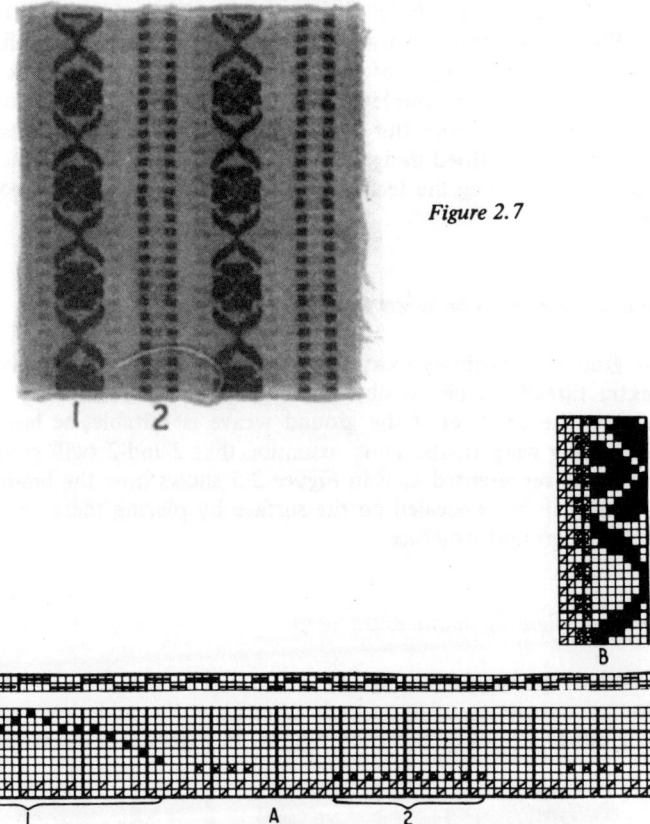

Figure 2.7

Figure 2.8

In *Figure 2.9* the spots are set out in full-drop style and are quite separate. In *Figure 2.10* the spots are also quite detached but this time they are combined with a small brocade figure forming distinct stems to the brocaded leaves.

The cloth represented in *Figure 2.6* consists of viscose rayon ground with the extra warp in cellulose acetate thus permitting the use of the cross-dyeing technique to achieve coloured figure on white ground. Where the extra warp is present it is drawn-in on the basis of one extra to one ground end and the designing technique in the extra thread stripe is exactly the same as explained in connection with *Figure 2.3*.

The cloth illustrated in *Figure 2.7* is a typical dobby stripe which can be produced on only 11 healds. The full draft for this design is given at A, *Figure 2.8*, with the order of denting indicated above it. Corresponding parts of the design in *Figure 2.7* and in the draft are marked 1 and 2. It will be noted from the denting plan that the ground ends are dented regularly two per dent with the figuring ends being added to some dents as they occur so that in some splits there may be four ends, in some three and in some others only the two ground ends. Regular order of denting for the ground ends is, of course, essential to maintain even spacing of the ends across the cloth. At B, in *Figure 2.8,* the lifting plan appropriate to the draft A is given from which the full design can be easily constructed.

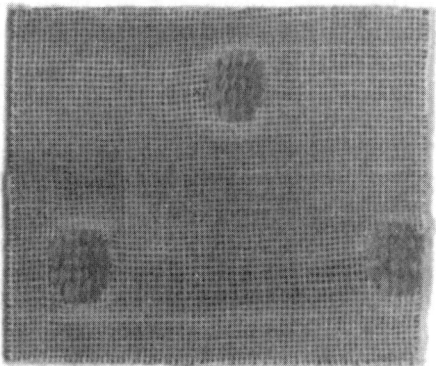

Figure 2.9

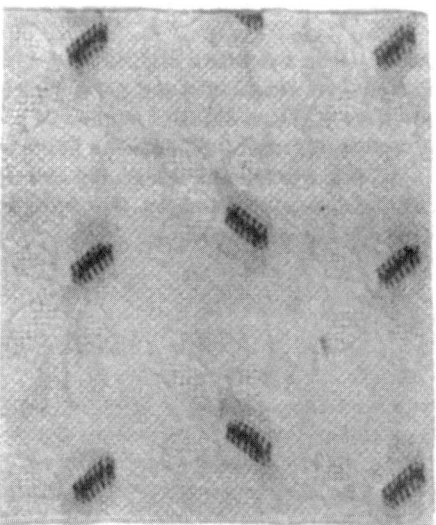

Figure 2.10

The fabric in *Figure 2.9* represents a dobby style clipped spot on an open ground cloth. As the extra yarn is cut away on the back of the fabric it could be easily plucked from the surface if it was insecurely anchored to the cloth. The necessary anchorage is achieved in this cloth in a two-fold way. Firstly, the ground ends are dented one per dent and where the extra ends occur they are crammed three per dent together with the ground end so that they are in effect

nipped fast between two adjacent ground ends and this in itself provides a
secure hold. Secondly, the extra ends are not permitted to float continuously
on the surface but are bound-in at intervals in such a manner that the dis-
continuity of the extra end float is not noticeable.

This is illustrated clearly in *Figure 2.11*, where at A full construction of one
spot is given with the denting order indicated above. B shows a weft section
through the structure illustrating the manner in which the longest extra warp
floats are bound-in. Due to the considerable degree of cramming of the extra
warp the binding points are invisible on the face and the spot appears quite solid.

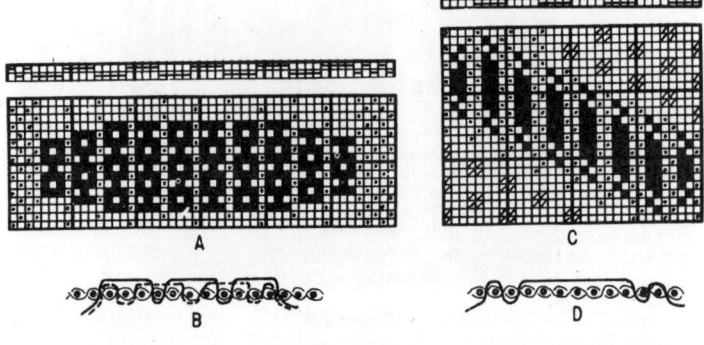

Figure 2.11

The cloth in *Figure 2.10* is a jacquard clipped spot and secure hold on the
extra warp is provided here by stitching-in the extra yarn in plain weave order
at the edge of the spot. This is also an efficient method of preventing the threads
from fraying out but it gives a blurred outline to the figure with some loss of the
clarity of definition. A fully worked-out design for one spot in this fabric is
given at C, in *Figure 2.11*, accompanied by the denting plan, with a weft
section at D showing clearly the fast bound float edge.

Figuring with two extra warps

Figure 2.12 shows a style in which two series of extra ends are introduced
continuously. A feature of the example is that the complete design extends
over 50 extra ends, whereas the order of interlacing repeats upon 25 ends.
This is due to the figure having been designed upon an odd number of ends,
which causes the colours to change positions in succeeding repeats. The warp
colours are also interchanged in the direction of the length of the design. In
weaving jacquard designs the method can be employed to obtain a width of
repeat that appears to require twice as many needles as are actually necessary—
e.g., a figure repeating upon 399 extra ends will produce an effect extending
over 798 extra ends. The system can be also used to produce a large repeat in
dobby weaving, and in *Figure 2.13* the complete draft is given for the design
shown in *Figure 2.12*, assuming that the ground weave is plain.

Figure 2.14 illustrates a simple arrangement (which is applicable to
elaborate designs) in which one extra is introduced continuously, as shown by

Figure 2.12

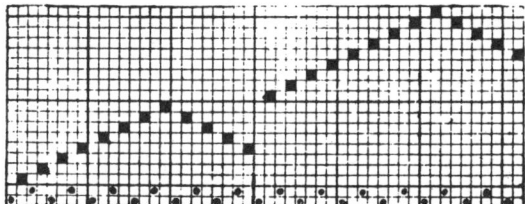

Figure 2.13

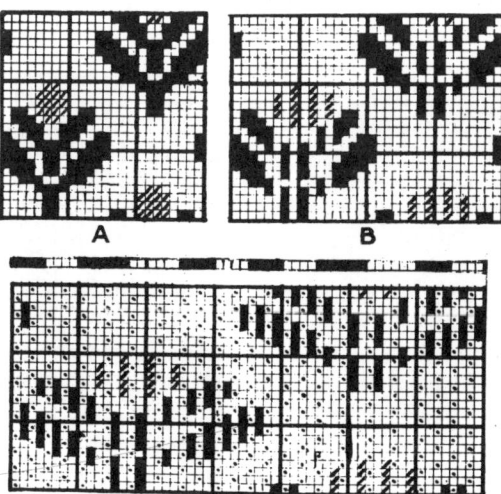

Figure 2.14

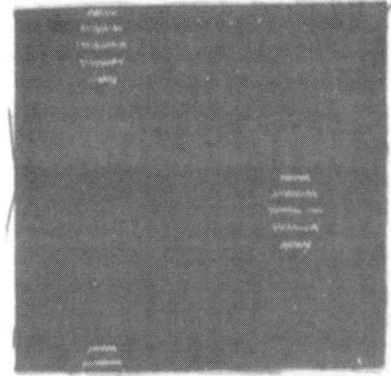

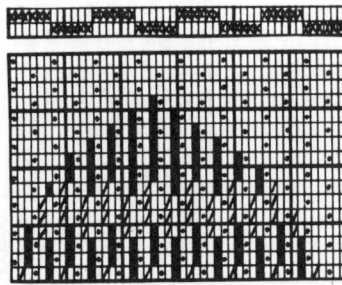

Figure 2.15   *Figure 2.16*

the solid marks, and a second extra intermittently, as indicated by the strokes, the former being used to form what may be termed the ground pattern, while the latter assists the first in producing a figure in two colours. The intermittent extra can also be used to form detached spots, etc. independent of the other. A shows a convenient method of first indicating the figure on design paper; at B the complete plan for the distribution of the extra ends is given; while C represents the full structure.

The fabric represented in *Figure 2.15* shows a detached spot figure, which is formed in two extra warps both of which are cut away between the spots. The firm interweaving of the extra ends in the figure, in this case, forms part of the effect. The spot is formed of coloured and white mercerised cotton on a self coloured woollen ground. A portion of the fully worked-out design for one spot is given in *Figure 2.16,* the full squares representing the coloured, the diagonal marks the white extra warp, and the dots the ground. The denting order is indicated above.

Figure 2.17   *Figure 2.18*

A more elaborate extra warp stripe in two colours is given in *Figure 2.17* for which a portion of the simplified design is shown in *Figure 2.18*. This cloth is also arranged on the basis of 1 ground end, 2 extra ends, which is the usual arrangement of the warp in two-colour extra warp structures. However, other orders of arrangement can also be employed such as for example: (a) 1 ground, first extra, 1 ground, second extra; or, (b) 2 ground, 2 extra. A two-colour extra warp fabric arranged on the basis of (a) is illustrated in *Figure 2.20*.

Extra warp planting

The design in *Figure 2.19* illustrates a system of arrangement termed 'planting', which enables a figure to be formed in a large number of colours without an addition being actually made to the series of extra threads. In the example five colours (represented by different marks) are employed, but it will be noted that

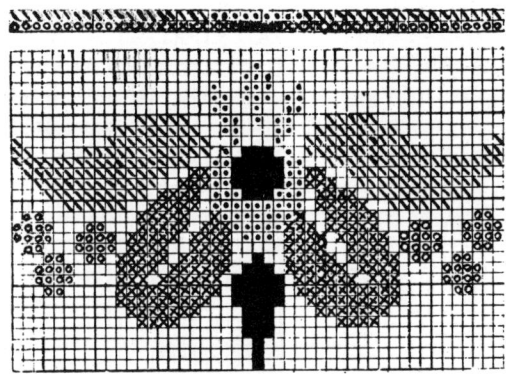

Figure 2.19

two colours only are introduced in any vertical line of the design. So far as regards the number of extra threads the arrangement is thus equivalent to a two-colour extra. The order in which the colours replace each other can be observed by following the spaces horizontally in the 'gamut' indicated above the design.

Stitching by means of special picks

When it is desired to retain the extra ends on the underside of a cloth without leaving them to float loosely between the figures, and when the ordinary method of stitching them between face floats of the ground ends is not feasible, the system, illustrated in *Figures 2.20* and *2.21* may be employed. The arrangement is applicable to any number of extras, and to either continuous or intermittent orders.

Figure 2.20 shows the face of the cloth, and it will be noted that the ground, which is plain weave, is at intervals quite free from the extra ends, yet, as shown at B, on the underside they are firmly bound-in.

Figure 2.20

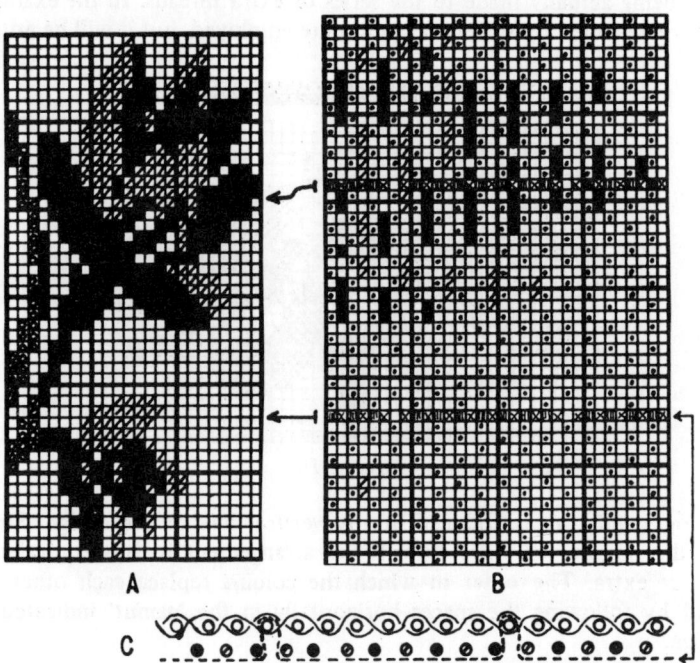

Figure 2.21

In *Figure 2.21* at A, a portion of the design, which is an extensively planted two-colour extra warp structure, is shown in a condensed form. The different marks indicate different extra warp colours and the arrows indicate the positions at which the special stitching picks are introduced. The last eight vertical rows of A are expanded into a fully worked-out design at B, which shows how the stitching picks operate. In this example the special picks are introduced after every 20 ordinary picks and if so desired they could be introduced more frequently. Whenever they are introduced the take-up motion must be rendered inoperative but unless the ordinary weft is very coarse they do not require to be different from the remainder of the weft. In most instances they are, in fact, the same as the ordinary weft and only perform a different function. It will be seen from B, in *Figure 2.21* and also from the warp section at C that on the stitching pick all the extra ends are raised, as are also the ground ends with the exception of those which are deliberately left down to provide a binding point for the special weft. In the example illustrated every eighth ground end is left down so that the stitching pick floats under eight extra and seven ground ends at a time. In the fully worked-out design the solid and the diagonal marks indicate the figuring lifts of the extra ends whilst crosses indicate the lifts of the extra ends on stitching picks; the ground weave is shown dotted and the lifts of the ground ends on stitching picks are represented by the vertically shaded squares.

In other forms of stitching alternate extra ends may be raised on alternate stitching picks so that the first special pick binds-in only the odd extra ends and the second only the even ones. The float of the stitching weft on the underside may also be made longer or shorter as required. Frequency of the stitching must not be too great, however, as in such cases indentations are liable to show on the face of the cloth.

EXTRA WEFT FIGURING

Extra weft-figured fabrics may be formed with one, two, or more extra weft picks in addition to the ground weft. Only one series of warp threads is used and the effect is obtained by floating the extra weft where desired on the face of the ground cloth produced by the interlacing of the warp with the ground weft in plain or in some other simple weave order. As already stated, weaving machines used for this purpose must have the capacity to insert more than one kind of weft.

When the extra weft is inserted continuously the take-up speed of the cloth is calculated in terms of ground picks only and the take-up mechanism can run continuously. When, as happens most frequently, the extra weft is introduced intermittently, i.e. when there are bands across the cloth where no extra material is required followed by bands where extra picks are inserted then an intermittent take-up motion is necessary. This, in effect, means that on extra picks the take-up device must be rendered inoperative.

Continuous figuring in one extra weft

A simple example is shown in *Figure 2.22* in which one extra weft is introduced continuously with the ground weft in the order of a pick of each alternately. The face of the cloth is represented on the left of *Figure 2.22*, and the underside on the right. The ground ends and picks interweave in plain order, while the extra picks float loosely on the back where no figure is formed on the surface. The condensed method of designing for the style is very simple since it is only necessary for the weft figure to be indicated on the paper (convention reversed), as shown in the corresponding design given at C in *Figure 2.23*. The card-cutting instructions are—cut two cards from each horizontal row, cut blanks for the extra picks, and cut the ground cards plain. The complete structure given at D in *Figure 2.23* shows the figuring picks arranged in alternate order with the ground picks, the former being indicated by the full squares and the latter by the dots. A warp section is given at D in *Figure 2.24*, which shows how the picks 2, 3, and 4 of D in *Figure 2.23* interweave with the ends 1 to 20. It will be appreciated that in all the designs in *Figure 2.23* the design convention has been reversed so that the marks indicate warp down and the blanks warp up. The reversal of the convention helps the designer to visualise better the shape of the figures formed on design paper.

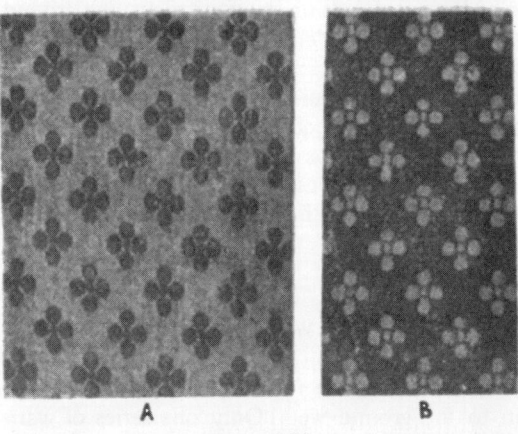

A   B

Figure 2.22

E in *Figure 2.23* shows the full development of the design C, assuming that it is produced in a loom with changing boxes at one side only, in this case two figuring picks alternating with two ground picks. Unless the figuring weft is heavy there is a tendency, in the 2-and-2 order of wefting, for the extra picks to show in pairs where the figure is formed, this being particularly noticeable if the ground picks interweave firmly underneath. Greater solidity of figure can be obtained by discontinuing the weave of the ground picks, beneath the extra weft floats, in the manner shown in the design E. The warp threads under the figure thus lie between the extra weft floats on the surface, and the ground weft floats on the underside, and no obstacle is offered to the pairs of figuring floats approaching each other. This is illustrated by the section given at E in *Figure 2.24*, which shows the interweaving of the picks 2, 3, and 4 of design E with the ends 1 to 20. With this arrangement it is not possible to repeat the ground cards, but they may be cut from the design painted solid, as at C. Two cards are cut from each horizontal row, the card-cutting instructions being—on extra picks cut all blanks solid; on ground picks, cut the marks, and cut the blanks plain.

Suitable weaving particulars for the fabric represented in *Figure 2.22* are—15/2 tex cotton warp, 32 ends per cm; and 15 tex cotton ground weft with 32 ground picks per cm; while the counts of the extra weft may be varied from the equivalent of 15 to 30 tex according to the desired prominence of the figure. In designing an extra weft figure in simplified form the count of the design paper is decided by the relative number of ends per cm to figuring picks per cm. In the 1-and-1 and 2-and-2 arrangements the number of extra

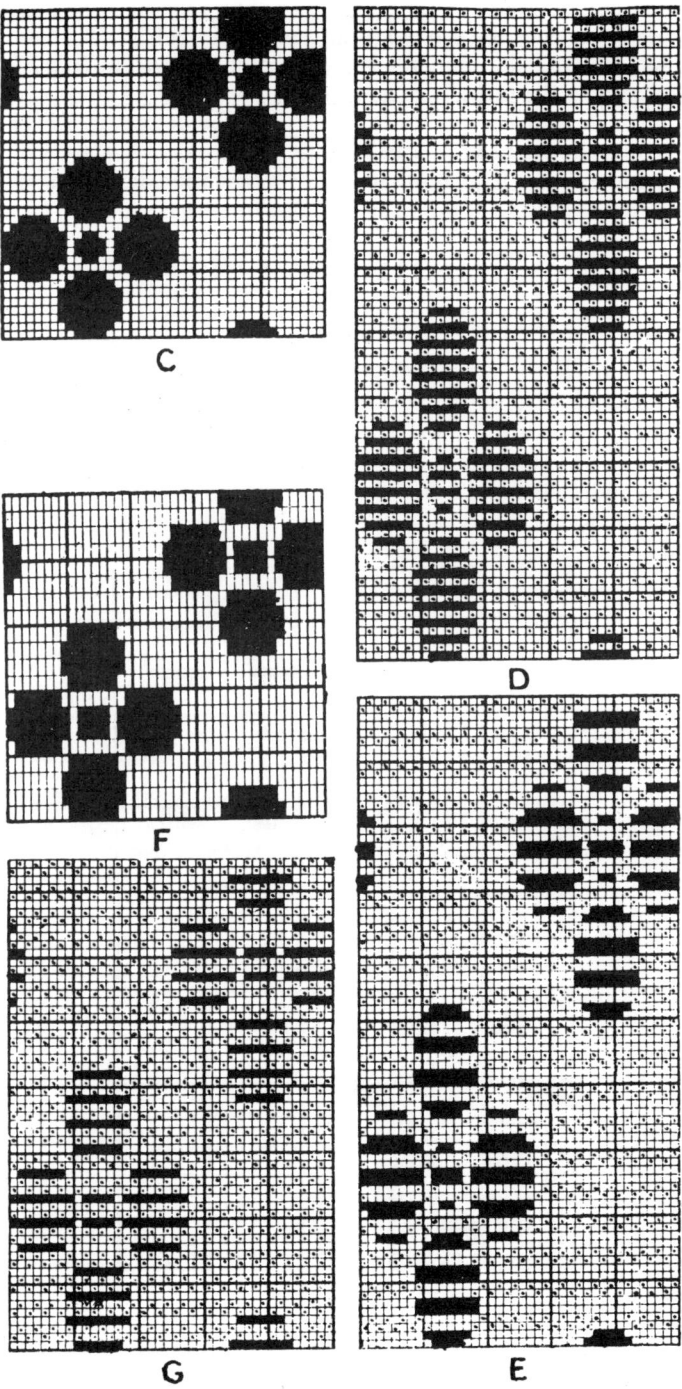

Figure 2.23

picks per cm are the same, therefore the count of design paper for the design C, with the foregoing particulars, is as 80 ends: 80 picks = 8 × 8.

The 1 extra, 2 ground order of introducing the extra weft is more economical than the 1-and-1, but with the same number of ground threads per cm the

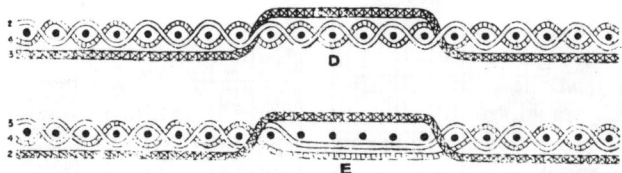

Figure 2.24

extra weft requires to be thicker, and the figure should usually be more massive. Assuming that the figure given at C in *Figure 2.23* is required to be produced in the 1 extra, 2 ground order, and that the ground threads per cm are as before, the extra picks per cm will be 40, and the count of the design paper as 80 ends: 40 picks = 8 × 4. To correspond with C the solid plan will then be as indicated at F, and the complete structure as shown at G, in *Figure 2.23*.

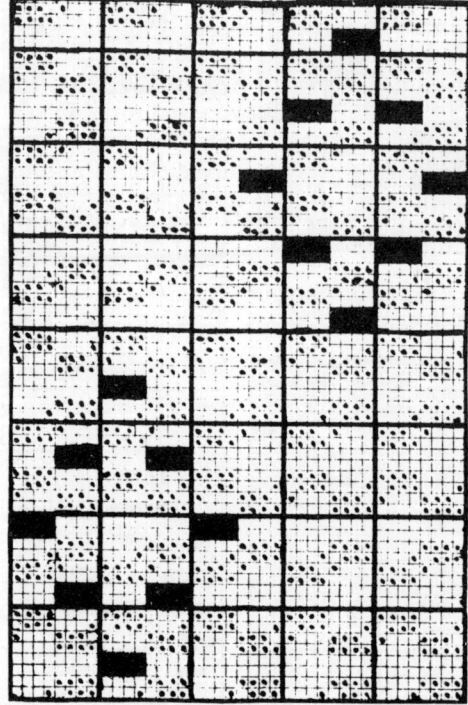

Figure 2.25

In the 2 extra, 4 ground order of wefting, a similar appearance could not be given to the figure shown at F in *Figure 2.23*, although the proportion of extra picks to ground picks is the same as in the 1-and-2 order, because the splitting of the figuring picks in pairs would be too pronounced. When the 2-and-4 arrangement is employed it is preferable to adapt the form of the figure to the wefting order, in the manner illustrated by the design in *Figure 2.25* in which the full squares show where the extra weft floats on the surface, while the dots represent the ground weave, which is a modified hopsack. *Figure 2.25* illustrates an important principle in extra weft spotting—viz. the selection of suitable positions for the figuring floats in relation to the ground weave. It will be noted in the design that the extra weft spots are formed in the centre of the warp floats in the ground, so that the best possible conditions are secured for showing the figuring floats prominently on the surface. It is necessary to avoid covering the figuring floats by adjacent ground weft floats (which would have occurred in the example if they had been placed four ends to the right or left) as much as possible.

Clipped spot effects

In the example given in *Figure 2.26* the figure is formed in extra weft on a plain transparent ground texture, which necessitates that the extra material be cut off on the underside. The face of the cloth is represented on the left and the underside on the right of *Figure 2.26*. In order to avoid the possibility of the severed picks fraying out, the extra weft is interwoven in plain order at the sides of the figuring floats, but as this causes the shape of the figure to be modified, the plain interweaving is shown extended completely round so as to produce an opaque outline between the weft figure and the thin ground texture. More plain weave is used than is really necessary to bind the figuring floats, the idea in this case having been to form distinct shapes upon which to develop the figure.

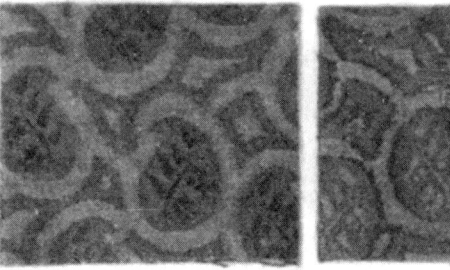

Figure 2.26

The condensed plan in *Figure 2.27* illustrates the method of indicating the design upon design paper. The figure is first painted in solid, then marks are inserted to stop excessively long weft floats, after which the plain binding is indicated round the figure. The complete structure of the portion indicated

Figure 2.27

Figure 2.28

between brackets is shown below the condensed design in *Figure 2.27;* while *Figure 2.28* shows how the last extra pick of the fully worked-out portion of the design and the ground pick on each side interweave with the ends 15 to 34.

Intermittent extra weft figuring

The fabric represented in *Figure 2.29* illustrates the principle of introducing the extra weft in intermittent order with the ground weft, for the purpose of producing a detached spot effect.

In this cloth an intermittent extra weft figure is combined with a figure formed by the ground weft—a useful method of obtaining an additional embellishment without adding to the production costs. In *Figure 2.30* at A, a portion of the fully worked-out design (convention reversed) for this fabric is given in which the ground weft float is indicated by the shaded squares and the dots whilst the extra weft float is depicted by the solid marks and the crosses. The background consists of an irregular 6-shaft warp satin upon which the coarse ground weft figure shows prominently. The dots indicate that the ground weft weaves plain with the warp under the extra weft figure area as shown by the warp section given at C. The plain weave interlacing of the ground elements under the extra weft figure is used often in an otherwise loosely bound cloth to achieve a reasonable degree of firmness in areas which without it would be very poorly bound. The extra weft figure is rendered in comparatively long floats as indicated by the solid marks, and as considerable distance separates one extra weft figure from another and the extra yarn is not cut away from the back of the cloth it has to be stitched in on the reverse side to prevent excessive floating—this is shown by the crosses and is also seen clearly in the warp section at C.

Figure 2.29

Generally the ground weft floats will close together and effectively conceal the binding points, but if the figuring weft is much thicker than the ground weft there is a liability of the stitches forcing the ground picks apart and showing on the surface, particularly if there is a strong colour contrast between the figuring weft and the ground. In such a case the stitches should be as infrequent as possible. In the example given the extra weft is bound-in in the 12-shaft irregular satin order, i.e. half as frequently as the ground weft. Where the stitch falls between two long floats of the ground weft it is concealed effectively as shown at C in *Figure 2.30*, but where it occurs at the point in which the ground weft float is short (as it will in the warp-faced background area) it may show on the face. If this is objectionable then the frequency of the extra weft stitching on the underside may be further reduced to, say, one third of the ground, thus binding it in the 18-shaft satin order.

At B, in *Figure 2.30* a condensed form of design is indicated for a portion of A. To achieve correct cloth construction from this the following card-cutting instructions would have to be given:

Cut two cards from each horizontal row;

1. Ground weft card—cut all blanks, cut solid marks and dots plain.
2. Extra weft card—cut all dots, blanks and shaded marks.

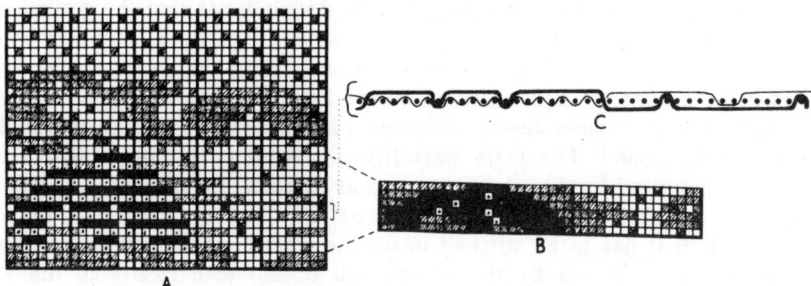

Figure 2.30

In *Figure 2.31* another example of an intermittent extra weft fabric is given. This is a fine tie cloth produced in polyester filament yarns to the following specification: Warp—120 dtex, 86 ends per cm; ground weft—170 dtex, 43 picks per cm; extra weft—same count as ground weft, inserted 1 extra, 1 ground where it occurs.

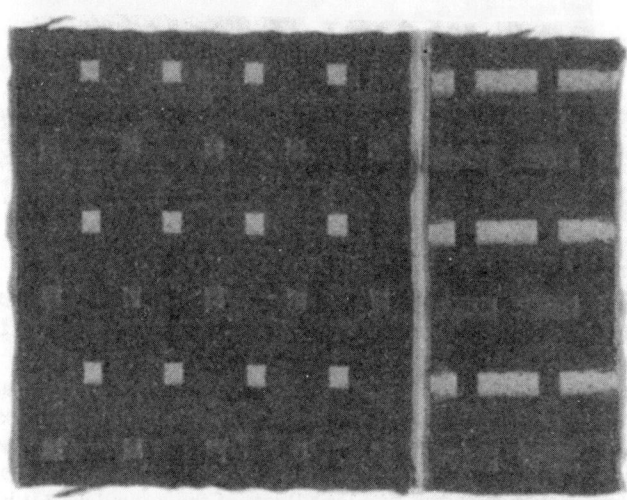

Figure 2.31

A portion of the design is worked out in full in *Figure 2.32* (convention reversed) showing the cord ground weave in dots, the extra weft float on the face in solid marks and the stitch marks of the extra weft on the reverse side in

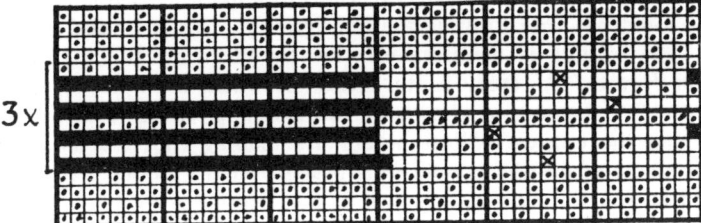

Figure 2.32

crosses. Each spot consists of 12 extra picks floating over the full width of the cord which is 24 ends wide. On the back, the extra weft floats under three cords before it again appears on the surface and is stitched in an irregular order.

Modification of ground weave

The continuation, under the figure, of a ground weave in which the weft passes over two or more ends, is sometimes not satisfactory, because the ground weft floats tend to cover up the figuring floats, and cause the edges of the figure to appear indistinct. In such a case the ground weave should be changed to warp surface under the figure, as shown in *Figure 2.32* where the ground weft instead of floating on the face, as it does in the ground weave area, is made to float on the back under the extra weft spot. The warp float, thus created, shows up the figuring spot distinctly.

Stitching by means of special ends

In fabrics in which the extra weft is considerably thicker than the ground elements and also of different colour, using the ordinary method of binding

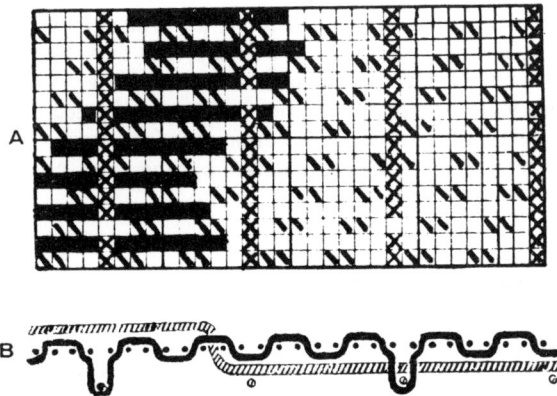

Figure 2.33

the float, as illustrated in the two preceding examples, is unsuitable because the stitches would be liable to force the finer ground picks apart and show on the surface. The binding is therefore effected by the warp and in the example shown in *Figure 2.33* every ninth thread of which is employed as an extra thread for the purpose, as shown in the design given at A in *Figure 2.33*. In the plan the crosses indicate where the binding ends are left down; the diagonal marks where the ground weft passes over the ground warp in 2-and-2 twill order; and the full squares where the figuring weft is on the surface. The binding ends are down on all the figuring picks, and are raised alternately on every fourth ground pick. B in *Figure 2.33* represents the interlacing of the second ground pick and the second figuring pick of A. In the ground portions of the fabric the figuring weft lies between the extra warp threads and the ground texture. The interweaving of the stitching warp with the ground picks is invisible on the surface of the cloth because it is of the same thickness and shade as the ground warp, and at each binding place it lies between two ground-warp floats.

The method is similar in principle to that illustrated in *Figures 2.20* and *2.21* in respect of extra warp fabrics but in this case warp instead of weft stitchers are employed.

Chintzing

In *Figures 2.29* and *2.31* two different colours of extra weft were used in forming different parts of the figure. The structure, however, was not changed in principle by this fact and would still be considered a one extra weft construction. Replacement of one colour of extra weft by another in succeeding horizontal rows of design is known as 'chintzing' and is simply a function of pattern change in the weft insertion and corresponds with planting in the warp direction (q.v.). It adds considerably to the variety of effect achieved without appreciably increasing the cost of production. Chintzing is often used also in true 2-thread extra weft styles, where two distinct series of extra threads are inserted and two differently coloured extra weft figures can, therefore, exist side by side. The differences between ordinary and chintzed effects are shown schematically in *Figure 2.34* in which at A, an ordinary extra weft effect is represented, at B a similar single extra weft effect is shown chintzed, with the green extra thread replacing the red one in a succeeding spot. C and D show two-colour extra weft figures, the former not being chintzed and in the latter, one of the two series of extra wefts is continuous (the grey) whilst the other is chintzed, the green threads being replaced by the pink ones in succeeding horizontal rows.

Continuous figuring with two extra wefts

Figure 2.35 shows the designing of a cloth in which a continuous figure is developed in two colours of extra weft, a pick of each being introduced regularly with each pick of the ground weft. Assuming that the number of ground picks per cm is the same as the number of ground ends, the figure may

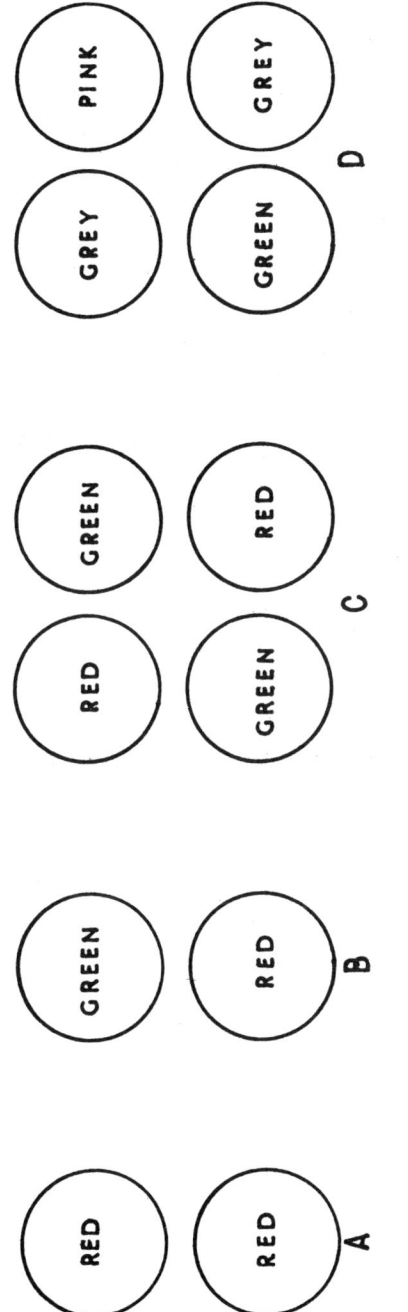

Figure 2.34

be painted solid in two colours on 8 X 8 design paper, as indicated at **A**. The full squares represent one colour of extra, and the crosses the other colour, each horizontal space on the paper being equivalent to three picks of which one is a ground pick. **B** in *Figure 2.35*, shows a portion of the complete structure which results from cutting each horizontal space of **A** as follows: First card, cut all but the full squares; second card, cut all but the crosses; third card, cut plain weave right across disregarding the marks. **C** shows the corresponding structure, assuming that the wefting order is arranged 2-and-2 to fit a loom with changing boxes at one end only.

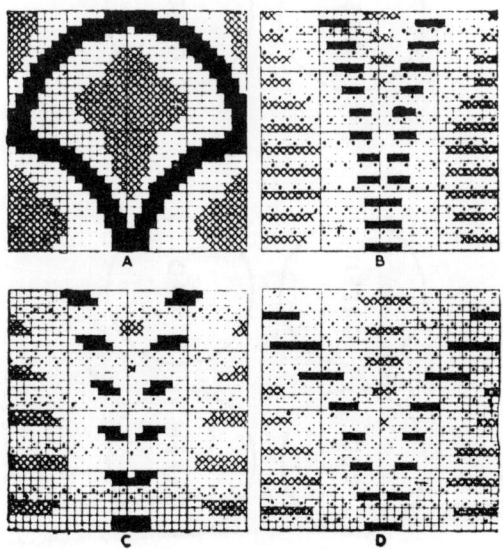

Figure 2.35

A more productive and economical method of introducing the two extras continuously consists of doubling each extra, and wefting in the order of 1 double pick—first extra; 1 ground; 1 double pick—second extra; 1 ground; as shown at **D** in *Figure 2.35*. The figure, however, is not so solid, and it is necessary to note that with an equal number of ground ends and picks per cm, 8 X 4 paper will be required in painting the design solid, since there are two ground picks to each double pick of each colour.

FIGURING WITH EXTRA WARP AND EXTRA WEFT

The combination of extra weft and extra warp threads gives very great scope in the development of designs, and for certain styles of ornament is more economical than when only one of the series of threads is employed. For instance, a fabric is represented in *Figure 2.36*, in which an all-over figure has been produced in extra weft and extra warp with a comparatively small consumption of extra material. In the corresponding design given in *Figure 2.37*, in which the marks represent warp, the lines below and at the side of the plan indicate the positions of the extra threads, which are arranged in warp and weft in the order of 1 extra and 6 ground threads. A special feature of the example is that the surface of the cloth is made perfectly plain by allowing each extra end and pick

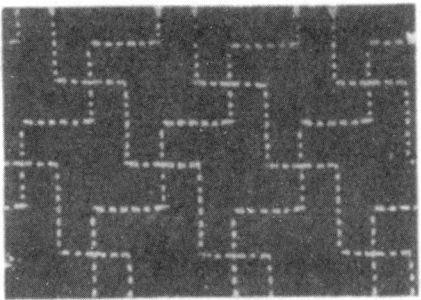

Figure 2.36

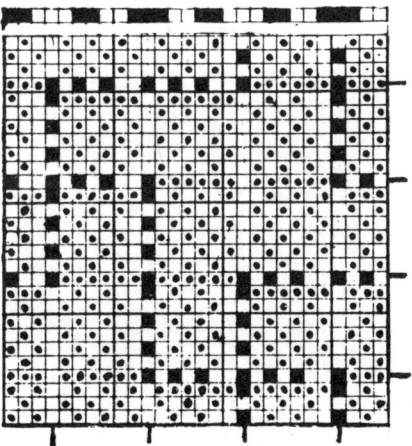

Figure 2.37

respectively to interchange with the preceding ground end and pick. Thus it will be seen that where an extra thread interweaves plain on the surface, as indicated by the solid marks, the ground thread which precedes it floats on the back, while where an extra thread floats on the back the ground thread is brought to the surface in plain order. If the ground threads had been interwoven in plain order throughout, the plain weave of the extra threads would have been the same as the preceding ground threads, and the former would have been partly concealed by the latter.

Extra warp and extra weft spot effects

A muslin fabric is represented in *Figure 2.38,* which shows detached figures formed by floating extra warp and extra weft threads in combination, and small spots produced by the extra warp alone. A convenient method of first indicating

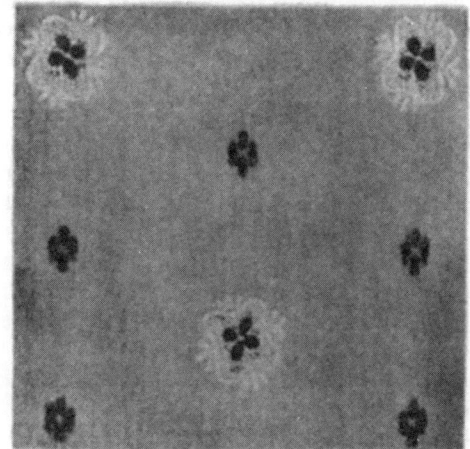

Figure 2.38

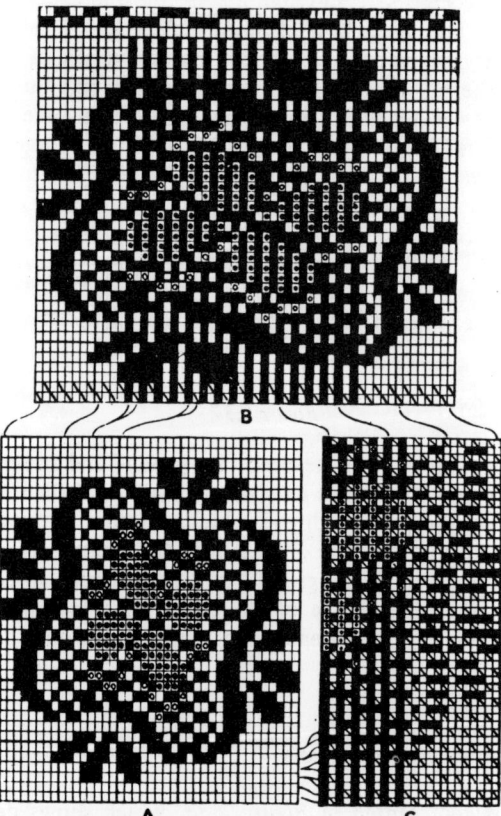

Figure 2.39

the combined weft and warp effect is illustrated at A in *Figure 2.39*, in which the solid marks indicate the weft figure and the dots the warp figure. The surplus warp and weft threads are sheared off the underside of the cloth, but the weft is so firmly interwoven in the figure that no stitches are required at the edges. In the warp figure, however, the ends are loosely interwoven, and they are therefore stitched with the ground picks, as shown by the circles in A, above and below the warp floats. Each horizontal space of A represents a ground pick and where weft figure is indicated, also an extra pick, while each vertical space represents a ground end, and where the warp figure is shown, an extra end also. The ratio of ground ends to ground picks per cm—in this case 24 to 18—gives the proper counts of squared paper in designing the simplified figure, and 8 × 6 paper is therefore shown at A. B is the same as A except for the inclusion of the extra ends which are down in places where they are not lifted to form figure or for stitching. The order of denting is indicated above B, in which it will be seen that twice as many ends are placed in each split where the extra warp is introduced as in the remaining parts of the design.

In the extra weft-figured portions of the design a figuring and a ground card are cut from each horizontal space, and in the remaining portions a ground card only. On the first two picks of B in *Figure 2.39*, the diagonal marks indicate the order of cutting by which the ground ends and picks are interwoven in plain order. The odd ground cards are cut like the first pick of B, and the even ground cards like the second pick (solid marks = weft up); but in addition, the extra warp lifts are cut where the latter are indicated by the dots and circles. On the extra weft-figuring cards, all but the solid marks and circles are cut. The complete weave of the first 24 horizontal rows and the last 24 vertical rows of B is represented at C in *Figure 2.39*. (The circles in C, which indicate the extra warp stitches, should be situated one pick later.) The ground ends and picks form plain weave throughout the cloth, and a feature to note is that where the extra warp figure is formed the extra weft lies between the plain foundation and the warp floats, the latter being thereby shown up very prominently.

IMITATION EXTRA THREAD EFFECTS

By careful and judicious arrangement special materials can be employed in such a manner in developing a design as to give the impression that extra threads are

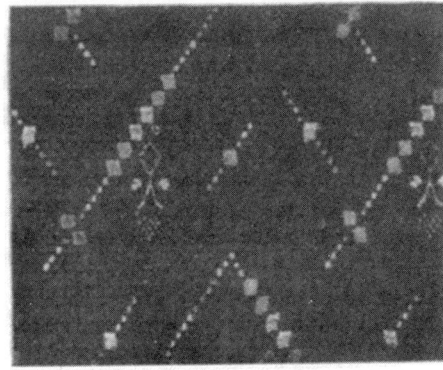

Figure 2.40

included in the cloth thus imitating successfully the more expensive extra warp or extra weft structures. The special figuring threads may be introduced, either in the weft or the warp, the latter method being usually the more convenient. In intermittent figured fabrics bright lustrous threads, either warp or weft, may be inserted where the figure is formed, and an ordinary class of yarn in the spaces between, both classes of threads being interwoven in the same manner in the ground portions of the cloth. Under certain conditions the difference between the yarns is scarcely observable in the ground, and a bright lustrous figure is produced with a minimum use of the more expensive yarn.

Imitation extra weft figures

The fabric represented in *Figure 2.40* and the corresponding design given in *Figure 2.41*, show how an ordinary fabric may be made to appear as if figured with extra weft. The figure and the order of wefting are arranged to coincide, and in the example the picks are inserted in the order of 16 dark and 16 light, to correspond with the arrangement of the figure, which in *Figure 2.41* is indicated in the same order by different marks (convention reversed). In order to produce the imitation extra weft effect, it is necessary for the ground of the cloth to be warp surface, in strong colour contrast with both wefts, and very

Figure 2.41

finely set; as for example, for the 8-thread satin ground weave—64 ends per cm of 15/2 tex cotton. Except for a slight shadiness, the weft intersections in the ground are concealed by the fine setting of the warp, and the figure is developed in two distinct colours on what is practically a solid-coloured ground.

Imitation extra warp figures

The foregoing principle can be employed in producing an imitation extra warp effect by using coloured ends in sections to correspond with the form of the design.

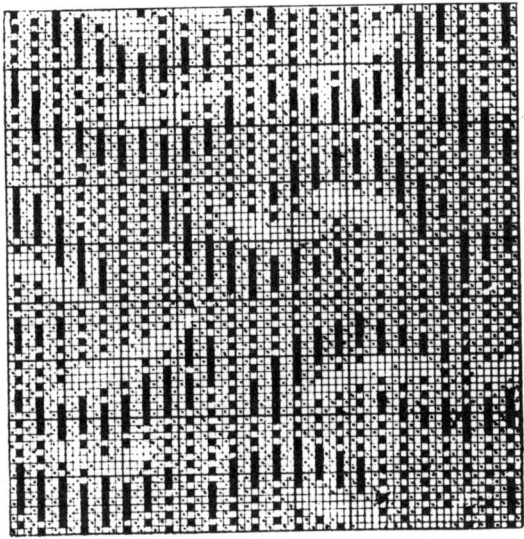

Figure 2.42

A method of figuring with special warp threads is illustrated in the design given in *Figure 2.42,* in which the marks indicate warp. The arrangement in the warp is 2 threads of 20/2 tex cotton to 1 special thread of 120/2 tex mercerised cotton, 20 threads per cm; while the weft is 30 tex cotton in the same colour as the ground warp, 22 picks per cm. The cloth is chiefly ornamented by floating the thick warp threads, as indicated by the solid marks, but weft figure is also formed as shown by the blank spaces in the design. Similar styles are woven in which the warp threads are arranged in the proportion of 3 ground to 1 special, or 4 ground to 1 special etc., the decrease in the proportion of the effect threads being usually compensated for by an increase in their thickness.

3

Backed Cloths

The backed principles of construction are employed for the purpose of increasing the warmth-retaining qualities of a cloth, and in order to secure greater weight and substance than can be acquired in a single structure which is equally fine on the surface. A heavy single cloth can only be made by using thick yarns, in conjunction with which it is necessary to employ only a comparatively few threads per unit space. A heavy single texture is therefore obliged to be somewhat coarse in appearance. By interweaving threads on the underside of a cloth, it is possible to obtain any desired weight combined with the fine surface appearance of a light single fabric.

When certain threads are inserted solely to give additional weight, the idea is to employ them in forming a back to a face fabric; and one of the advantages of the backed construction is that the extra weight can be obtained in an economical manner, since material which is inferior to the face yarns may be used on the underside. Backed cloths are constructed on both the backed weft and the backed warp principle; a cloth consisting in the former case of two series of weft threads and one series of warp threads, and in the latter case, of two series of warp threads, and one series of weft threads.

The construction of these fabrics on design paper is best carried out methodically in several stages:

(1) Mark out on design paper face threads and back threads in the order in which they are inserted, e.g. 1 face, 1 back, as shown in *Figure 3.1* at A and B for the warp-backed and weft-backed structures respectively.

(2) Insert the face weave on face threads only using normal convention for warp backing and reversed convention for weft backing (C and D in *Figure 3.1*).

(3) Insert the back weave on back threads only—normal and reversed convention as before—taking care to place a mark of the back weave between two long floats of the face weave thus concealing the binding marks of the back weave by the covering float on the face (E and F in *Figure 3.1*).

In reversible structures the binding marks of the face weave should be equally well concealed on the back which is in fact achieved in the examples given in *Figure 3.1* by a suitable choice of the face and the back weaves which are 4 up, 1 down twill face and back, for the warp-backed structure and 1 up, 4 down

42

twill face and back, for the weft-backed structure. Interlacing diagrams and cross-sectional views which correspond with E and F are given at G and I, and H and J respectively.

Many other weaves can be employed for the purpose of producing backed cloths and several examples of each are given in the succeeding pages. At this

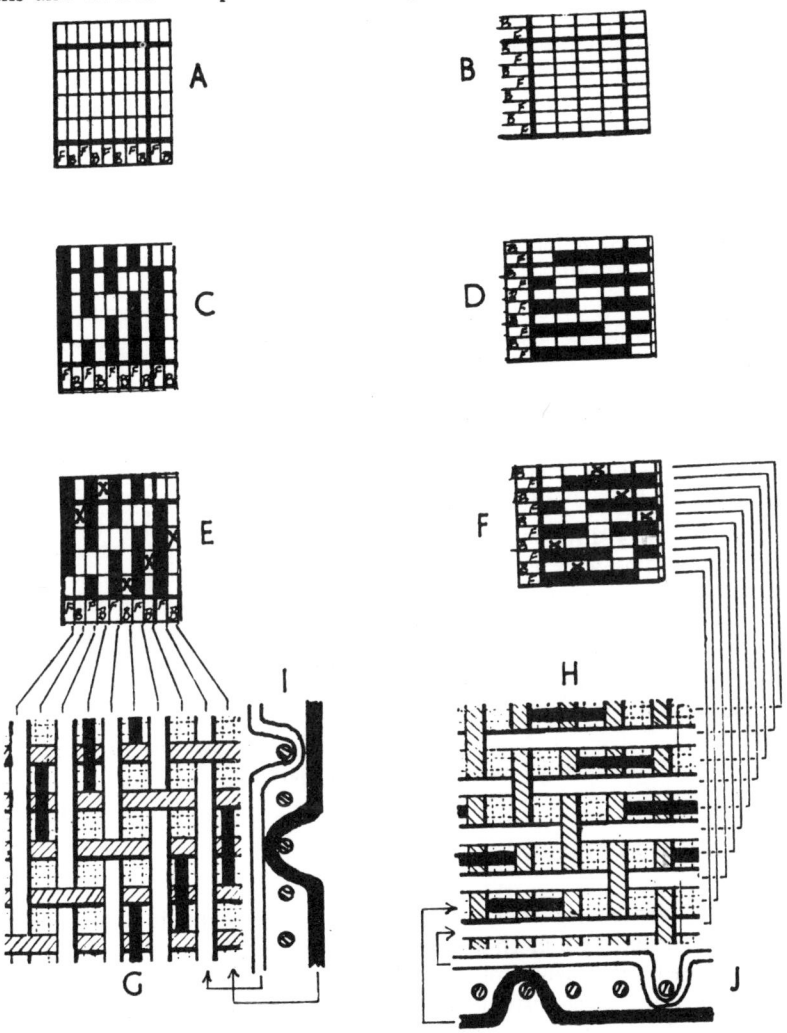

Figure 3.1

point it may be noted that warp-faced weaves are more suitable for warp backing and weft-faced weaves for weft backing whilst certain square-faced weaves can be successfully applied to both structures. In order to achieve a well-covered face in a backed cloth, correct settings are very important as without sufficient density of the face threads the binding marks of the back weave cannot be covered no matter how cleverly they are placed. As most of these fabrics are milled and raised exact rules of setting cannot be given because the eventual density of thread spacing will depend on the amount of shrinkage achieved during finishing. However, as a general guide it can be stated that in 1 face, 1 back orders of backing the percentage cover of the face yarns could be up to 12 per cent less than in a similar well-constructed single cloth if comparatively short floated weaves are used with firm backing, and about 6 per cent less if longer floated weaves with loose backing are employed. In 2 face, 1 back orders the reduction in the setting compared with a similar single cloth must be less than for 1 face, 1 back orders and under conditions similar to those stated for the latter could be 8 per cent and 4 per cent less respectively.

WEFT-BACKED CLOTHS

Weft-backing is sometimes used in preference to warp backing because a softer and more lofty handling cloth can be obtained owing to the weft containing less twist and being under less tension than the warp. Similar conditions can be obtained in a worsted face cloth by using woollen weft, and, as a weaker yarn can be employed as weft than as warp, the use of cheaper material can compensate to some extent for the increase in the cost of weaving due to the insertion of the backing threads weft way.

The standard orders of arranging the picks in weft-backed cloths are: (1) 1 face to 1 back; (2) 2 face to 1 back; (3) 3 face to 1 back; (4) 2 face to 2 back; (5) 4 face to 2 back. The last two arrangements are used in place of the first two when a different kind of backing weft from face weft has to be inserted in looms with changing boxes at one side only. The first and the second systems are most commonly used, the former being employed for fine cloths in which the face and back wefts are similar in thickness. The latter has the advantage of cheapness of production compared with the first method, as there are only half as many backing picks per cm, so that the cost of weaving is less, and generally the backing yarn is thicker and may be of a lower and cheaper quality. The 2 face, 1 back method of backing, however, produces a less attractive cloth because the underside appears coarser.

Reversible weft-backed weaves

Weft-backed designs, in which the same weft face weave is formed on both sides, as in the example given at F and H in *Figure 3.1*, are a distinct class that is chiefly used for heavily milled cloths which are composed of woollen weft and cotton warp. It is customary to use much thicker weft than warp in this structure, and to insert more picks than ends per unit space, so that the milled and raised finish that is applied to the cloth causes the weft entirely to conceal the warp. A number of designs are given in *Figure 3.2*, which form a weft sateen weave on both sides. The backing weave is shown by crosses placed on the backing picks, and the face weave by solid marks. As sateen weaves form a smoother surface than twills, they are more suitable than the latter for the heavily milled woollen weft and cotton warp structures. The designs A, B, C, and D are arranged in the order of 1 face pick, 1 back pick, and they respectively form the 4, 5, 6, and 8-thread weft sateens on both sides of the cloth. In the

design E in *Figure 3.2,* the 4-thread weft sateen on both sides is arranged to suit
a 2-and-2 order of wefting; in the design F the 6-thread weft sateen is similarly
arranged with 4 picks face to 2 picks back; whereas the design G shows the
4-thread weft sateen on both sides arranged 2 picks face to 1 pick back. On the

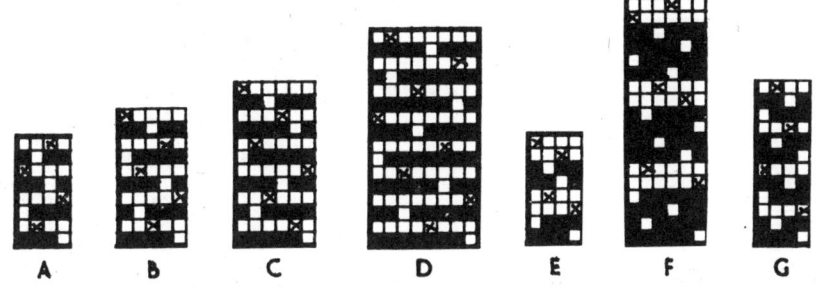

A   B   C   D   E   F   G

Figure 3.2

same principle certain twill weaves which have floats of more than one end at a
place may be made reversible, as for instance, 2 up, 4 down face, 4 up, 2 down
back; 2 up, 6 down face, 6 up, 2 down back; 3 up, 5 down face, 5 up, 3 down
back etc.

Suitable weaving particulars for the design A are: 38/2 tex cotton warp
16 ends per cm, 160 tex woollen weft, 24 picks per cm; and for the design B;
30/2 tex cotton warp, 22 ends per cm, 120 tex woollen weft, 32 picks per cm.
The cloths are shrunk from 20 to 30 per cent in width.

By employing differently coloured wefts for the face and back a cloth is
produced in which the two sides are differently coloured, since each weft is
retained on one side, or a cloth may be woven coloured on one side and white
on the other. In the latter case, in order to avoid the liability of the coloured
weft showing through on the white surface, the coloured yarn may be slightly
finer than the white yarn. By interchanging two different wefts elaborate
designs for dressing gowns, coats, bedspreads, rugs, etc., are woven.

Methods of weft-backing standard twill and hopsack weaves

The examples given in *Figure 3.3* illustrate different methods of weft-backing
the 2-and-2 twill. In the design H the picks are arranged 1 face, 1 back, and the
weave on the underside is 1-warp and 3-weft twill and the cloth is thus as firm
on the back as on the face. The section I in *Figure 3.3* shows how the picks
1 and 2 of H interlace. In the design J the picks are arranged 1 face, 1 back, but
the back weave is 8-thread weft sateen. The cloth in this case is looser and softer
on the back than on the face, as long weft floats are formed on the underside,
as shown at K, which represents how the picks 1 and 2 of the design J interlace. In L and M in *Figure 3.3* there are 2 face picks to 1 backing pick, but
each design, although used in practice, is defective in that the back weave
stitches occur only on alternate ends, so that the odd and even ends are liable
to vary as regards take-up. The design M is additionally defective because the
ties occur only on alternate face weft twill lines which show more prominently

than the unstitched twills. In L the stitches produce a very firm back which appears like plain weave. The design N is imperfect, but it shows the best method of weft-backing the 2-and-2 twill when the picks are arranged 2 face and 2 back. In the design O the picks are arranged 3 face to 1 back, and the

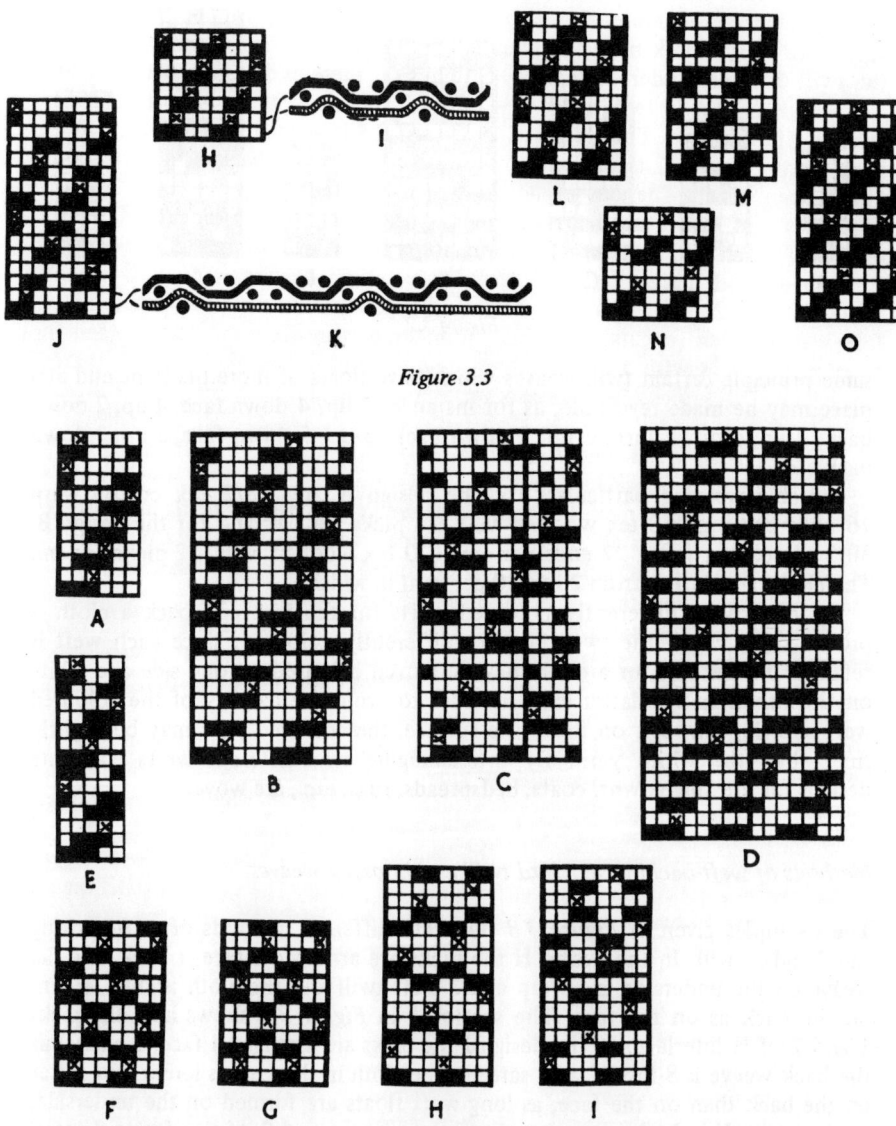

Figure 3.3

Figure 3.4

stitches are arranged in twill order in the reverse direction to the face twill so that there is a possibility of a cross twill appearing in the cloth. Suitable weaving particulars for design J in worsted yarns are—37/2 tex warp, 26 ends per cm, 45/1 tex weft, 46 picks per cm.

The designs A and B in *Figure 3.4*, which are wefted in the order of 1 pick face, and 1 pick back, show the 3-and-3 twill face weave. In A the backing weave is 1-and-5 twill, and in B 12-thread sateen, and the 3-and-3 twill is extended to two repeats in each direction; each back weave mark in the latter could also have been passed over two ends at a place. In weft-backing the 4-and-4 twill considerable variety of firmness of stitching can be obtained, as each back weave mark may be extended over 1, 2, or 3 ends at a place in either the twill or sateen order of backing. Stitches on these ends can be woven 1 up, 1 down, 1 up, but in twill order this causes the back weave to be much firmer than the face weave. The design C in *Figure 3.4* is wefted 1 pick face, 1 pick back, and shows a 12-thread sateen-derivative face weave backed in corresponding sateen order. The design, D, in which the face weave is 3-and-2 twill, illustrates that in a 1-and-1 arrangement of the threads three repeats each way of a twill on an odd number of threads are required when a sateen back weave is formed. The design E shows the 3-and-2 twill backed with weft in the proportion of 2 face picks to 1 back pick, and this example illustrates that a face weave which repeats on an odd number of threads must be extended to two repeats to fit with a 2-and-1 arrangement. Thus the design contains 10 face picks and 5 back picks, and repeats on 5 ends, and the back weave is 5-thread sateen.

The designs F, G, H, and I in *Figure 3.4* show different methods of weft-backing a 2-and-2 hopsack weave. F and G are wefted in the proportion of 2 face picks to 1 back pick, the backing weave in the design F being in alternate order. In G the stitches are in 4-satinette order, which is a better arrangement, because a smoother and softer back is produced at the same time that a tie is placed on every warp thread. H in *Figure 3.4* is wefted in the proportion of 2 picks face to 2 picks back, and a stitch is placed on each warp thread. Each design F, G, and H, is so arranged that a backing pick is placed between two face picks that are in the same shed, and not between two picks that cut with each other, so that it is possible to place each stitch with a face weft float on both sides. In the design I the 2-and-2 hopsack weave is backed with weft in the order of 1 face, 1 back, and in this case therefore it is only possible to arrange one half the stitches with a face weft float on both sides. A mark, which is covered on one side only by the face weft, should precede the covering pick, as shown in I, because it is better concealed by the subsequent beating up of the covering pick than if the latter preceded the tie. The order of stitching in design I is the 8-thread sateen which yields a soft back, but the regular distribution of the back weave marks is liable to produce a twill effect on the face.

Warp-face weaves backed with weft

The designs given in *Figure 3.5* illustrate warp-face weaves that are backed with weft, in which it is only possible to cover each tie on one side. In each example the face weave is shown alongside with the positions of the back weave marks indicated between the squares, and it will be seen that the back weave is looser than the face weave. The designs A and B both show the 4-thread warp twill backed with 8-thread sateen, but the former, in which the ties follow the face weft floats, is given simply to illustrate incorrect placing of the back weave marks, the correct method being indicated at B. In the design C the face is 5-thread warp satin, and the back 10-thread weft sateen; the arrangement of the picks is 1 face, 1 back, in each case. The design D is also 5-thread warp satin face, but the picks are in the proportion of 2 face to 1 back, and the backing weave is extended 5-sateen.

The type of design given in *Figure 3.5* is employed for a class of piece-dyed coatings in which a worsted face warp largely predominates in quantity over the face weft, while thick wollen weft is used for the back. The cloth is milled and

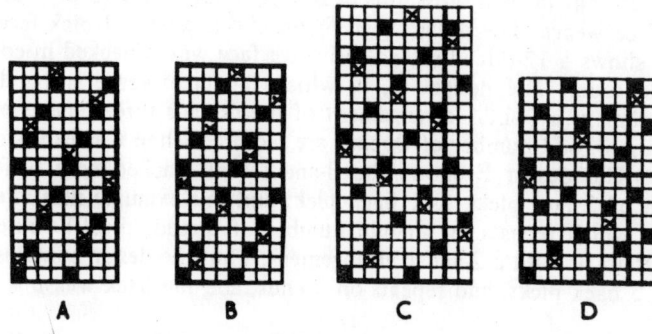

Figure 3.5

raised on the underside, and is therefore soft and full to the feel. Suitable weaving particulars for the design C are: Warp, 35/2 tex worsted, 40 ends per cm; face weft 30/2 tex cotton, backing weft 105 tex woollen, 38 picks per cm. In any weft-backed cloth in which, in order to secure greater weight and substance, thicker backing weft than face weft is used, fewer face picks than ends per cm may be employed with the result that the angle of twill face weave is steeper than 45°, but a very fine twill can be obtained in a heavy cloth.

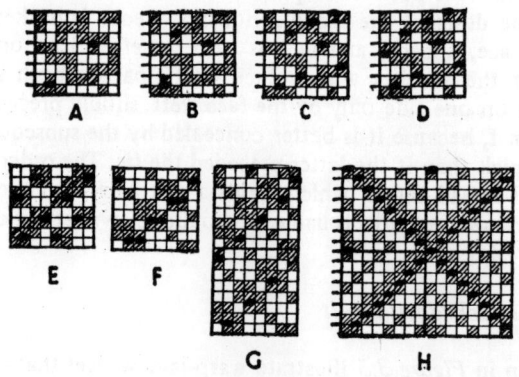

Figure 3.6

The ties for face weaves that are regular in construction are, as a rule, easily arranged, but before constructing a backed design it is convenient, in many cases, to indicate the face weave lightly, and to scheme the distribution of the ties by inserting back weave marks between the squares. The ties for irregular face weaves are sometimes difficult to arrange, and in *Figure 3.6* a convenient method of working is illustrated in stages at A, B, C, and D. The positions of the backing picks, in the order in which they are inserted with the face picks, are indicated, as shown by the marks alongside the face weave given at A; and the first ties are marked between the face picks where only one tying position is available. Second, the ties are marked on the ends which afford only one suitable tying position, as shown at B on the third and seventh ends. Third, the ties are marked in the remaining positions, care being taken to indicate one for each backing pick, and, if possible, to so distribute them that one tie is placed on each end. In the plan C seven ties are correctly indicated, only that of the fourth backing pick being omitted. This tie can only be covered on both sides by placing it on the second or eighth end, on both of which, however, a tie has already been indicated. In such a case, unless there is a strong contrast in colour between the face and back wefts, it is better to place the stitch with a face weft float on one side only, as shown at D, and thus have the ties properly distributed, than to stitch twice on one of the ends. The complete design may then be readily made, but it is quite convenient to peg the dobby lags straight from a face plan, constructed as shown at D.

Special examples of weft-backing

As previously shown, in 2 face, 1 back weft-backing it is frequently impossible to place a tie on every warp thread, and it is only in certain weaves that perfect distribution can be obtained. Twill weaves that repeat on an odd number of threads, are examples in which every thread may be stitched, as shown in the design E, *Figure 3.4*. A face weave, such as that given at E in *Figure 3.6*, may be stitched on every end, by placing two ties on each backing pick, as shown; while F is an example in which a similar result is obtained by floating each tie over two consecutive ends. A weave such as F, however, can be stitched on every end by extending it to two repeats, as shown at G. The design H in *Figure 3.6* illustrates the principle of weft-backing a diamond weave with the same number of stitches on each end.

WARP-BACKED CLOTHS

The arrangement of two series of warp threads to one series of weft threads enables a considerable saving in the cost of weaving to be effected, as compared with the weft-backed principle, because of the reduction in the number of picks per cm; a more solid appearance can be given to the cloth by the formation of stripe patterns on the underside, which is impossible in weft-backed textures; while owing to the greater strength warp way the cloths are superior from a structural point of view. On account of the greater strain in weaving, however, such a low quality of backing yarn cannot be used as in weft-backing; drawing in the warp is more costly because there are more ends; the drafts are usually more complicated, and a greater number of healds are required in producing similar effects.

The standard orders of arranging the ends in warp-backed cloths are: 1 face to 1 back, 2 face to 1 back, and 3 face to 1 back (there is no necessity to arrange the ends in even numbers); while in some cases a backed weave is combined in stripe form with a single weave.

Reversible warp-backed weaves

The design A in *Figure 3.7* is a standard reversible weave, in which the face weave and the order of stitching produce a 3 up, 1 down twill on both sides so that the cloth is perfectly reversible except for the difference in the direction of the twill lines, as is illustrated by the corresponding diagrams at E and G. The view, given at E, represents the structure as viewed from the face side, and that shown at G, as viewed from the back, assuming that the cloth has been turned over vertically, while D and F show how the first two ends interlace. The design shown at E in *Figure 3.1* is similar except that it results in a 4 up, 1 down twill on both the face and the back of the cloth. Warp weaves on both sides of the

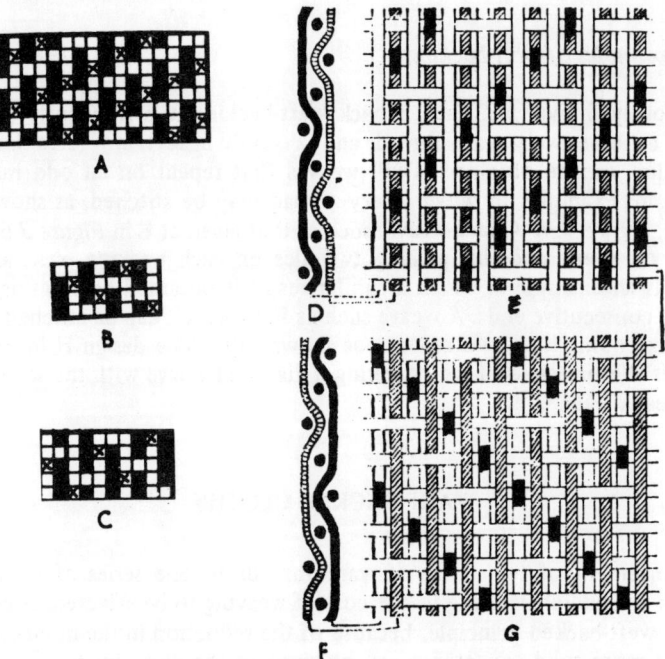

Figure 3.7

cloth are constructed in a similar manner, and B and C in *Figure 3.7* respectively represent the 4-thread and 5-thread satin weaves made reversible. The cloths may also be made reversible as regards the colouring, or different colour patterns may be formed on the two sides by employing different schemes of colouring for the two series of warp threads.

Beaming and drafting warp-backed structures

Figure 3.8 shows various methods backing a 2-and-2 twill face weave, and also illustrates different systems of drafting warp-backed structures. The designs Q and S are arranged 1 face end, 1 backing end. The order of stitching in Q results in a 3-and-1 twill back and R represents the interlacing of the first and second ends of the design. The twill on the underside makes the cloth as firm on the back as on the face. In the design S a loose satin weave is formed on the underside, as shown at T, which indicates how the first and second ends of S interlace.

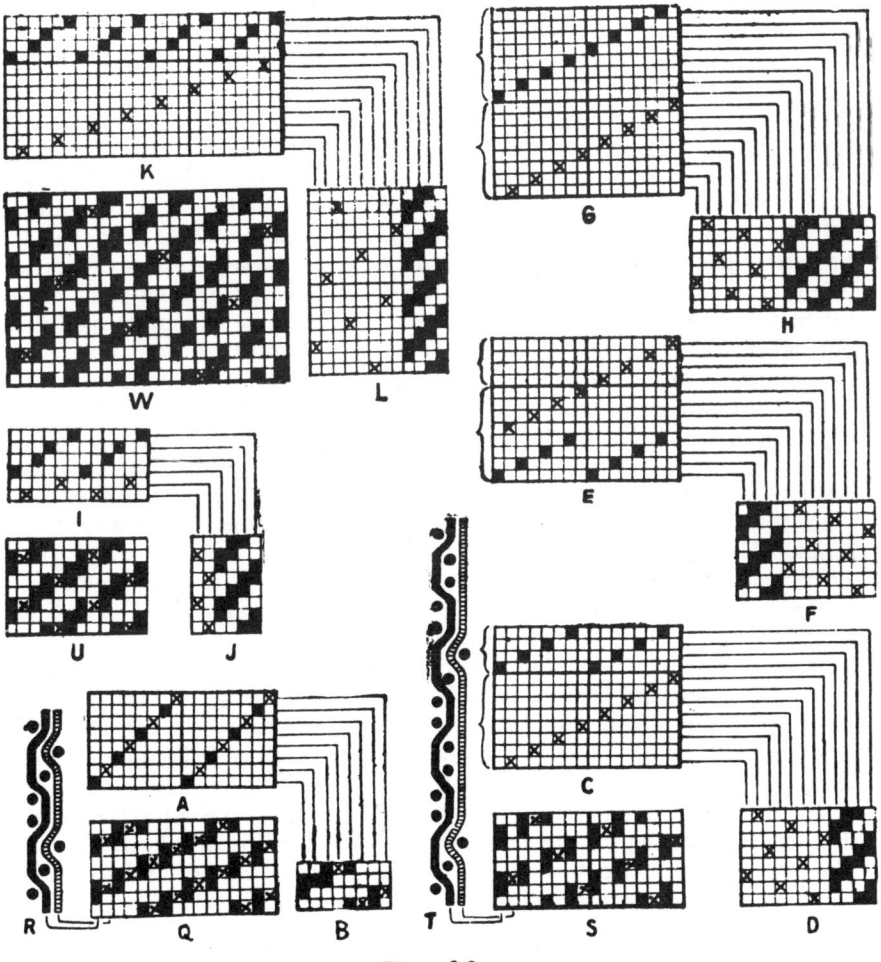

Figure 3.8

In the designs U and W in *Figure 3.8* there are two face ends to each backing end, but a commencement is made with one face and in order that the backing ends may be readily dented in the reed with a face end on each side. In the design U an alternate order of stitching on the odd face picks is employed, and in W a sateen order is similarly arranged, the ties thus occurring on only half the face picks.

For the design S the following are suitable weaving particulars in a worsted cloth: Face and back warp 45/2 tex, 48 ends per cm; weft 45/1 tex, 24 picks per cm. The design U might be woven in a woollen cloth with 66 tex face warp, 96 tex backing warp, 24 ends per cm; weft 66 tex, 16 picks per cm.

In beaming the warp for a warp-backed cloth all the threads may be placed on one warp beam so long as the face and back yarns are similar, and the face and back weaves are equal in firmness, as in the design given at Q in *Figure 3.8.* Preferably however, the two series of threads are placed on separate beams, in order that they may be independently tensioned, and there is then no restriction as to the comparative firmness of the weaves or thickness of the threads. A weave such as that shown at S may require the face warp to be as much as 8 per cent longer than the back warp.

In drafting warp-backed designs simple patterns may be drawn straight over, as shown at A in *Figure 3.8,* which is the draft for the design given at Q, while lifting plan B is exactly the same as Q. In the draft A the healds which carry the backing ends are intermingled with those upon which the face ends are drawn, and a similar order of drafting upon 16 healds can be employed for the design S, the latter then forming the lifting plan. In cases, however, where there is difference in thickness or material between the face and backing ends, or if different warp patterns for the two sides of the cloth are employed, or if the face weave requires a special draft, it is better to draw each series through a separate set of healds. Whether the weaker yarns (usually the backing yarns) should be drawn over the front or the back healds is dependent on the shed geometry of a weaving machine and both methods are employed contingent upon the specific circumstances in operation at a given machine. The two positions of the face and backing healds are illustrated at C and E in *Figure 3.8,* which show two methods of drafting the design S upon the smallest number of healds, while the respective lifting plans are given at D and F. In the draft C the backing healds are shown in front of the face healds, and in E behind them.

The use of only four healds for the face weave as shown at C and E in *Figure 3.8,* gives very little scope for producing different weaves in the same draft, whereas if eight face healds are employed, as shown in the draft given at G, any face weave that repeats on four or eight threads may be woven; also the wires are less crowded on the shafts so that the healds will last longer, and there is less friction on the warp. H shows the corresponding lifting plan for the design S.

The draft and lifting plan for the design U in *Figure 3.8* are given respectively at I and J, and for the design W at K and L; the backing healds in each case being placed in front of the face healds. A comparison of drafts I and K shows that fewer healds are required for a firm back than for a loose back.

Methods of warp-backing standard weaves

The examples in *Figure 3.9* show standard methods of placing back weave ties between the floats of the face structure. The twill order of stitching in A and B coincides with one repeat of the 3-and-3 twill face weave (in the latter design the backing ends are stitched on two consecutive face picks), whereas the sateen order of stitching in the design C requires that the face weave be extended over two repeats in each direction.

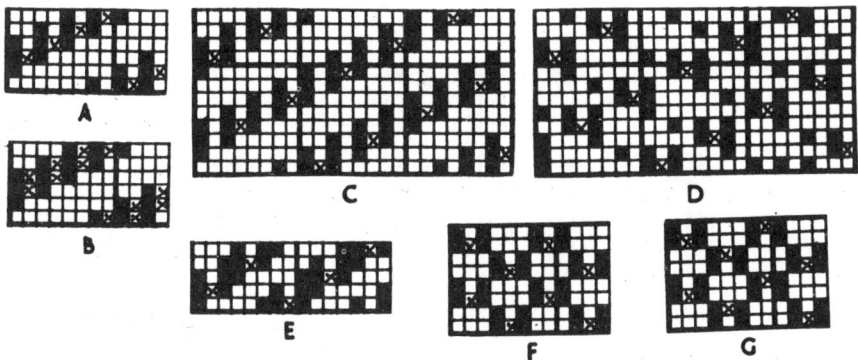

Figure 3.9

The design D in *Figure 3.9* shows a sateen-derivative face weave backed in sateen order. E shows a 5-thread twill extended to two repeats horizontally to fit with a 2 face, 1 back arrangement of the ends. The 2 up, 1 down twill is a standard fine coating weave, and in 1 face, 1 back arrangement the same twill on the back produces a firm structure, and a 9-sateen back a softer handling cloth. In 2 face, 1 back arrangement the 3-thread twill can be formed on both sides running in the same direction when the cloth is turned over. The design F shows the 2-and-2 hopsack face weave and the alternate order of back stitching, while G shows the same weave backed in 4-sateen order, which is a better arrangement than that shown at F. In warp-backing the 2-and-2 hopsack weave in 1 face, 1 back order the back weave may form 4-thread twill or satin, or an 8-thread satin back may be formed, but in each case one-half the stitches are covered by a face float on one side only.

Method of selecting warp ties for irregular weaves

A convenient simplified system of arranging the ties in warp-backing an irregular face weave is illustrated in stages at I, J, and K in *Figure 3.10*. The face weave is marked in, and the positions of the backing ends—in the order in which they are arranged with the face ends—are indicated below the face plan, as shown at I. The ties are first indicated between the face ends in the places where only one tying position is available, as shown by the marks between the squares of J. Then, as shown at K, the remaining ties are indicated in the positions which will give the most regular and uniform distribution. Afterwards, the draft and lifting plan may be constructed directly from the face plan, as shown at L and M.

N in *Figure 3.10* shows a type of design in which, in a 1 face, 1 back order of warp-backing, certain of the stitches of the backing ends—in this case the fourth and eighth—can only be covered on one side by a face warp float. In a 2 face, 1 back order of backing the design, however, it is possible to avoid placing a backing thread between the face threads that cut with each other, so that proper positions for the ties can be readily found, as shown at O.

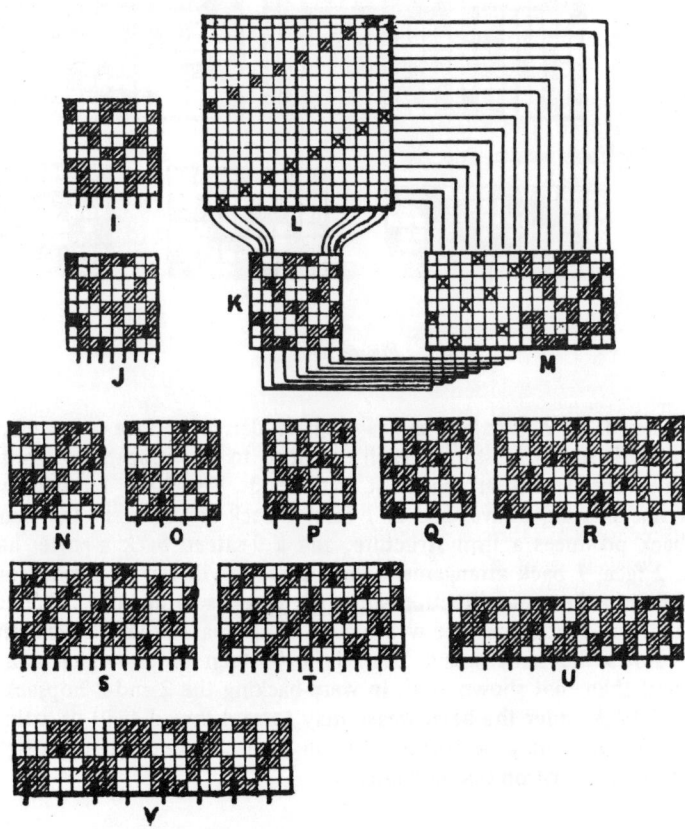

Figure 3.10

P, Q, and R in *Figure 3.10* show different methods of backing the Mayo weave with warp in the proportion of 2 face to 1 back. In the design P the ties are placed on alternate picks only, but Q shows the face weave in a different position relative to the backing ends which permits each to be tied twice so that a stitch is placed on every pick at the same time that the back weave is made much firmer. R shows another method of tying the Mayo weave on every pick; the face weave in this case is extended over two repeats and each backing end is only stitched once, hence the backing weave is as loose as in the design P. The design R, however, requires twice as many backing healds as either P or Q.

The designs S and T in *Figure 3.10* show two arrangements of the stitches in a 1 face, 1 back warp-backed weave, which is composed of 2-and-2 twill and twilled hopsack. In both designs the stitches are correctly placed as regards being covered on both sides by face warp floats, but in S the distribution is not so good as in T. Further, a complicated draft of the backing ends is required in the former, whereas in the latter the draft of these ends is quite regular.

The design U in *Figure 3.10* shows a stripe face weave, composed of 3 up, 2 down twill and Venetian, which is arranged in 2 face, 1 down order. In the twill section of the face weave the ties are arranged in sateen order with the picks, and in the Venetian section in twill order, so that in this case both the face and the backing threads require to be specially drafted.

The design V in *Figure 3.10* shows a stripe weave composed of 3-and-3 hopsack and 3-and-3 twill derivative that is backed with warp in the proportion of 3 face to 1 back. The 3 face, 1 back arrangement of the threads is particularly suited to the face weave, and the ties are so distributed that only two backing healds are required.

INTERCHANGING FIGURED BACKED FABRICS

Interchanging weft-backed fabrics

These cloths are chiefly used for blankets, dressing-gowns, and rugs. The weave is the same in every part of the cloth, and a weft surface is produced on both sides. The design is due to the manner in which differently coloured wefts are interchanged from one side to the other, a dark figure on a light ground on one side corresponding with a light figure on a dark ground on the other side. This is illustrated by the fabric represented in *Figure 3.11*, in which the reverse side of the cloth is shown in the bottom left-hand corner. A portion of a ·

Figure 3.11

similar design is given in *Figure 3.12*. Generally, the wefts should be brought about equally to the surface on both sides in order that one side will not appear darker than the other, this being particularly the case when the cloth is seen on both sides at the same time. A raised finish is applied alike to both the back and face, and when woollen weft is used the shrinkage in width ranges from 15 to 30 per cent. The warp is almost invariably cotton, and suitable weaving particulars for a heavy fabric in a 4-thread weave are: 60/2 tex cotton warp, 9 ends per cm, and 350 tex woollen weft, 19 picks per cm; and for a lighter cloth: 39/2 tex cotton warp, 14 ends per cm and 120 tex woollen weft, 28 picks per cm. The felted and raised finish causes the cotton ends to be entirely concealed, and gives a full soft feel to the cloth. Cheap cloths are made entirely of cotton, the flannelette class of weft being used, which is generally inserted in even picks, and the following weaving particulars are suitable: Warp, 38 tex, 19 ends per cm; weft, 49 tex, 34 picks per cm.

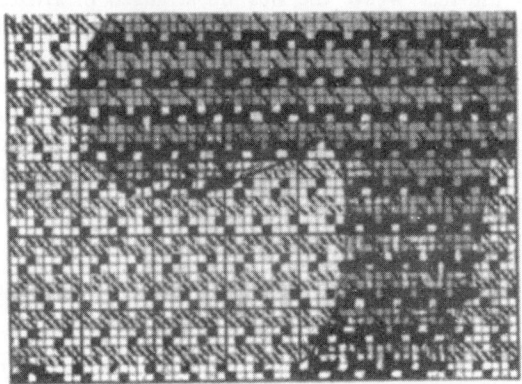

Figure 3.12

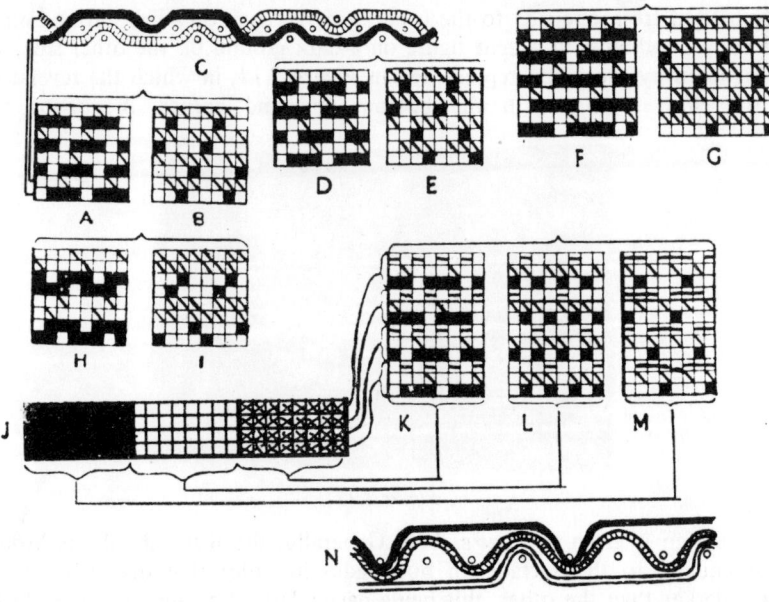

Figure 3.13

The weaves for the figure and ground are constructed upon the same principle as the weft-backed reversible designs illustrated in *Figure 3.2*, and as shown in *Figure 3.13*, in which the most commonly used reversible plans are given. Both

A and B in *Figure 3.13* in which the marks indicate weft, show the double-face 1 up, 3 down weft twill weave, but in A the odd picks are on the surface and the even picks on the back; whereas in B the odd picks are on the back and the even picks on the surface. If, therefore, the picks are arranged 1 dark, 1 light, weave A will produce a dark surface and a light back, and weave B a light surface and a dark back. By combining the two weaves in sections the wefts interchange between the face and back, and a design in two colours is formed, as represented in the diagram C, which shows the interlacing of the picks 1 and 2 of A and B in combination.

D and E in *Figure 3.13* show the 4-thread and F and G the 5-thread weft sateens made reversible in the same manner as A and B, while H and I illustrate the construction of the reversible 4-sateen weave to fit with a 2-and-2 order of wefting. Other weft-face twill and sateen weaves can be similarly arranged, but, as a rule, a sateen produces a smoother surface, and is therefore more suitable for the raised finish than a twill weave upon the same number of threads. As in all weft-backed cloths the chief point to note in each weave of a pair, is that the interlacing points of the back picks occur between face-weft floats.

Figure 3.12 illustrates the method of painting out a design in full. The reversible 4-thread weft sateen weaves are combined; and the order of wefting is 2 dark, 2 light. The figure is indicated lightly in a wash of colour, then, in order to produce a dark figure upon a light ground, the weave H in *Figure 3.13* is indicated in the figured portions, and the weave I in the ground. As the design shows the complete interlacing of the threads, one card is cut from each horizontal space, and the cutting instructions are: Cut all but the weave marks.

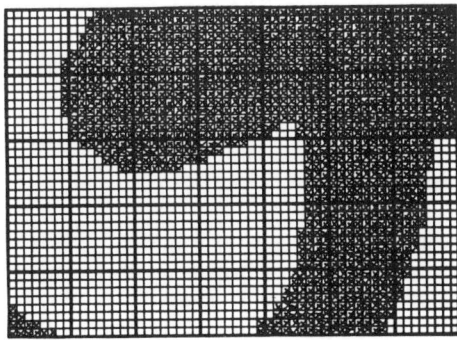

Figure 3.14

It is a tedious process to paint out a design in full, and in most cases a simplified painting would be made in preparation for cutting on one of the advanced card-cutting systems outlined in Chapter 1. Such a design is shown in *Figure 3.14* and corresponds with the fully worked-out design in *Figure 3.12* as far as the shape of the figure is concerned. It will be noted that in simplified designing for interchanging weft-backed structures it is only necessary to paint the figure solid in one colour and then the ground equals paper, i.e. unpainted area. The simplified design would then be programmed into the area selector and to make the construction identical with that in *Figure 3.12,* weave H from *Figure 3.13* would be placed on the weave selector during the cutting of the paper area, i.e. the unpainted ground, the weave selector pattern would be that represented by I in *Figure 3.13.* Thus, the two areas, i.e. the ground and the figure would be rendered in cloth exactly as shown in *Figure 3.12.* The same design could be produced in other structures if instead of H and I other pairs of weaves shown in *Figure 3.13* were used successively on the weave selector whilst the paint and the paper areas were selected as before on the area selector.

The simplified design in *Figure 3.14* could be easily condensed weft-wise by 2, or even by 4 without an appreciable loss of definition because in these structures, due to the raised finish, only rather solid designs are applicable as any fine design lines are liable to become lost under the surface nap.

In the interchanging weft-backed structures only two colours can be brought to the surface in each horizontal line but more than two colours can be obtained by chintzing in succeeding portions of the design. Even without chintzing it is possible to obtain a third effect by weave 'mixing'. Thus, in a pick-and-pick order of wefting a third effect can be formed by combining two weaves such as D and E in *Figure 3.13* with a third weave, such as H or I, while in a 2-and-2 order of wefting, two weaves, such as H and I, can be combined with a third weave, such as D or E. In each case two of the weaves produce solid effects, whereas in the third the weft colours are intermingled and a subsidiary pattern is formed which can be used to give variety to a design.

In certain low qualities of the woollen-weft cloths the structure is strengthened by the insertion of extra cotton picks at intervals which interweave in plain order with the warp threads. The arrangement may be 4 picks of wool to 1 pick of cotton, or 10 to 2, 12 to 2, etc., plain cards being combined with the figuring cards in the required order. The appearance of the cloth is not altered, but the presence of the cotton picks prevents any tendency of the woollen picks to slip. The production of a loom is, of course, reduced by the insertion of the extra cotton picks.

Treble-wefted interchanging backed fabrics

The weft-backed structures are made to a limited extent with three figuring wefts, which enables an effect to be woven in three colours; while increased weight can be obtained combined with greater firmness, as in the centre the wadding threads may be interwoven more frequently than on the face and back. A figure in two colours on a ground in the third colour may be formed on both sides of the cloth, or one of the wefts may be used to form a solid colour effect on one side of the cloth, while the other two wefts interchange so as to form a figure on the other side.

The simplified plan J in *Figure 3.13* illustrates a method of indicating a treble-wefted design, arranged a pick of each alternately, in which each weft interweaves on the face, in the centre, and on the back so as to produce a figure in two colours upon a ground in the other colour, on both sides of the cloth. The weave on the face and back is 4-sateen, and in the centre plain. The design J is condensed by 3 weft-wise, therefore, three cards must be cut from each horizontal row of J. The complete weaves to correspond with different portions of J are shown separately at K, L and M in *Figure 3.13* in which it will be seen that in K the first weft floats 3-and-1 on the face, the second weft floats 3-and-1 on the back, and the third weft weaves plain in the centre. In N the first weft weaves plain in the centre, the second weft floats 3-and-1 on the face, and the third weft floats 3-and-1 on the back. In O, the first weft floats 3-and-1 on the back, the second weft interweaves plain in the centre, while the third weft floats 3-and-1 on the face. A figure formed by the first and second wefts on the face is similarly formed by the second and third wefts respectively on the back, while the third weft forms the ground on the face and the first weft the ground on the back. The plain centre weave gives the cloth great firmness, and may be too firm for a heavily-wefted cloth, and in such a case another weave may be used, such as 2-and-2 twill, or 2-and-2 weft rib. The floats in the centre require to be shorter than those on the face and back in order that they will be invisible on both sides. The warp section N in *Figure 3.13* shows the interlacing of the first three picks in portion K.

Interchanging warp-backed figured fabrics

This class of structure can be produced in a similar manner to interchanging weft-backed fabrics by employing different colours in the warp. The construction of the weaves that are combined will be illustrated by turning *Figure 3.13* one-quarter round, and taking the marks to indicate warp.

BACKED CLOTHS WITH WADDING THREADS

In this construction the object is to obtain increased weight—as compared with ordinary backed cloths—by introducing a thick cheap yarn between the face texture and the backing threads, with neither of which it is usually interwoven. In weft-backed cloths the wadding threads are introduced in the warp, and in warp-backed cloths in the weft, each type thus consisting of two series of warp and two series of weft threads. The threads may be arranged either 1 ground, 1 wadding or 2 ground, 1 wadding, and each order may be used in conjunction with 1 face, 1 back, or 2 face, 1 back orders of backing. The 2 ground, 1 wadding order allows very thick wadding yarn to be used, and along with 2 face, 1 back order of backing is largely employed for worsted face cloths with low woollen or cotton wadding threads, particularly when the wadding yarn is in the weft.

Weft-backed and warp-wadded designs

The system of constructing weft-backed and warp-wadded designs is illustrated in *Figure 3.15*, in which the weave marks indicate weft. In order that comparisons may be made, the 4-and-4 twill weave is shown arranged on the ordinary weft-backed principle at A, while the construction of a wadded design to correspond is illustrated in stages at B and C. The arrangement of the threads is 1 face pick, 1 backing pick, and 1 ground end, 1 wadding end. At B the marks of the face weave are inserted where the ground ends and face picks intersect, as shown by the solid marks, while the backing weft stitches, which are represented by the crosses, are indicated on the ground ends between face weft floats. C shows the completion of the design, the wadding ends being marked down, as shown by the dots, on the face picks. On the backing picks the wadding ends are left blank so that they are raised, and therefore lie between the face texture and the backing picks, as shown at D, which represents the interlacing of the picks 1 and 2 of C. E shows the appearance of the design C when only one kind of mark is used to represent weft up. All the wadding ends work alike so that one heald only can operate them.

A different arrangement of the threads is given at F in *Figure 3.15,* which shows the weft-backed 2-and-2 hopsack weave, wadded with warp in the proportion of 2 ground to 1 wadding end, while the picks are in the proportion of 2 face to 1 back. As before, the weave marks on the face ends are exactly the same as in the ordinary weft-backed design, and the wadding ends are marked down on the face picks. The section through the weft given at F, which represents the interlacing of the first two ends of F, shows how the wadding ends lie between the face fabric and the backing weft.

Warp-backed and weft-wadded designs

The system of constructing warp-backed and weft-wadded designs is illustrated by the examples given at H to M in *Figure 3.15* in which the weave marks indicate warp up. The design H in *Figure 3.15* shows the 4-and-4 twill arranged on the ordinary warp-backed principle, while I represents the construction of the wadded design in the same structure. The arrangement of the threads is 1 face, 1 back in the warp, and 1 ground, 1 wadding in the weft, the positions of the backing ends and wadding picks being indicated by the lines on the fringe of I. The face and back weaves are shown by the solid marks and crosses, then the face ends are lifted on the wadding picks, as shown by the dots, whereas the backing ends are left down. The wadding picks therefore lie between the face and the backing ends, as shown in the diagram given at J, which represents the interlacing of the first and second ends of I. K represents the appearance of the design I assuming that only one kind of mark is used to indicate warp up. The drafting of the design is the same as for the warp-backed design H.

The design L in *Figure 3.15* shows a 2-and-2 twill backed with warp in sateen order and illustrates a method of interweaving the wadding picks with the backing ends, which is sometimes practised, in order to produce a firmer cloth. The diagonal strokes which precede the crosses in L indicate the lifts of the backing ends over the wadding picks; it is necessary for these lifts to be made either immediately before or immediately after the backing-warp stitches (the former, which is shown at L, being preferable), in order to avoid breaking the backing-warp floats on the underside of the cloth. Diagram M represents the interlacing of picks 2 and 3 of L and shows how the wadding picks lie between the face and the backing ends, except where they pass under the latter at a

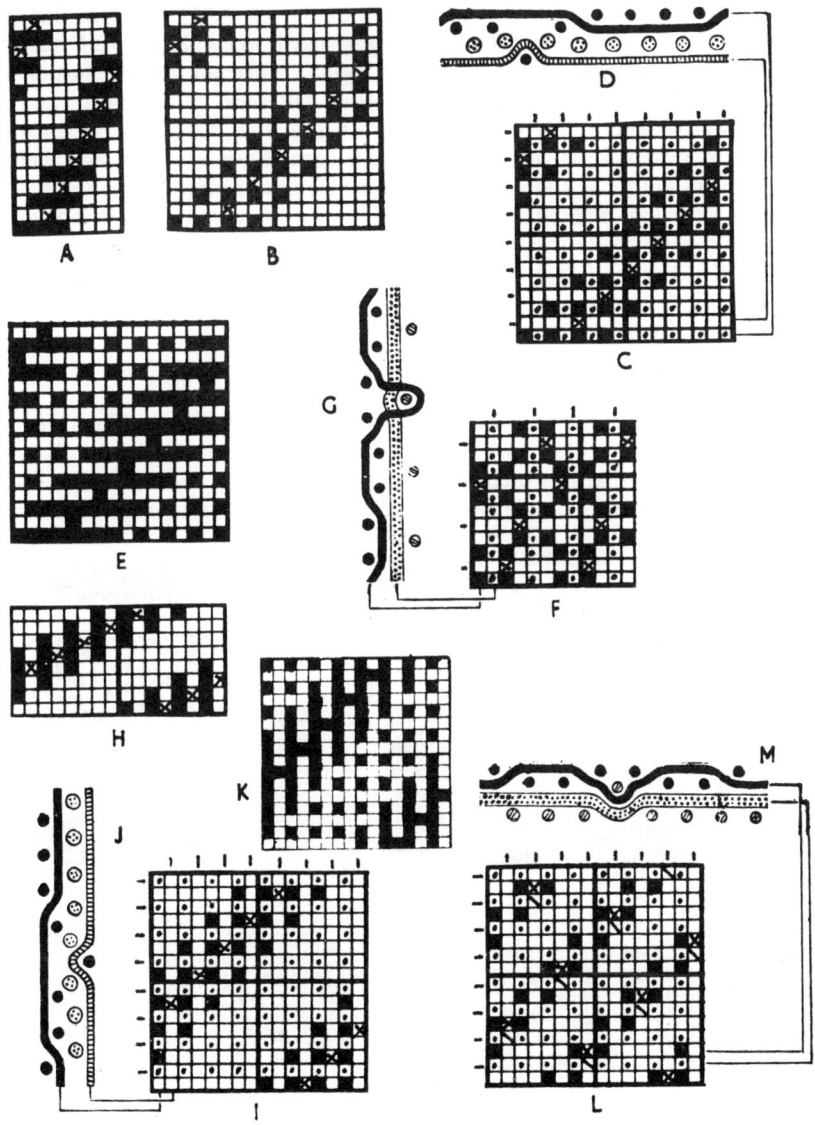

Figure 3.15

stitching place. In the design L the wadding picks and backing ends interweave in 8-sateen order with each other and really form a loosely-woven wadding fabric in the centre of the cloth.

IMITATION BACKED CLOTHS

Nearly any ordinary weave can be so modified as to produce a structure which very closely resembles a weft or a warp-backed texture, but in which each thread interweaves regularly on both sides of the cloth. The system has the advantages that a heavy single cloth is produced which has a fine surface appearance, and is elastic and soft in the handle, while the threads sustain an equal amount of friction in the manufacture and wear of the cloth, and in imitation warp-backing only one warp beam is required. An inferior quality of yarn cannot, however, be introduced on the back, since each end and pick is interwoven on both sides, while colours cannot be so effectively applied to the surface as in proper backed cloths. The principle of construction is, however, very useful, particularly for piece-dyed fabrics. The designs may be made in imitation of either the 1 face, 1 back, or the 2 face, 1 back order of backing.

Imitation weft-backing

The method of constructing imitation weft-backed designs is illustrated in *Figure 3.16*, in which the marks indicate weft up. In modifying the 2-and-2 twill weave, given at A, to imitate a 1 face, 1 back order of wefting, the repeat of the imitation weave is made one thread less, or one thread more than twice

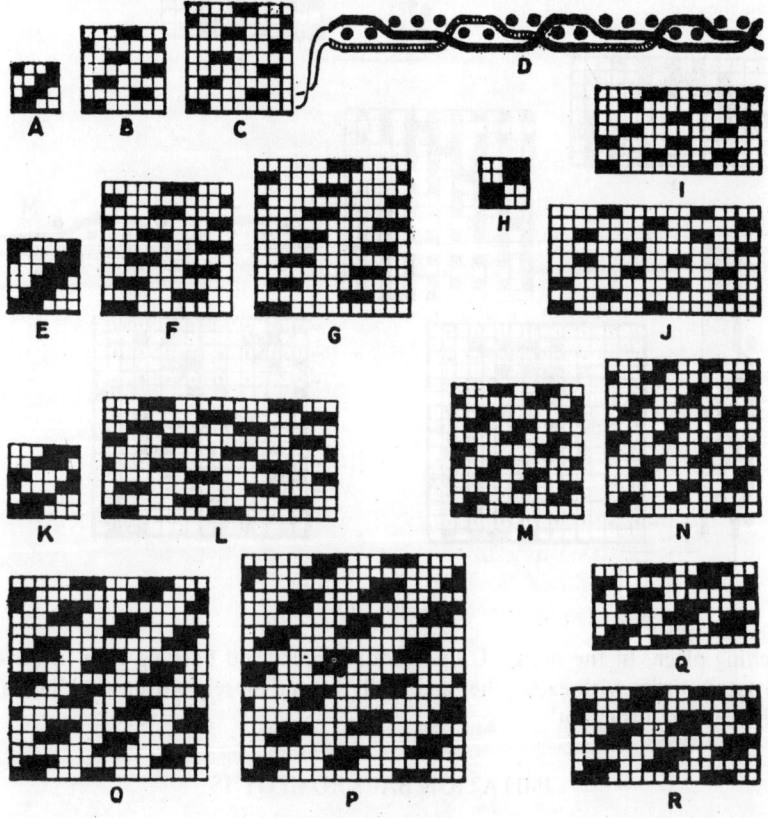

Figure 3.16

the number of threads in the repeat of the twill. Thus, in B and C, both of which are imitations of a weft-backed 2-and-2 twill, the respective repeats are on 7 and 9 ends and picks. In constructing the weaves a line of marks of the original twill is inserted on alternate horizontal spaces of B and C, but as the repeats contain an odd number of picks the twill marks fall first on the odd and then on the even horizontal spaces. Diagram D represents the interlacing of picks 1 and 2 of C, and shows that the odd and even picks form separate twill lines behind which the even and odd picks respectively float. The design should be woven with about twice as many picks as ends per unit space in order that the picks will be beaten up very close together, and so cause the twill lines to appear as solid as in an ordinary single cloth. The long weft floats on the underside give the appearance of a loose or sateen back weave, and complete the resemblance to a weft-backed structure.

E, F, and G in *Figure 3.16* illustrate the modification of a 3-and-3 twill on the imitation 1 face, 1 back principle; the larger weave, of course allowing of finer setting than the smaller. In the same manner, H, I, and J in *Figure 3.16* illustrate the modification of 2-and-2 hopsack weave, and K and L of a 3-and-3 twill derivative. In constructing imitation designs that are not twill weaves, a series of floats of the original weave is inserted on the odd horizontal spaces, then a second series is run in on the even spaces, at such a distance from the first series as will give the nearest resemblance to the weave when the picks are beaten close together. In some cases, as shown at J and L it is necessary to insert several lines of the floats in order to complete the design, each line being placed in the same relative position to its neighbours as the first two lines are to each other.

In re-arranging twill weaves in imitation of the 2 face, 1 back order of weft-backing, the repeat is made one thread less or one more than three times the number of threads in the repeat of the twill. For instance, a 2-and-2 twill imitation weave may be made on 11 or 13 threads as shown at M and N respectively in *Figure 3.16;* and a 3-and-3 twill imitation weave on 17 or 19 threads, as represented at O and P respectively. A design in imitation of a twill that repeats on an odd number of threads is complete on twice as many threads in one direction as the other, as shown at Q and R in *Figure 3.16*. These are weft-backed imitations of a 3-and-2 twill and repeat respectively on 14 ends by 7 picks, and 16 ends by 8 picks. Assuming that the 3-and-3 twill modifications given at G and P are woven in 50/2 tex worsted warp and weft—25 ends and 50 picks per cm will be suitable for the former, and 25 ends and 38 picks per cm for the latter, and in each case the face twill will run at 45° angle. Sometimes a firm back is produced by stitching the weft on the underside between weft-face floats; thus the first pick of weaves G and P may be stitched on the ninth end, and the other picks in corresponding positions to the twill lines of marks.

Imitation warp-backing

Imitation warp-backed designs are constructed on the same principle as imitation weft-backed effects, and if *Figure 3.16* is turned one-quarter round and the marks are taken to indicate warp up, the example will illustrate imitation warp-backed structures. In the designs given in *Figure 3.17*, the marks indicate warp, and in order that comparison may be made, the construction of an imitation warp-backed 2-and-2 twill is illustrated at A, B, C, and D, which correspond with the examples similarly lettered in *Figure 3.16*. In this case a line of marks of the original twill is inserted alternately on the odd and even vertical spaces of B and C, so that the odd and even ends form separate twill lines with long floats at the back, as represented in the diagram given at D.

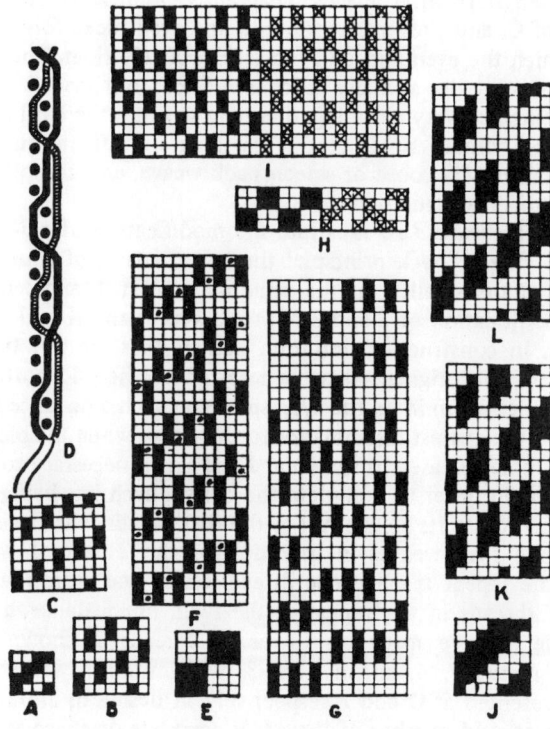

Figure 3.17

F and G in *Figure 3.17* illustrate the arrangement in two ways of the 3-and-3 hopsack weave E in imitation of 1 face, 1 back warp backing; while I shows a 1 face, 1 back imitation of the stripe weave given at H.

The designs K and L in *Figure 3.17* show 2 face, 1 back imitation warp-backed modifications of the 4-and-3 twill given at J, the former containing one pick less, and the latter one pick more than three times the number of picks in the twill.

The 1 face, 1 back imitation warp effects should have about twice as many ends as picks per unit space, while in the 2 face 1 back styles the proportion of ends to picks should be about 3 to 2. The warp on the underside may be stitched between face warp floats so as to produce a firm back, as shown by the dots in weave F.

4

Figured Pique Fabrics

Figured pique fabrics, also known as fancy toilet cloths, belong to the same group of structures as welts and piques (see *Watson's Textile Design and Colour*). As in latter fabrics, a structural unit in figured piques consists of three ends, arranged: 1 ground, 1 stitching, 1 ground. The ground ends, which are comparatively slack, weave plain with ground weft and the figure is formed by the taut stitching ends, brought from a separate heavily tensioned beam, which by lifting over ground weft in selected places form indentations in the slack, plain weave ground cloth. The areas of the ground fabric which are not stitched down form prominent ridges or 'blisters' between the indentations due to excess length of the ground ends brought about by the light tension applied to the ground warp beam. Simple geometric figuring may be produced with the aid of dobby shedding motions but in most instances the designs are large and elaborate requiring the employment of a jacquard shedding system. A specially developed harness mounting with healds and working comber-boards on which these fabrics were at one time extensively produced is shown in Appendix I. Modern figured pique cloths are generally less stiff and 'boardy' than the traditional constructions which were almost exclusively made for bed covers, dress shirt fronts, etc.

A simple style in the figured pique structure is represented in *Figure 4.1* for which the condensed plan is given at A in *Figure 4.2*. Each mark of A indicates the lift of a taut stitching end over two face picks, by which the slack face cloth is drawn down and indentations formed in the surface; the blank diamond spaces between the marks correspond with the raised portions of the cloth. In weaving the design A in a dobby machine four healds would be employed for the plain face ends and nine healds for the stitching ends, as shown in the draft given at B in *Figure 4.2*.

Classification of the structures

The cloths are classed as 2-pick, 3-pick, 4-pick, 5-pick, and 6-pick stitch according to the number of picks per stitch; and they are also described as 'loose-back' and 'fast-back' according to whether the stitching ends are floated loosely or are interwoven on the underside. The bulk of loose-back toilets are made on the

Figure 4.1

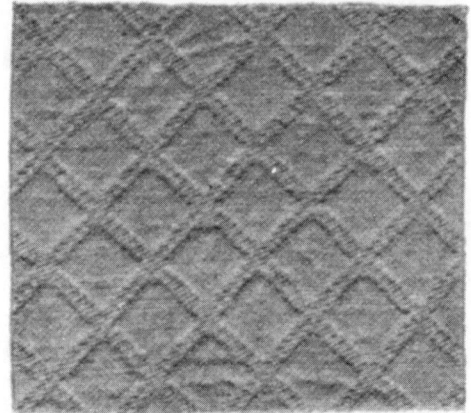

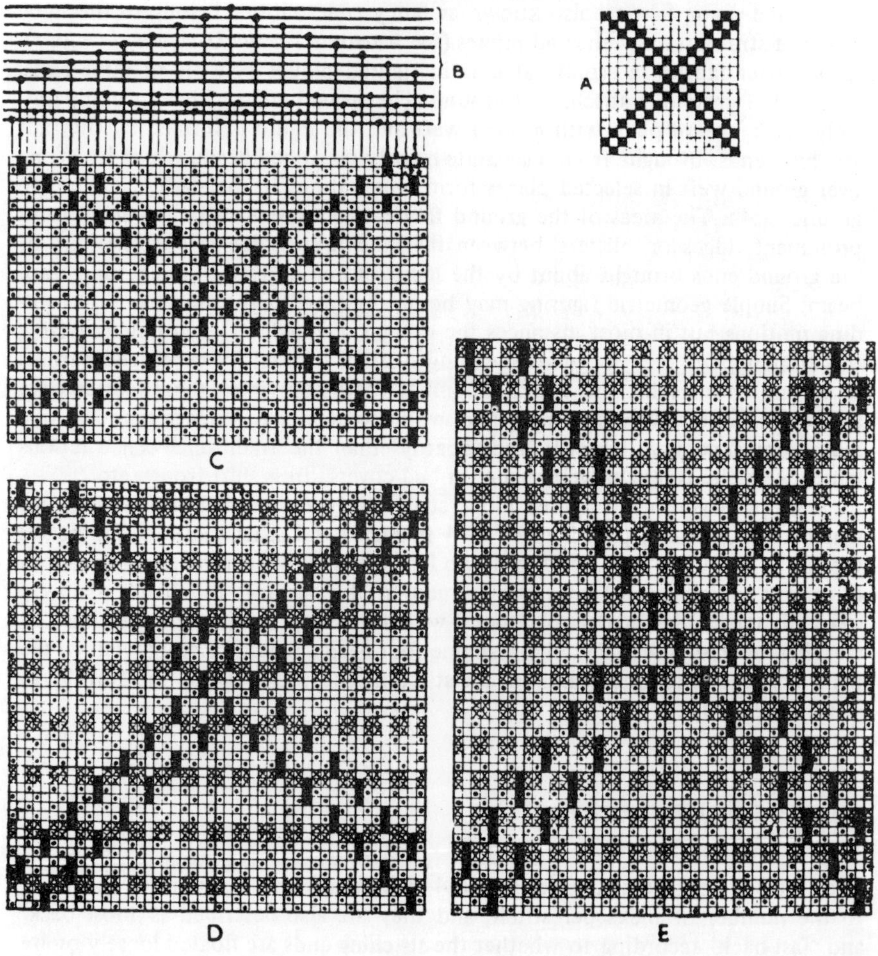

B

A

C

D

E

Figure 4.2

2-pick and 3-pick basis, whereas fast-back cloths are made four, five or six picks per stitch.

Loose-back piques

In these the stitching ends are not interwoven below the slack plain ground cloth, and in *Figure 4.2* full weaves to correspond with the plan A are given at C, D, and E, which are arranged two, three, and four picks per stitch respectively. In C no wadding picks are inserted, and the stitching ends are up for two picks, forming the lifts shown in solid marks, while the ground ends are raised in alternate order so as to produce the plain weave, as represented by the dots.

In most of these cloths, however, wadding weft, which usually is soft spun and thicker than the ground weft, is introduced and it lies between the tight stitching ends and the unstitched portions of the ground fabric, thus giving greater prominence to the latter while making the cloth more substantial. In the 3-pick cloths the picks are usually arranged in the order of 4 ground to 2 wadding, as shown at D in *Figure 4.2,* in order that a loom with changing boxes at one end only may be used. In this case, the stitching ends are up for three picks at a time, whilst the ground ends weave plain with the ground picks as indicated by the dots, and are lifted over all the wadding picks, as indicated by the crosses in D and E. In a 4-pick structure the picks are arranged 2 ground and 2 wadding, as shown at E in *Figure 4.2,* and each stitching end is up for four picks at a time. The system of designing for elaborately figured loose-back piques, which is exactly the same as for fast-back cloths, is illustrated in *Figure 4.4.*

Half fast-back piques

This class of structure (also termed 'stocking-back') comes between the loose-back and fast-back varieties. The stitching ends are interwoven in the plain order

Figure 4.3

with the wadding picks at rather infrequent intervals in order to avoid the formation of loose warp floats below the figure. In one order of lifting two plain sheds are formed by the stitching ends in every six groups of picks, and in another order in every four groups of picks. Thus E in *Figure 4.2* would be made half fast-back in the first order of lifting by raising the even and then the odd stitching ends on picks 12 and 24 in every 24 picks, and in the second order of lifting, on picks 12 and 16 respectively in every 16 picks. In applying the first order of lifting to D in *Figure 4.2* the even and odd stitching ends would be respectively raised on picks 9 and 16 in every 18 picks.

Fast-back piques

Fast-back pique fabrics are usually woven with four, five, or six picks to each stitch, the latter two constructions, however, due to high production cost, are now encountered rather infrequently. A portion of a figured pique fabric is shown in *Figure 4.3,* and a condensed design to correspond in *Figure 4.4.* In this structure the tight stitching ends are interwoven on the underside in plain order with alternate wadding picks, so that the two plain fabrics are formed—one above the other—in every part of the cloth. In the raised figure areas the two fabrics are quite separate from each other, and the inoperative wadding picks lie between them, but in the ground they are firmly united by the lifts of the stitching ends over the ground picks.

Method of designing

In preparing a design the raised figure may be indicated by a wash of colour, in the manner represented by the shaded squares in *Figure 4.4,* and plain weave marks are inserted round the figure in order to separate it from the ground, the solid marks indicating the lifts of the stitching ends, The order in which the stitching ends are required to be interwoven with the ground weft is then indicated in the ground of the design, and various small weaves, such as those shown at G, H, and I in *Figure 4.4,* may be employed. In the example shown in *Figure 4.3* the ground texture is as firm as it is possible to make it, as the stitching ends are raised in alternate order. In condensed designing represented at F in *Figure 4.4* the alternate lifts of the stitching ends appear as plain weave markings and, therefore, only need to be indicated at the left of the design, as shown in *Figure 4.4,* on the understanding that the remainder of the ground area is exactly the same. The degree of condensation achieved in *Figure 4.4* is by 3 warp-wise, each vertical row representing one stitching end and two adjacent ground ends, and by 4, 5 or 6 weft-wise, each horizontal row representing four, five or six picks depending on the class of structure which it is intended to produce. In the cloth, each 2 × 2 area of figure and ground indicated at J in *Figure 4.5* is expanded as shown at K and L respectively, assuming that a 4-pick structure is produced.

It will be appreciated that the condensed design shown at F in *Figure 4.4* has been constructed so that the solid marks show exactly the places at which the stitching ends are floated on the surface of the ground cloth thus binding it in. The tight, alternate stitching shown at the edges of the shaded figure areas is necessary to achieve prominent puckering of the figure; whether it is also continued in the remaining ground areas is a matter of choice. In the example given it has been done, thus achieving the maximum prominence of the raised figure,

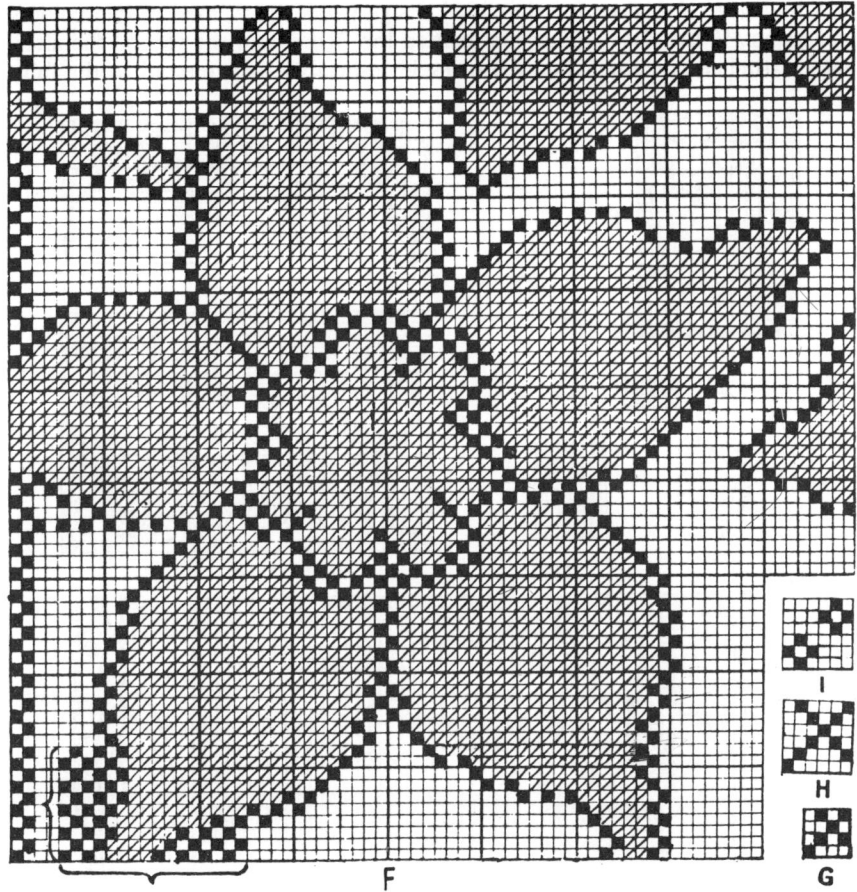

Figure 4.4

but if desired, any one of the motifs G, H, I in *Figure 4.4* could have been used instead to provide small, auxiliary corrugations or 'blisters' in the ground area. It will be noted that all the above motifs cut correctly with the alternate lifts of the stitching ends at the edge of the figure which is necessary to prevent the occurrence of disorderly long and short floats of the stitching ends on the surface.

In simplified designing for modern card-cutting systems the somewhat laborious build up of the structure indicated in *Figure 4.4* for the purpose of explaining the constructional features of these fabrics is unnecessary and where the ground area is the same only one colour of paint would be used to denote the figure as previously and the paper, or unpainted area, would then represent the ground. If the degree of condensation was the same as before then for a 4-pick structure each 2 × 2 area of figure (i.e. paint) would be worked out as K, and of ground (i.e. paper) as L, in *Figure 4.5*. If the ground, apart from the tightly stitched surround to the figure, consisted of a different order of stitching such as G, H or I in *Figure 4.4*, then the simplified design would be painted in two colours and paper. If the first colour represented the figure, and the second the alternately stitched surround then these would respectively correspond to the fully

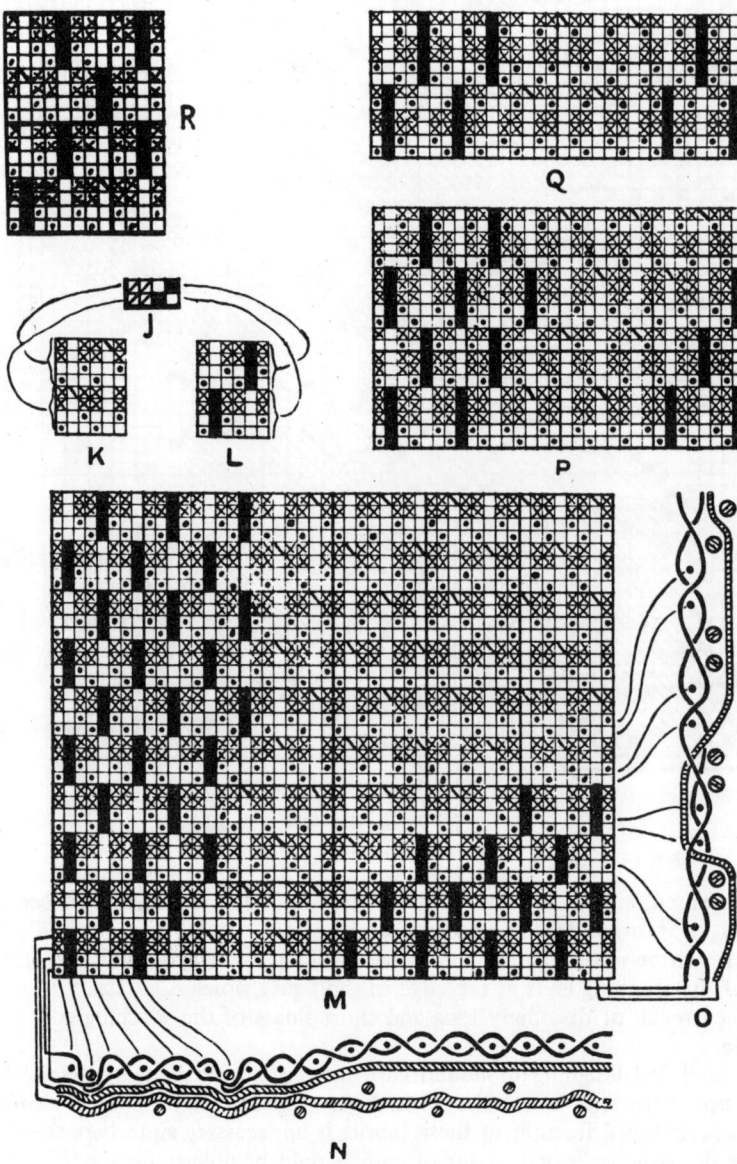

Figure 4.5

worked-out plans at K and L in *Figure 4.5* as before, but the paper would represent the new order of stitching in the ground and would have to be worked out in detail. This is shown at R in *Figure 4.5* which shows a fully worked-out area equivalent to a 4 X 4 area of paper in the simplified design assuming that the motif G has been used for the ground.

Four-pick structure

As stated before, the plans K and L in *Figure 4.5*, each of which repeats upon six ends and eight picks. respectively show the weaves of the puckered and bound portions of a cloth which is woven with four picks per stitch. The warp is arranged the same as in all other pique structures, and the weft is inserted in the order of 2 fine ground picks and 2 thick picks. The raised area K is formed as follows: The ground ends are raised in alternate order on the two fine face picks, so as to form the plain ground or face cloth (represented by the dots), and all together on the thick picks (shown by the crosses). The stitching ends are raised in alternate order on the fourth and eighth picks (indicated by the diagonal strokes), the plain back weave thus being formed by the second of each pair of thick picks, known as the backing pick, while the first of each pair forms the wadding. In the bound weave L *Figure 4.5*, the order of lifting of the ground ends is the same as in K, but the stitching ends are raised in alternate order over a group of four picks at a place, as shown by the solid marks.

The complete weave, in the 4-pick structure, is given at M in *Figure 4.5* of the vertical rows 5 to 20, and the horizontal rows 1 to 10 which are enclosed by brackets at F in *Figure 4.4*. From a comparison of M in *Figure 4.5* with E in *Figure 4.2* it will be seen that the fast-back and loose-back structures are alike, except that in the embossed or puckered portions of the fast-back cloths the tight stitching ends are interwoven in plain order with the even thick picks.

The interweaving of the picks, 1, 2, 3, and 4 with the ends 1 to 24 of the design M, is represented at N in *Figure 4.5*, the bound structure being shown on the left, and the embossed effect on the right. O shows the interweaving of the three last ends of M with the picks 9 to 16, the embossed and bound structures being represented in the upper and lower portions respectively. The plain ground threads are shown in solid black, and the threads are connected by lines in order that the sectional drawings may be readily compared with the plan M.

The following are the weaving particulars of a typical bedcover cloth with four picks per stitch: Ground warp, 15/2 tex cotton; stitching warp, 24/2 tex cotton; ground weft, 20 tex cotton; wadding and backing weft 60 tex cotton; 43 ends and 57 picks per cm. The stitching ends contract from 5 to 8 per cent, and the ground ends from 15 to 20 per cent, while the shrinkage in width is about 12 per cent. Since each vertical space of the design paper is equivalent to three ends, and each horizontal space to four picks, the count of the design paper is in the ratio of (108 ÷ 3) to (144 ÷ 4)—or 8 X 8. Other cloths are frequently made with single yarns, and the ground threads range from 15 to 28 tex, the stitching warp from 25 to 34 tex, and the wadding weft from 30 to 72 tex; the ends and picks per cm in the plain ground cloth vary from 20 to 32.

Five-pick and six-pick structures

The structure of a 5-pick cloth is illustrated at P in *Figure 4.5*, which corresponds with the picks 1 to 16 and the ends 1 to 32 of M. In this structure in each series of 10 picks the fifth and tenth form the backing picks, and these, in looms with bo·es at one side only, consist of the same kind of weft as the face picks, so that the complete order of wefting is 2 picks fine, 2 picks thick wadding, 2 picks fine, 2 picks thick wadding, and 2 picks fine. Each stitching lift operates for five picks; the ground ends are operated in alternate order on picks 1, 2, 6, and 9, and then are raised together on picks 3, 4, 5, 7, 8, and 10; for the purpose of interlacing on the back, the stitching ends are raised alternately on picks 5 and 10. To correspond with the foregoing particulars of a 4-pick cloth the structure represented at P would be woven with about 75 picks per cm, of which 30 fine picks would form the face, 15 fine picks the back, and 30 thick picks the wadding.

Six-pick cloths are constructed as shown at Q in *Figure 4.5* which correspond with the picks 1 to 8, and the ends 1 to 32 of M. In this case also the face and backing picks are alike, and each group of six picks is arranged 2 picks fine, 2 picks thick wadding, and 2 picks fine, the last pick being a backing pick. There are three face (or ground) picks per stitch compared with two face picks in the 5-pick cloth.

The system of designing, illustrated in *Figure 4.4* is applicable for both the structures P and Q if each horizontal row is taken to represent respectively five and six picks.

5

Stitched Figuring Weft Constructions

In extra weft fabrics (see Chapter 2) the figuring weft is usually displayed in full float, i.e. it is not broken on the surface by any regular binding lifts of the warp. As a result its colour stands out clearly and the figure formed by it is sharply defined, but, in certain uses the existence of an unbound float on the surface may result in faults due to plucking, cutting, etc. This is liable to occur particularly when the ground cloth is very open or when the fabric demands frequent and severe laundering or other form of cleaning treatment. The damage is also more prone to take place when the figuring weft is considerably thicker than the ground materials so that in such a case it will tend to stand proud from the level of the base cloth. In such circumstances the figuring weft requires to be stitched at frequent intervals and several standard methods have been developed to bind the figuring material firmly yet unobtrusively. Two such structures dealt with in this chapter are the figured book muslin and the patent satin. In both of them the effect is achieved by displaying a comparatively coarse figuring weft on the surface of a much finer plain ground fabric. The figuring float is bound-in by the use of fine stitching or ground warp which, being of the same colour as the figuring weft, does not disturb the solidity of the figure. In the book muslin structure the figuring weft is clipped outside the bound-in portions, whilst in the patent satin it is continuous, interchanging between the face and the back, and thus forming a reversible cloth. The two structures, although quite different in appearance and in the manner in which the extra weft is bound-in, are constructed according to somewhat similar principles. At one time they were produced on special harness mountings described in Appendix I and, indeed, the term 'book muslin' derives from the harness of that name.

FIGURED BOOK MUSLIN FABRICS

In the book muslin structure a light plain foundation texture is ornamented with extra weft, in the manner represented in *Figure 5.1*, which shows the appearance of a typical fabric as viewed from opposite sides. The extra weft is thick and soft spun, and in the figured portions of a cloth it is, as a rule, inserted in 2-and-2 order with the plain ground picks. As the cloth is woven, the figuring picks are floated loosely on the surface between the parts in which they are incorporated in the ornament, but in the finishing process the loose floats are cut away so that an opaque figure appears upon a semi-transparent ground. The cloths are practically reversible, but, as shown in *Figure 5.1*, the uncut side—represented in the lower portion of the figure—is neater in appearance than the cut side, which is shown in the upper portion, as on the latter side the severed ends of the figuring picks impart a rough edge to the figure. The textures are used as window curtains, and in small designs for skirtings, blouse, and dress fabrics.

Structure of the cloth

In *Figure 5.2*, A shows a simplified design which is represented in full at B, while C illustrates the interlacing of the threads to correspond with B as viewed from the cut side (the face side as the cloth is woven). The design A is condensed by 2 warp-wise, each vertical row representing two ends, and by 4 weft-wise, each horizontal row representing two ground picks and two figuring picks. On the ground picks plain weave is produced as indicated by the diagonal marks and the dots in design B. On the extra picks the odd ends are left down, but the even ends are raised where figure is required to be formed—i.e., the marks in the design B in *Figure 5.2* indicate warp up.

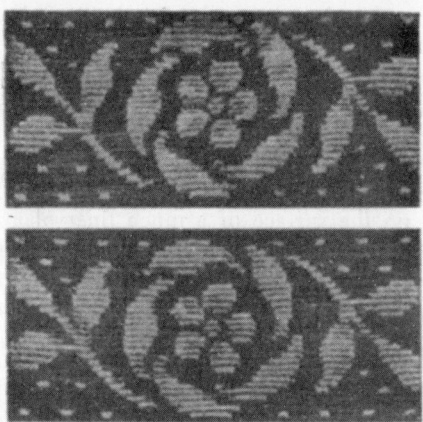

Figure 5.1

In the most usual structure of the figure the extra picks float over only one end at a place on both sides of the cloth, as shown in the spot on the left of B and C in *Figure 5.2*. On the uncut side of a cloth a longer float than over one end is not produced, because all the odd ends are depressed on the figuring picks. On the cut side the even ends may be operated in any desired order, but if long figuring floats are made they are liable to be cut away in the shearing process along with the floats which extend between the figures. It is therefore, customary in a figure to leave down not more than one even end at a place, which gives a float of three ends in the cloth on the cut side, as shown in the central spot in B and C, *Figure 5.2*. The weave development of a figure is thus limited to floats of one and three in the cloth, but a further variation of the structure is made by interweaving both figuring picks of a pair in one portion and only one pick in another portion, as shown in the spot on the right of B and C in *Figure 5.2*. An effect is obtained which, being between the semi-transparent ground and the opaque figure, is useful in shading a design.

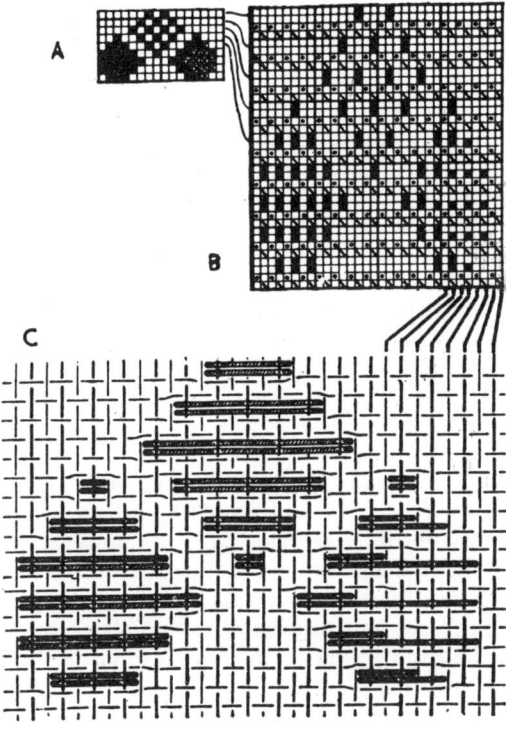

Figure 5.2

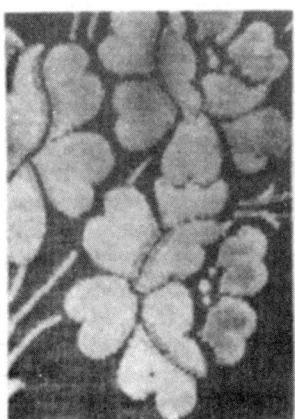

Figure 5.3

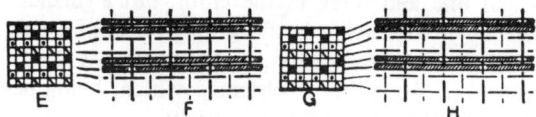

Figure 5.4

The three methods of interweaving the figuring picks, illustrated at B and C in *Figure 5.2* are extensively used, but as both picks of a pair are inserted into the same shed, adjacent pairs are distinctly separated ,from each other by the plain ground picks. The running of the figuring picks in pairs is clearly shown in *Figure 5.1*, and generally, this is considered a feature of the structure. Sometimes, however, this formation is avoided, and in *Figure 5.3* a fabric is represented in which a fuller and more solid figure is obtained by floating the picks of each pair alternately in 3 up, 1 down order. The 3-and-1 floats may be arranged as shown at E and F in *Figure 5.4*, in which the last figuring pick of one pair is in the same shed as the first pick of the next pair, or as indicated at G and H, in which they run continuously in alternate order.

Method of designing

The plan given at K in *Figure 5.5*, which corresponds with a portion of the design represented in *Figure 5.3*, will serve to illustrate several features in the preparation of designs for these fabrics. It is a rule to separate two portions of figure—between which the light ground texture is required to show distinctly—by at least two vertical spaces of the design paper. Otherwise the weft floats between the parts will not be long enough to be engaged by the shears, and by being retained in the cloth will make the two portions of figure appear to join up. Frequently, in the finer set cloths three consecutive vertical spaces are left blank to ensure that the light ground texture will show clearly between the separate parts of a figure.

The design K is shown painted in four colours and paper indicating the existence of five different structural areas. The blanks (paper) indicate the ground area in which no extra weft material is incorporated, whilst the solid marks, the alternate diagonal marks and dots, the diagonal marks alone and the dots alone represent respectively four different orders in which the extra picks are bound-in. The detailed weaves for each of the five structural areas are shown at K1 and K5 in *Figure 5.5*. It will be noted that the bulk of the figure is produced in the weave K3 which results in the formation of alternate 3 up, 1 down figuring weft floats exactly as shown at F in *Figure 5.4*. Floats of one thread only are made in the remaining portions of the design—by both figuring picks where the solid marks are indicated, K2; by the odd figuring picks where the diagonal marks only are shown, K4; and by the even figuring picks where only the dots are inserted, K5. The two last orders of marking are for shading the figure and it will be seen that where two shaded effects are made close together, one is formed by the odd picks (the diagonal marks) and the other by the even picks (the dots). The object of this is to get as great a length of float as possible between the separate parts of the figure, so that the floats will be

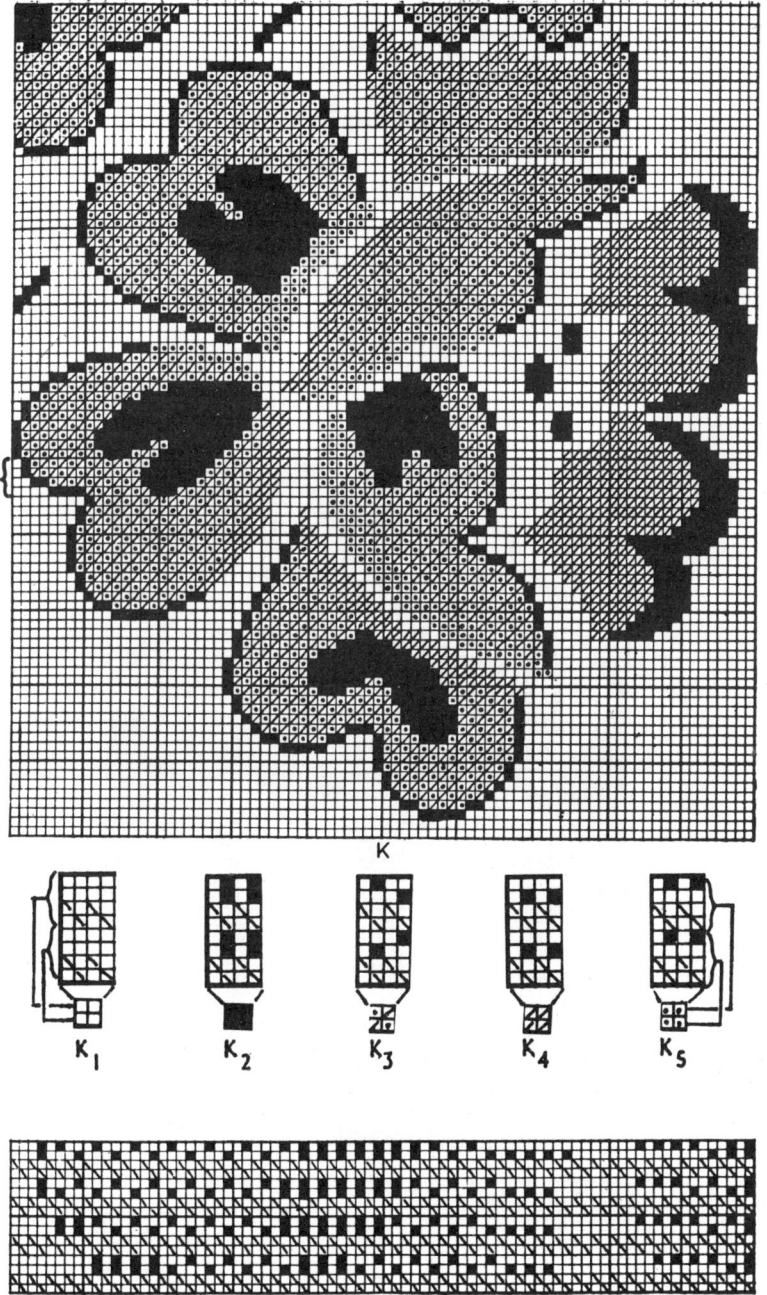

K

K₁ K₂ K₃ K₄ K₅

L

Figure 5.5

77

Figure 5.6

Figure 5.7

effectively cut away during shearing; while the arrangement also helps to equalise the lifts of the harness. In order that comparisons may be made, the complete weave of the horizontal rows 37 to 40 (indicated by the bracket at the side of K) and the vertical rows 1 to 40 is given at L in *Figure 5.5*, assuming that the weaves K1 to K5 are employed.

Ground weave variation

In the cloth shown in *Figure 5.6* a useful variation of the plain ground texture is illustrated in the form of a 5-and-1 imitation leno weave, which is used in these fabrics to a considerable extent. The corresponding complete design, condensed as before is given at M in *Figure 5.7*; the solid squares result in the same structure as K2 in *Figure 5.5*, and the diagonal marks represent additional lifts of the even ends on the even ground picks. The complete weave of the horizontal rows 12, 13, and 14 (indicated by a bracket), and the vertical rows 1 to 25, is given at N, while the plan O, which corresponds with the last six ends of N, with the figuring picks omitted, shows the imitation leno weave that is formed in the ground.

Other modifications of the ground weave may include the introduction of extra cord ends which may be interwoven with the ground structure continuously or only intermittently. However, as the introduction of extra ends requires modification of the harness, discussed in connection with extra warp figuring, it is only rarely attempted. More frequently cord effects are achieved by substituting thick ends for the fine ground ends at intervals which, when combined with a suitable extra weft figure, results in pleasing stripe effects without any need to disturb the existing harness arrangements.

Weaving particulars of book muslins

The number of ends per cm with which the cloths are woven traditionally ranges from 18 to 24 and the number of ground picks from 16 to 22; the warp yarns range from 8 to 10 tex cotton, and the ground weft yarns from 6 to 8 tex cotton; while from 30 to 38 tex soft spun figuring weft is used. The average number of figuring picks per cm varies according to the order in which they are inserted, and the cloths are classed as 'full cover', in which the extra picks are inserted continuously with the ground picks, or one-half, two-thirds, two-fifths cover, etc., in which the extra picks are inserted intermittently. The proportion of the cover can be obtained by finding the number of horizontal spaces of the design upon which the extra figuring picks are indicated in relation to the total number of horizontal spaces in the repeat. The design paper should be ruled in the same proportion as the ends per unit space are to the ground picks. Thus, for a cloth that counts 24 ends × 21 ground picks per cm 8 × 7 paper is suitable. In most cases the yarns are white, but occasionally a coloured figure is made upon a white ground and sometimes, by chintzing, a figure is woven in white and a colour, or in two colours.

More recently the book muslin structure has been adapted for such uses as hangings and fancy shirtings, and as a result the transparent ground has given way to a much more solid structure and the cloth looks different although it uses exactly the same principle of binding the extra weft as the more open structures. Shirtings have been woven in cotton or cotton/polyester blended yarns with 24 ends and 26 picks per cm using 28 tex warp and 34 tex ground weft yarns.

PATENT SATIN STRUCTURE

The 'patent satin' structure has developed from the Mitcheline quilt which was similar, but wefted pick-and-pick, as opposed to the two ground, two figuring wefting of the former. As the structure was employed mainly for bed covers it required large design repeats and this led to the development of a special harness mounting including healds and working comber-boards. This is described in Appendix I. The structure, being originally dependent on regular lifts of the special shed-forming elements, is very rigid and consists essentially of a stout plain weave ground fabric upon which a coarse figuring weft is floated to form the figure. The extra weft is continuous and when not floating on the surface is

Figure 5.8

displayed on the back thus producing a perfectly reversible construction. The coarse figuring weft is stitched at frequent intervals on the face and on the back by a fine stitching warp. The term 'satin' is a misnomer as no satin weaves are employed and it probably derives from the somewhat flat appearance of the figure.

The structural unit in this type of fabric consists of three ends, arranged 1 ground, 1 stitching, 1 ground, and four picks, inserted 2 ground, 2 figuring and this corresponds to the usual degree of condensation of 3 warp-wise and 4 weft-wise normally employed in design painting. A small portion of a design, condensed as indicated above, in which the solid marks represent the figuring weft on the face is shown in *Figure 5.8*, whilst the appearance of a patent satin cloth is illustrated in *Figure 5.9*. The quality of these fabrics varies considerably and the following are the weaving particulars of a medium quality cloth: Ground warp 34 tex cotton, stitching warp 18 tex cotton, 16 ground and 8 stitching ends per cm; ground weft 15 tex cotton, figuring weft 74 tex soft spun cotton, 18 picks of each weft per cm. The ground warp contracts about 2 per cent and the stitching warp, which is placed on a separate beam, from 20 to 25 per cent, while the contraction in width varies from 10 to 15 per cent.

Figure 5.9

The cloths are mostly woven grey and then bleached, but sometimes the ends which form the ground are all coloured or are arranged in stripes of white and colour; a white figure then being formed upon a coloured or a striped foundation.

Method of designing and structure of the cloth

Taking the order of wefting as 2 picks fine and 2 picks coarse the order of shedding is as follows: The ground ends lift in alternate order on the two fine picks, and form the plain weave represented by the dots in the plans A and B in *Figure 5.10*. The stitching ends lift in 2 up, 2 down order alternately and produce the weave shown by the crosses in A and B. The figuring lifts of ground ends occur on the coarse picks and the ground threads in pairs (one on each side of a stitching thread) form the design that is required by lifting in one portion

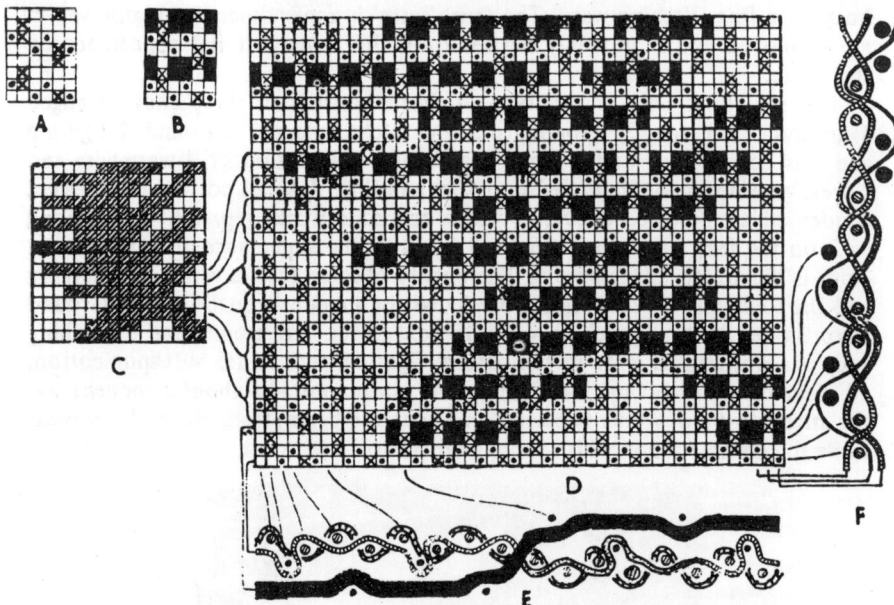

Figure 5.10

and remaining down in another portion of the cloth. The figure is formed by the floats of the thick weft, and in weaving the cloth right side up the marks in the design given in *Figure 5.8* indicate ground ends down, or weft float on the face. Taking the marks in A and B in *Figure 5.10* also to indicate weft, the former shows the full ground weave and the latter the full figure weave in the cloth. The stitching ends bind the figuring weft to the plain ground cloth very firmly by working alternately 2 up and 2 down and changing places between the two fine picks and, again, between the two coarse picks.

Variations are not produced in a design by altering the structure of the cloth, but a subsidiary effect can be achieved by shading either as shown on the left of the small plan given at C in *Figure 5.10,* or, as indicated on the right of E. The complete weave, to correspond with the horizontal rows 1 to 10 of C is given at D in *Figure 5.10.* Taking the marks to indicate weft, the warp section E shows how the picks 1, 2, and 4 of D interlace with the ends 1 to 24, while F represents the interlacing of the last three ends of D with the picks 1 to 16. In E and F the figuring picks and the stitching ends are shown in solid black, in order that they may be clearly distinguished, and connecting lines are indicated to enable the threads to be easily traced.

The count of design paper for a cloth with 24 ends and 32 picks per cm is in the proportion of $(24 \div 3)$ to $(32 \div 4) = 8 \times 8$. On account of each horizontal space of the design representing two thick picks, the figure in coarse cloths has a steppy outline, and in order to avoid this, sometimes the figuring lifts of the ground ends are different for each one of the two coarse picks in which case the degree of condensation of the design is reduced to 2 weft-wise.

6

Damasks and Compound Brocades

DAMASKS

Damasks were originally produced as all silk fabrics. Later, cotton and linen threads were used and in the bleached state the cloths were chiefly employed for table napery. Today cotton and linen are still used to produce the traditional types of damasks but other materials, such as viscose rayon in filament and staple form, polyester and others, find their place in the production of curtainings, fine upholstery cloths and similar fabrics. In a true damask figured fabric, a weft sateen figure is formed upon a warp satin ground, or *vice versa,* and the structure is described as reversible. The term damask, however, is also applied to cloths in which the figured portions are developed in diverse ways upon a sateen or satin ground, the texture being then known as a one-sided damask. *Figure 6.1* illustrates a sateen figure with warp rib and weft float

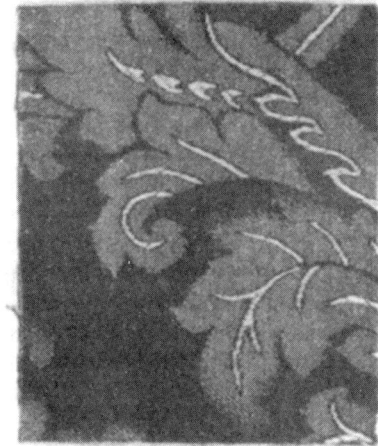

Figure 6.1

typical in a one-sided damask upon satin ground. The weaves which are most frequently used are the 5-thread and the 8-thread sateens and satins but other opposite pairs of sateens and satins and warp-faced and weft-faced twills are also occasionally employed.

As the design repeats in finely set damask table cloths and other fabrics are very large, several jacquard systems were developed to increase the figuring scope of the machines and to simplify the design painting and card cutting. The main types of the special systems were the pressure harness, the bannister harness and the self-twilling jacquards which were also sometimes referred to as the scale jacquards. The self-twilling machines are still occasionally used and, therefore, the principle of their operation is described later; however, nowadays most damasks are produced on ordinary, high-capacity, fine pitch jacquards and the use of modern card-cutting machinery permits the design to be painted in simplified or condensed form without the need to introduce any weave binding marks. For such jacquards, apart from the limited number of weaves employed, the designing does not differ from that described with reference to figured fabrics in general in *Watson's Textile Design and Colour.*

As stated above, one of the objects of the special mountings was to increase the figuring capacity of the jacquard and thus to obviate the need for mounting several machines in tandem above the loom or, at least, to reduce the number of jacquard machines per loom. This was achieved by making one needle control (in a variety of ways) two or more consecutive ends and one card to act for two or more successive picks. Thus, a 600s jacquard in which one needle controlled three ends and one card of a set of 400 acted for three picks would produce a design repeating upon 1800 ends and 1200 picks. In addition, a design would need to be painted over an area of 600 X 400 only, thus achieving a considerable saving in the cost of design preparation and, as in the special systems the binding weaves were introduced automatically by the mechanism itself, the most laborious part of design preparation, the insertion of binding weaves, was also saved. In a modern fine pitch machine the above figuring capacity is easily attained and as, instead of paste-board cards, the machine uses a continuous paper roll, the weight and the cost of a long pattern roll is considerably reduced and the operation of card lacing is completely eliminated. This, combined with the higher production rate of fine pitch jacquards and the use of modern card-cutting machinery has resulted in a marked decline in the use of the special mountings some of which, such as the pressure harness, were only suited to handloom weaving.

The self-twilling jacquard

This type of jacquard is still used occasionally in the coarse pitch version and is offered by some jacquard makers also in the fine pitch execution for extra large figuring capacity sometimes necessary for the very wide fabrics favoured presently. Several systems of operation exist but the basic principles in most of them are similar and will be appreciated by reference to the schematic diagrams in *Figure 6.2.*

In *Figure 6.2* the diagram X1 shows the connection in an 8-row machine of three figuring hooks B to each needle A, the capacity of the machine in this case being trebled. Each figuring hook B is made at the lower end in the form of a loop D which is rather longer than the depth of the shed, and when in their lowest position the hooks rest upon bars E, one of which extends right through each long row of hooks. The bars E offer no obstruction to the lifting of the figuring hooks by the jacquard. At each side of the long row of figuring hooks, a special row of strong twilling hooks H—shown in diagram X2—is provided, each

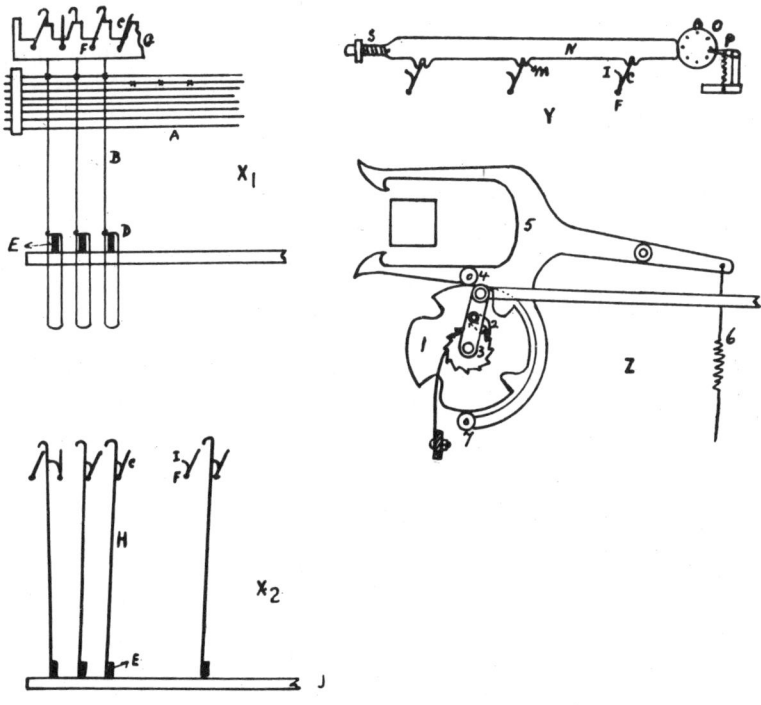

Figure 6.2

of which forms a loop round a bar E so that the bars are supported at both ends. A plate J at each side serves as a rest for the twilling hooks H when the latter are in their lowest position, and the plates are pierced with guide holes through which the straight lower ends of the hooks pass. The heads of the figuring hooks B are turned towards the card cylinder and are over the knives C in the ordinary manner, but those of the twilling hooks H are in the normal position held clear of the knives.

When in their ordinary position the swivelling knives C are inclined towards the figuring hooks B in the usual way. Each knife C, however is capable of being swivelled upon its fulcrum F in such a manner that its upper edge is clear of the corresponding figuring hooks, as shown by the knife 2 in diagrams X1 and X2. At the same time the lip I of the knives which is in line with the twilling hooks, pushes the latter over the preceding knife, as shown by the hook 1 in diagram X2. (It will be noted that the arrangement necessitates the use of one blade more than there are hooks in a short row.)

The rocking movement of the knife C is obtained from a small revolving cylinder O in which projections or studs are fixed, each of the latter acting upon the end of a control bar N, as shown in the diagram Y in *Figure 6.2*. The upper edge of each knife C fits within a recess M formed on the underside of a control bar N. As many bars are provided as there are threads in a repeat of the binding weave, and each, by means of the recesses M, control every fifth or every eighth knife, according to whether the binding weave repeats upon five or eight threads. The pressure of a stud upon O moves a control bar N (each of which is acted upon for one pick in every five or eight picks as the case may be) so that the knives to which the latter is connected assume the vertical position shown by knife 2 at X1 in *Figure 6.2*. This knife is thus put out of engagement with the corresponding figuring row of hooks, as shown in diagram X1, and into position for engaging the preceding row of twilling hooks, as shown in diagram X2. When the griffe rises on the following pick the figuring hooks in the long row 2 (and every fifth or eighth row) will thus be automatically left down. The twilling hooks 1, and every fifth or eighth twilling hooks, however, will be raised and lift up the corresponding bars E, and as each bar supports a long row of figuring hooks the rows 1 etc., will be automatically lifted. The arrangement causes one long row of figuring hooks to be left down, and one long row to be raised to each repeat of the binding weave, quite independently of the figuring cards; and each hook that is left down is next to a hook that is raised. That is, the jacquard lifts the ends in solid groups in forming the design, except that one in each repeat of the binding weave is left down through the action of the twilling motion, while of the ends that are left down by the jacquard the same propor-tion is raised. A spiral spring S is used to return each twilling bar to its normal position after the pressure of the stud has been removed.

It will be clear from the foregoing that the increase in the figuring capacity of the machine depends on the number of hooks per needle. This is usually two to four but sometimes the number of hooks controlled by one needle varies so that two consecutive needles control two hooks and the third, three, the fourth, two, and so on.

Apart from a desire to achieve a specific degree of multiplication this may be necessary to obtain a number of long rows of hooks which is a multiple of the binding weave repeat. For example, a machine with eight needles per row in which each needle controls three hooks will result in 24 long rows of hooks which is perfect for the 8-shaft binding weaves but not suited for the 5-shaft binding weaves for which the number of long rows of hooks should be 20, 25, etc., and thus to achieve the required number the needles in a row of eight may control in succession, 3, 2, 2, 3, 3, 2, 3 and 2 hooks.

A reduction from the full capacity of the machine can be effected by taking out of action in each row a number of hooks equal to, or a multiple of, the number of threads in the repeat of the binding weave and the same sets of cards can be used to produce identical designs in different qualities. Smaller reduc-tions, equal to half the size of one repeat of the binding weave, can also be made if necessary by spreading the number of hooks taken out of action between two successive short rows of needles but such fractional changes are rarely required.

The saving of cards and card lacing in machines which do not use continuous paper roll is of some importance and a device which can present the same card several times in succession is illustrated at Z in *Figure 6.2*. The card cylinder can be rotated forwards when the upper catch engages the lantern as the cylinder slides out, or backwards when the lower catch similarly engages the lantern. This is determined by the cord connection 6 which ensures that in the former case bowl 4, and in the latter case bowl 7, acts against the cam face 1. The cylinder will not turn when either of the bowls act against the projecting segment of the cam 1 but will turn when the bowl falls into a recessed segment, because it is only in this position that either the upper or the lower hook (depending on whether the cards are turning forwards or backwards) of the turning catch 5 can

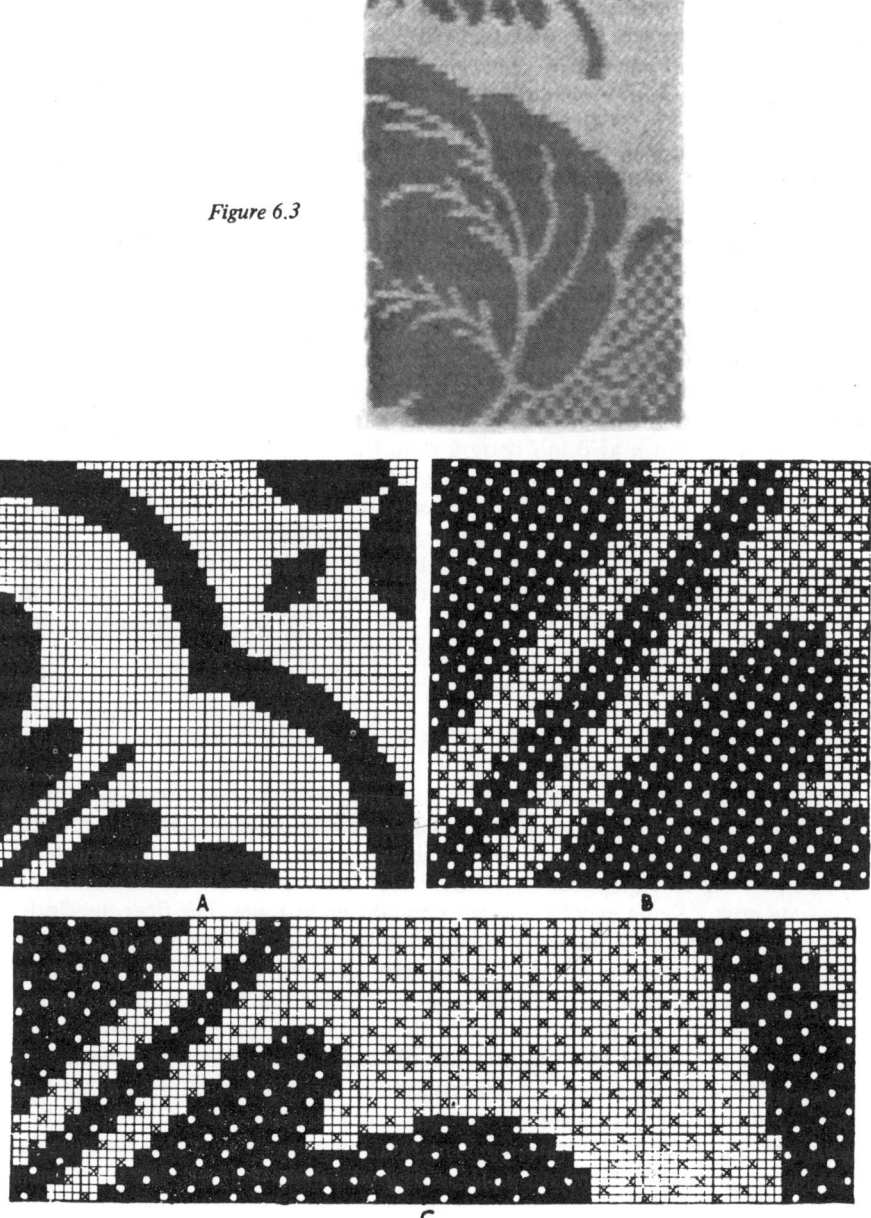

Figure 6.3

Figure 6.4

engage the cylinder. In the illustration the cam is turned one-twelfth of a revolution each pick by the pawl 2 engaging the ratchet 3 which has 12 teeth. As the projecting segment of the cam 1 acts upon bowl 4 (or 7) for two picks and the recessed segment for one pick the same card will be presented on three picks in succession before turning. To change the number of picks per card the cam must be changed and if each projecting segment, which renders the turning action inoperative, acts for only one pick, the same card will be presented for only two picks in succession, but when it operates for three picks before a recess occurs then four presentations will result before the card is turned.

The designing for this system is very simple and consists of painting the figure solid and leaving the ground blank. If the paint corresponds to the warp-faced portion of the design then it will be cut solid and the special mechanisms will introduce the binding weaves automatically as explained above, also enlarging the design vertically and horizontally according to the number of hooks per needle and picks per card. As each step of the original design is multiplied the figure results in a characteristically steppy outline as shown in *Figure 6.3*.

In the example given at A in *Figure 6.4* the design is painted solid in the usual way, and if the needles are connected to two and three hooks alternately, and the twilling motion produces 5-sateen and satin binding weaves while each card acts for three picks, the full weave of the first 20 vertical and 16 horizontal rows will be as shown at B in *Figure 6.4*. In the same manner, C shows the full weave of 16 horizontal rows of the design A, assuming that each needle is connected to two hooks, the twilling motion produces 8-sateen and satin binding weaves, and each card acts for two picks. Taking the marks to indicate warp, the blanks in the figure represent ends that are left down on account of the knives being made vertical and missing the figuring hooks, while the crosses in the ground indicate the lifts produced by the bars that are raised by the twilling hooks.

Diversification of effect in damasks woven on self-twilling machines

In the twilling jacquard systems the figure and the ground can only be woven in opposite sateen and satin weaves so that no variety of weave development can be produced. It is possible, however, to diversify the ornamentation of a fabric, as shown at D in *Figure 6.5* in which a few shading effects are indicated to illustrate how a figure may be brought up in different tones. The finer the cloth, and the fewer the ends to each needle and picks to a card, the nearer the effects are to those that can be woven in an ordinary jacquard; and it should be kept in mind that very fine lines of figure may be indicated in the solid design, since each small square represents a group of threads. E in *Figure 6.5* shows the full weave of a portion of the design D, assuming that the hooks are connected to the needles in the order of 3, 3, 2; successive cards act for 3, 3, and 2 picks; and 8-shaft binding weaves are formed.

In designing for the twilling jacquard the count of the design paper is in the proportion of—

$$\frac{\text{ends per cm}}{\text{hooks per needle}} \qquad \frac{\text{picks per cm}}{\text{picks per card}}$$

For example, assuming that a cloth contains 38 ends and 56 picks per cm, and that 32 hooks are connected to each short row of 12 needles, and there are 10 picks to three cards—

$$\frac{38 \text{ ends}}{32 \div 12} \qquad \frac{56 \text{ picks}}{10 \div 3} = 12 \times 14 \text{ design paper.}$$

Figure 6.5

The best qualities of damask cloths are generally woven with more picks than ends per unit space, and the following are typical weaving particulars: Warp—33 tex linen, 36 ends per cm; weft—24 tex linen, 52 picks per cm. Five-shaft binding weaves are usually employed in the lower qualities of cloths.

FIGURED WARP RIB BROCADES

The construction of simple warp ribs has been explained in Chapters 3 and 6 of *Watson's Textile Design and Colour*. In figured warp ribs the ends which form the rib are generally composed of a lustrous material, and they are floated on the surface in the manner illustrated by the fabric represented in *Figure 6.6*.

Figure 6.6

Figure 6.7

The following are suitable weaving particulars: Warp, 2 figuring ends (in a decked mail) of 30/2 tex mercerised cotton, 1 binding end of 20/2 tex cotton, 16 double-figuring ends, and 16 binding ends per cm. Weft 24 tex cotton in the same colour as the binding ends, 22 picks per cm.

A in *Figure 6.7* shows the complete weave of a portion of the design represented in *Figure 6.6*, the solid marks indicating the lifts of the double-figuring ends and the diagonal marks those of the binding ends. The ground of the cloth is plain weave, and the figuring warp floats are arranged to fit with the plain, which, as shown in *Figure 6.6*, causes the figure to appear very prominently, but with a stepped outline. It will be seen in A, *Figure 6.7*, that all the figuring ends are raised on the even picks; the binding ends are raised on the odd picks in the ground, but where the figure is formed they are lifted in alternate order. The figuring lifts of the figuring warp occur on odd picks and it is only these lifts that need to be indicated in simplified designing as shown at B, in *Figure 6.7*. The simplified design is thus condensed by two in each direction as each vertical row equals two ends and each horizontal row two picks. The detailed weave for the ground area, represented by the unpainted paper at B, is given at C, and that for the figure area, represented by the solid marks at B, is shown at D.

Figure 6.8

The combination of weft figure with a warp rib figure is illustrated by the fabric in *Figure 6.8*. The weft effect is indicated in a different colour on the simplified design and when the weft float is short all the ends in the weft figure area are simply left down which is the case in the fabric given in *Figure 6.8*. When, however, the weft floats created in the above manner are too long they may be stitched at intervals by the binding ends in the manner illustrated at E in *Figure 6.13*. As the binding ends are usually of the same colour as the weft, the stitches do not detract from the solidity of the figure produced by the weft.

Rib designs produced in two colours of warp

The design given in *Figure 6.9* illustrates a class of figured warp rib which is different in structure from the foregoing. In this case all the ends are identical in thickness and quality but differ in colour, the arrangement being an end of pink and light green alternately. In the ground differently coloured horizontal tie lines are formed, and both colours are employed in producing the figure.

Figure 6.9

In the design the full squares represent the figure produced by the pink warp, and the crosses that formed by the light green warp; while the shaded squares show how a subsidiary effect is obtained by floating both colours of warp together, the, intermingling of the colours giving a grey appearance to this portion of figure. Since both colours of warp are used separately for figuring, the warp threads should be finely set, and equally lustrous. The weft requires to be even and smooth, and of a neutral shade, while fewer picks per cm than ends may be inserted—as for example, for plain ground weave 24 picks per cm of 20 tex cotton, and 32 ends per cm of 280 dtex viscose filament rayon.

Figure 6.10

In the fabric represented in *Figure 6.10,* the warp threads are arranged alternately in two colours, as in the preceding illustration, but in this example a portion of the figure is formed by the weft, the colour of which is in contrast with the warp colours. Thus, as is shown in the corresponding design given in

Figure 6.11, the figure is formed by floating the odd ends in one section, as represented by the crosses while the weft forms the surface in a number of abstract shapes in the repeat. As the settings in this cloth are comparatively low

Figure 6.11

the weft areas are bound in a 5-sateen order. Further variety is added by introducing figure areas in which both warps work together and are bound in a 5-shaft warp satin order as shown on the right of *Figure 6.11*.

Methods of ornamenting warp rib structures

In order to further illustrate the diversity of effect that can be produced in the rib structures, a fabric is represented in *Figure 6.12*, which shows a method of figuring by means of brightly coloured weft. Also a number of rib weaves are

Figure 6.12

shown separately in *Figure 6.13*, two or more of which can be used in combination. It is assumed that in the plans two figuring ends alternate with a fine binding end, and a fine pick with a thick pick, the figuring ends in the ground passing under the former and over the latter, as shown at A. In some cases the rib ground is given a very rich appearance by employing brightly coloured filament weft for the fine binding picks. Where the filament weft floats over the double-figuring ends, it shows a fine line of bright colour between the ribs forming by the warp.

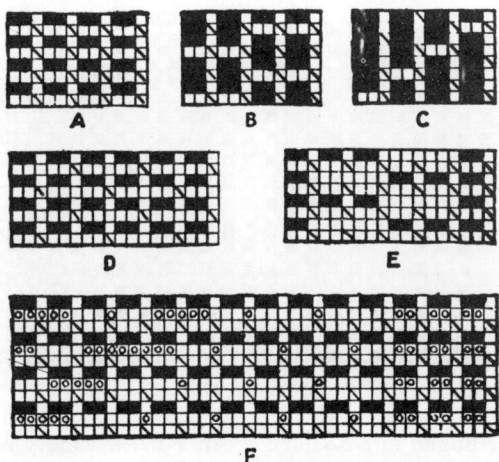

Figure 6.13

A, B, and C in *Figure 6.13* show the weaves that are used in the ground of *Figure 6.12,* and this and other variations of the rib structure can be employed along with, or instead of, the ordinary plain rib. C illustrates the usual method of figuring with the rib ends; each float should be arranged to commence and terminate with a thick pick.

The plan D shows how brightly coloured binding picks can be floated on the surface so as to diversify the form of a design; while, as represented at E, the thick picks can be similarly floated. The plan F illustrates the introduction, between the fine binding and the rib picks, of special figuring picks, the floats of which are represented by the circles. These extra picks can be used to spot the rib ground, as shown on the right of F, and to form a weft figure, as indicated on the left; both of these systems of interweaving are represented in *Figure 6.12.* All the ends are lifted on the extra picks, except where the circles are indicated in F, *Figure 6.13,* and the centre portion of the plan shows how the picks are bound-in on the underside of the cloth.

MULTI-WEFT BROCADES

Very ornate designs are often produced for hangings, corsets, and fine upholstery cloths in which two or three wefts are employed in conjunction with one series of warp threads. There is no special ground weft, as in extra weft fabrics, and all the wefts are floated on the surface as required in producing the figure, but each also assists in making the ground structure.

Two-weft brocades

Figure 6.14 represents a cloth in which the figure is formed in two colours of weft-woven pick-and-pick upon a warp satin ground produced by the two wefts interweaving with the warp. The following are suitable weaving particulars for an upholstery cloth: Warp and weft, 110 dtex filament polyester; 72 ends and picks per cm.

Figure 6.14

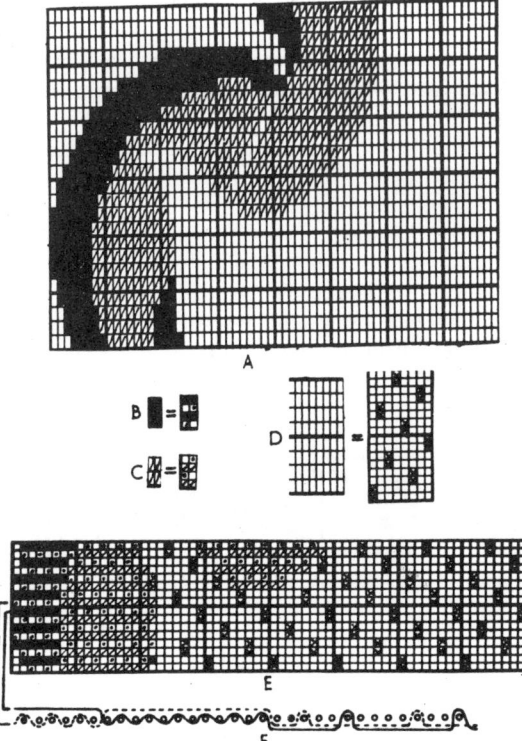

Figure 6.15

A in *Figure 6.15,* which corresponds with a portion of *Figure 6.14,* illustrates the system of constructing the squared paper design. The figure is painted in solid in two colours to represent the different wefts, and the ground is left blank; while it is convenient also to use white in binding the floats of the figure. Each horizontal row corresponds to a pick of each colour, and the count of the design paper is therefore in the proportion of the number of ends per cm to the number of picks per cm of each colour, or 8 × 4 with the foregoing particulars. As the design is condensed by 2 weft-wise, two cards are cut from horizontal row. Assuming that the solid marks represent one colour of weft, say, black and the diagonal marks the other colour of weft, say, grey and that the order of wefting is one grey, one black, the detailed weaves for each differently painted portion of A in *Figure 6.15* will be as indicated at B, C, and D (convention reversed). The effect of the superimposition of the detailed weaves is illustrated at E in *Figure 6.15,* which shows the full weave of the horizontal rows 5 to 12 of A. It will be seen that the ground weave is an 8-shaft satin—two picks in a shed, while where one weft is floated on the surface, the other weft forms plain weave underneath. The warp threads are usually so finely set in these cloths that the weft intersections in the ground have scarcely any effect upon the solidity of the warp colour. It will be noted from E that where the plain weave is produced odd lifts are cut for the odd picks of each colour and even lifts for the even picks. The disposition of the two wefts in the different parts of the design is clearly shown at F which is a warp cross-section of the structure cut through picks 8 and 9 of E.

If it is necessary to insert the weft colours in 2-and-2 order, on account of the loom being provided with changing boxes at one end only, the detailed weaves remain the same as before for each different portion of the design but are re-arranged to coincide with the 2-and-2 order of wefting.

Two-weft brocade ground weaves

The figure areas in two weft brocades are constructed according to a standard system described above but the ground weaves vary considerably. A number of ground weaves, different from the one shown at D and E in *Figure 6.15,* which

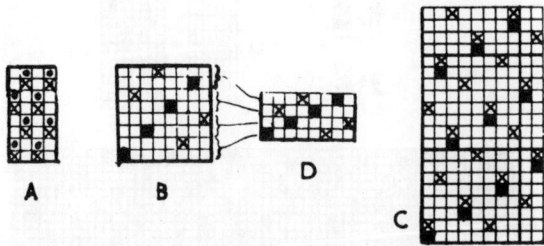

Figure 6.16

are employed in the pick-and-pick brocade fabrics are given, using the reversed convention, at A to C in *Figure 6.16.* A shows a fine 2-and-2 warp rib ground which is occasionally employed in these fabrics. B represents an ordinary 8-shaft warp satin and this is used extensively in the lower quality fabrics in which the extended satin weaves such as the one represented at D in *Figure 6.15* would result in excessively long warp floats due to a comparatively few picks per cm. At C the ground weave indicates that the successive picks operate alternately in 5-shaft and 10-shaft satin orders, the idea in this case being to interweave one weft more firmly than the other.

The longer float of the odd picks causes them to stand out behind the even picks on the underside of the cloth, and as they interweave with the warp in the same shed as the even picks they are prevented by the latter from showing on the surface. The method enables a weft which is thicker, or in stronger colour contrast with the warp than the other, to be thrown chiefly to the back in the ground, so that the solidity of the warp colour is affected as little as possible. Other satins can be arranged in the same manner as the 5-shaft satin.

For the piano card-cutting machines the ground weaves have to be painted and if the design is given in a condensed form ingenious methods of indicating the ground are used so that the weaves are properly reconstituted during cutting. This is shown at D in *Figure 6.16* where the 8-shaft satin ground given at B is shown in a design condensed by 2 weft-wise. Different marks are used to denote the different colours of weft and to achieve the correct satin binding in the cloth the following card cutting instructions are appropriate:

Cut two cards from each horizontal row—

First weft—cut blanks and crosses;

Second weft—cut blanks and solid marks.

The result of this order of cutting is the ordinary satin ground as shown at B. Other ground weaves can be similarly condensed and reconstituted by means of properly formulated instructions. Fortunately, the tedium of painting large areas of ground in an elaborate manner is avoided when modern jacquards and card-cutting machines are used.

Pick-and-pick weave shading

A fabric is represented in *Figure 6.17* in which different degrees of light and shade are formed by means of weave shading, in a pick-and-pick order of wefting. The warp is blue, while the weft is arranged 1 pick grey, and 1 pick white. Similar weaves to those employed in the cloth are given in full in *Figure 6.18,* in which the solid marks represent the grey weft floats, and the dots the white weft floats. A portion of white weft figure, under which the grey weft interweaves in plain order, is produced by section A. In section B a bluish surface is formed by the white weft interweaving in plain order with the warp, the grey weft floating on the back in 7-and-1 order. Section C is in contrast to section B, as the grey weft interweaves in plain order with the blue warp, so that a different mixed colour surface is formed, under which the white weft floats in 7-and-1 order. In section D, above a plain foundation formed by the interweaving of the white weft with the blue warp, the grey weft is floated in gradually increasing lengths of float, so that the bluish surface gradually merges

Figure 6.17

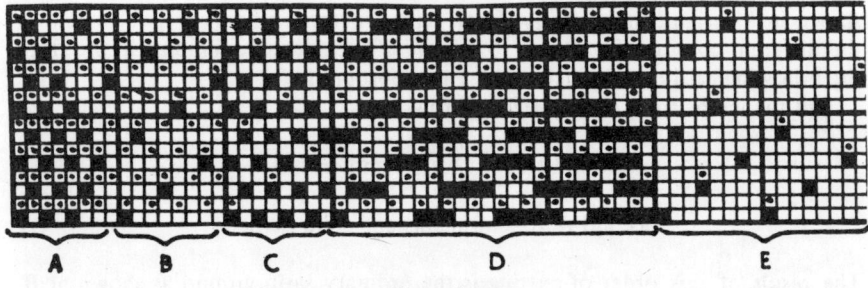

A   B   C   D   E

Figure 6.18

into a grey surface. Section E shows the weave which is used in forming a solid blue ground, the blue warp interweaving in 8-satin order with the grey weft and in 16-satin order with the white weft.

Pick-and-pick reversible structures

A perfectly reversible structure can be produced by figuring with the two wefts in such a manner that when one forms a float on the face the other forms a corresponding float on the back. The ground weave which in these structures is usually a warp rib is also the same on both sides. This type of construction is particularly suitable for curtaining fabrics.

Where a weft figure is formed the picks do not interlace and for this reason the floats which are produced must be comparatively short. To keep the length of float of each weft the same, every face float on the right-hand side of the figure is taken down whilst the back weft float is brought up. This is illustrated in *Figure 6.19* which also shows the typical 2-and-2 rib which surrounds the weft figures and the 4-and-4 warp rib in the ground. One weft is represented by the solid marks and the other by the crosses using the reversed convention. The warp section shown below the design has been taken across ends 33 to 64 and shows the operation of the seventh and the eighth picks.

As the warp in the weft figure areas simply lies straight between the face and back floats of the weft without any interlacing it is advisable to break the figure

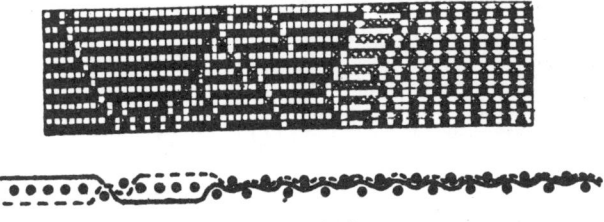

Figure 6.19

in the length by frequent introduction of small areas of ground. Diagonal or horizontal disposition of the figure is particularly helpful in this respect whilst long vertical shapes may lead to the formation of slack ends.

Three-weft brocades

Figure 6.20 represents a fabric in which the figure is produced in three colours of weft which are inserted, a pick of each in succession. Two of the wefts are inserted continuously and the third one is chintzed to add further variety to the design. In this construction when one weft is floating on the surface, the second one weaves plain underneath and the third one floats on the back where it is loosely stitched. The ground is usually a warp satin to the formation of which each of the three wefts contributes although occasionally warp-rib grounds are also used.

Figure 6.20

A in *Figure 6.21* shows a simplified design for this type of structure which has been condensed by 3 weft-wise. A different colour is used to indicate the float of each of the three wefts on the surface and the ground is left blank. The detailed weaves for each differently painted area are shown at B in the corresponding columns.

From an examination of B it will be seen that the first figuring colour floats in 15-and-1 order on the underside, except where it forms figure, while the third figuring colour floats in 7-and-1 order in the ground and under the figure formed by the first colour. Plain weave is formed by the second colour under the figure formed by the first and third colours, and by the third colour

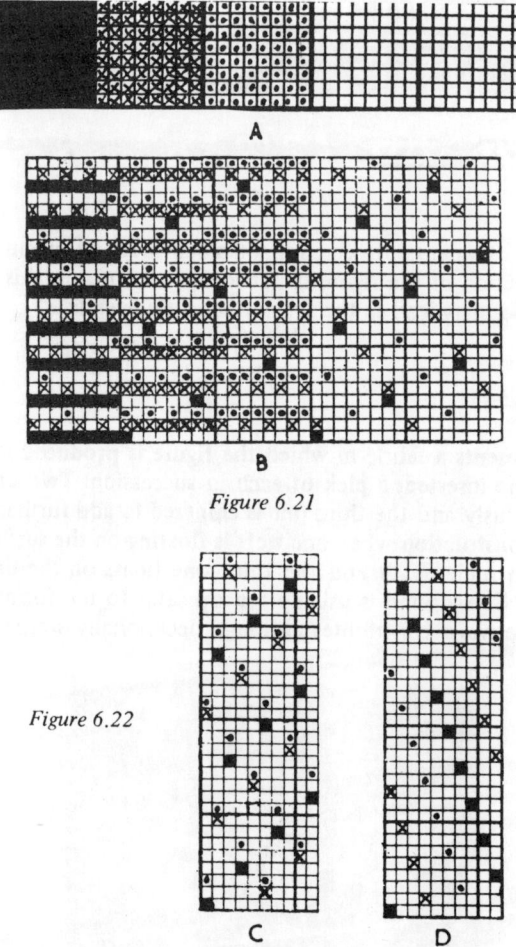

Figure 6.21

Figure 6.22

under the figure formed by the second colour. In the ground 8-satin weave is formed by the second and third colours, and the first colour is stitched on alternate binding points in the same shed as the second colour. This tends to throw the first colour of weft, which is in a strong contrast with the warp, more distinctly to the back thus preventing it from spoiling the solid colour ground effect achieved by the warp satin weave.

Examples of other ground weaves used in conjunction with three weft figuring are given at C and D in *Figure 6.22*. At C the first two wefts operate in a 10-shaft satin order and the third one in a 5-satin order stitching-in together with the second weft on alternate binding points. At D all the three wefts contribute equally to the formation of the ground each making a stitch in succession in a regular 10-satin order.

MULTI-WARP BROCADES

In these constructions two or more different series of warp threads are used to produce a figure in conjunction with a common weft. As opposed to extra warp fabrics in which one warp is specifically allocated to produce the ground structure, in multi-warp brocades all the warps are floated on the face in turn to produce the figure and all of them also assist in the formation of ground weaves.

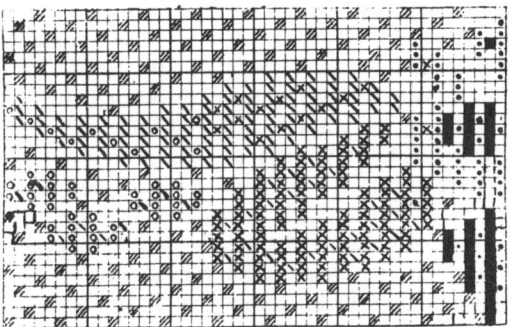

Figure 6.23

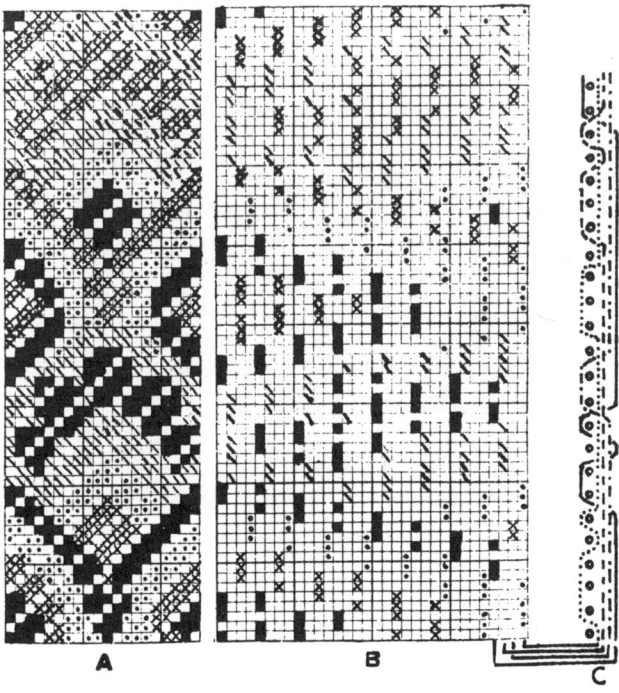

Figure 6.24

The cloths produced in these styles are mainly used for curtainings and are, as a rule, woven in coarse yarns and low settings with the warp frequently entered two ends per mail for better cover.

In *Figure 6.23* a fully worked-out portion of a design suitable for an end-and-end structure is given to show the principle of this construction. In the ground areas, represented by the shaded squares, the weave is a 5-thread weft sateen and as the colour of the weft is usually in considerable contrast with the warp it shows off the brightly coloured warp figure very distinctly. Where one warp floats on the face the other floats on the back where it is loosely stitched with the weft. A feature of this type of construction is the extensive planting (q.v.) in the warp, and in the structure represented in *Figure 6.23* although only two warps are employed per vertical row of design use is made of five different colours of warp, denoted by the different marks, as the original colours are supplanted by other ones in succeeding portions of the design.

A four-colour warp figure is illustrated in *Figure 6.24,* in which there are two features to note—viz. (1) The surface of the cloth is entirely covered by the warp figure, and there are no ground ends, the necessary firmness of structure being obtained by interweaving each colour, where forming figure, in 3-and-1 twill order; (2) variety of effect is obtained by the interchange of the colours in succeeding repeats. Thus, an examination will show that while the complete design is on 64 picks, the figure in the upper half is exactly like that in the lower half, except that the colours, represented by the full squares and crosses, interchange. A in *Figure 6.24* shows how the figure may be conveniently indicated by first making the different colours solid, and then inserting the twill marks over the design; while B shows the fully worked-out weave of the first eight vertical rows of A.

Figure 6.25

Figure 6.25 shows a fabric constructed according to the principles outlined above in which, however, only two series of ends are employed.

7

Stitched Double Cloths

Double cloths are fabrics in which there are at least two series of warp and weft threads each of which is engaged primarily in producing its own layer of cloth, thus forming a separate face cloth and a separate back cloth. The two layers may be only loosely connected together in which case each may be readily identified as a different entity or they may be so intricately stitched or tied together that they appear to form a complex single structure. The purpose of the construction may be entirely utilitarian, such as the improvement of the thermal insulation value of a fabric in which a fine, smart face appearance is necessary; or, it may be aesthetic in intention for which purpose the existence of two series of threads in each direction improves the capacity for producing intricate effects dependent upon either colour, or structural changes.

Classification of double cloths

Most of the double cloths can be classified under well defined headings and the following list gives the principal structural types with the simple schematic diagrams in *Figure 7.1* illustrating the basic principle of each construction.

(1) Self-stitched double cloths. These fabrics contain only the two series of threads in both directions and the stitching of the face cloth layer to the back layer is accomplished by occasionally dropping a face end under a back pick, or, by lifting a back end over a face pick, or, by utilising both of the above systems in different portions of the cloth. This type of structure and the three different methods of stitching are illustrated at A, B and C in *Figure 17.1*.

(2) Centre-stitched double cloths. In these fabrics a third series of threads is introduced either in the warp or in the weft direction whose entire function is to stitch the two otherwise separate layers of cloth together. The centre threads lie between the face and the back cloth and for the purpose of stitching oscillate at regular intervals between the face and the back thus achieving the required inter-layer cohesion as shown at D in *Figure 7.1*.

(3) Double cloths stitched by thread interchange. These structures are similar to the first category inasmuch as they do not contain an additional

series of stitching threads. However, they are distinguished from the self-stitched fabrics by the fact that the stitching of the face and the back cloth is achieved by frequent and continuous interchange of some thread

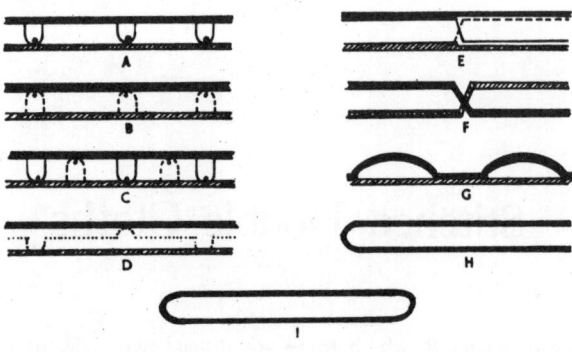

Figure 7.1

elements between the two cloth layers. Thus, in some portions of the cloth the face ends may be made to interweave with the back picks and the back ends with the face picks as illustrated schematically at E in *Figure 7.1*. The point at which the threads interchange represents the stitch point.

(4) Double cloths stitched by cloth interchange. In this class of constructions the principle of the interchange is taken one stage further than in the third category and complete cloth layers are made to change places as shown at F in *Figure 7.1*. As stitching between the two fabrics occurs only at the point of cloth interchange the degree of cohesion in this type of cloth depends on the frequency of the interchange.

(5) Alternate single-ply and double-ply construction. In some fabrics the constituent thread components are occasionally merged together into a heavily set single cloth and occasionally are separated into distinct layers to form figure areas of open double cloth on the firm single cloth ground. Usually, the effect depends upon a degree of distortion as the crammed single cloth areas tend to spread out, thus affecting the appearance of the double cloth 'pockets'. A cloth of this type is illustrated at G in *Figure 7.1*.

In addition some cloths are produced on the double cloth principle of construction but due to the deliberate absence of stitching between the layers become single cloths upon their removal from the loom. Two such constructions, the double width and the tubular cloth are shown respectively at H and I in *Figure 7.1*.

SELF-STITCHED DOUBLE CLOTHS

The self-stitched double cloth is composed of two series of weft and two series of warp threads; one series of each kind forming an upper or face fabric, and the other, an under or back fabric. It is necessary for the face picks to be arranged in definite order with the back picks, and the face ends with the back ends. The two series of ends require to be drawn through the healds or harness in such a manner that one series may be operated quite independently of the other series.

Separate weaves are required for the two fabrics, which may be either alike or different from each other. Then by interweaving the face picks only with the face ends according to the face weave, and the back picks only with the back ends according to the back weave, two distinct fabrics are formed one above the other. The method in which this is accomplished is illustrated in *Figure 7.2.*

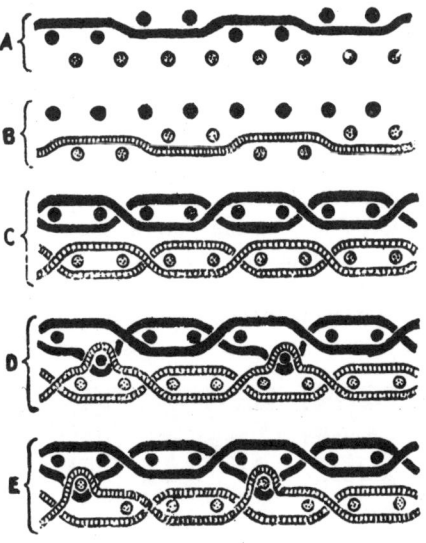

Figure 7.2

The threads are arranged 1 face, 1 back in warp and weft, and a 2-and-2 weft rib weave is employed for both the face and back textures. A represents the position of the warp threads when the first face pick is inserted. All the back ends are left down in order that they will be out of the way of the face weft, and half the face ends are raised in forming the face weave. B shows the position of the warp threads when the first back pick is inserted. In this instance all the face ends are raised in order that they will be clear of the back weft; also half the back ends are raised in forming the backing weave. By allowing each series of weft picks thus to interweave only with its own series of warp threads, two fabrics are produced which are quite separate and detached from each other, as shown at C. If, however, a proportion of the face warp threads be left down when a back pick is inserted, as shown at D in *Figure 7.2*, or if a proportion of the back warp threads be raised when a face pick is inserted, as indicated at E, the threads of one fabric interweave with the threads of the other fabric; and although there are still two distinct fabrics formed one above the other, they may be so closely united that separation of the two layers is impossible. The tying or stitching together of the two fabrics forms one of the principal features of double cloth construction. If a cloth is not soundly stitched, the two fabrics are liable to become separated from each other during wear, particularly if the back fabric is heavier than the face. Diversity of design and colouring can be applied to both sides of a double cloth, and a more perfect structure is obtained than in the case of single fancy cloths or backed cloths.

Relative proportions and thicknesses of the face and back threads

These are decided mainly by the weight to be added to the face texture, but the order of arrangement of the weft threads is determined partly by the weft insertion of the loom. The most common varieties of double cloths are arranged in warp and weft 1 face, 1 back, as shown at F in *Figure 7.3*, and 2 face, 1 back, as shown at G. For looms with boxes at one side only, and when the back weft is different from the face weft, similar effects may be obtained in many weaves by changing the wefting to 2 face, 2 back and 4 face, 2 back, respectively, as shown at H and I. Cloths which require a very fine face are sometimes arranged 3 face, 1 back in warp and weft, as shown at J. The threads may also be arranged in a mixed order, as, for example, 1 face, 1 back in the warp, and 2 face, 1 back in the weft, and *vice versa,* as shown at K and L respectively, or 2 face, 1 back in the warp, and 2 face, 2 back in the weft, as shown at M. Irregular arrangements such as 5 face to 4 back (shown at N), and 7 face to 5 back (shown at O), are also employed, and these are occasionally useful as they admit of relative proportions of face and backing threads being used which cannot be obtained in any of the regular bases.

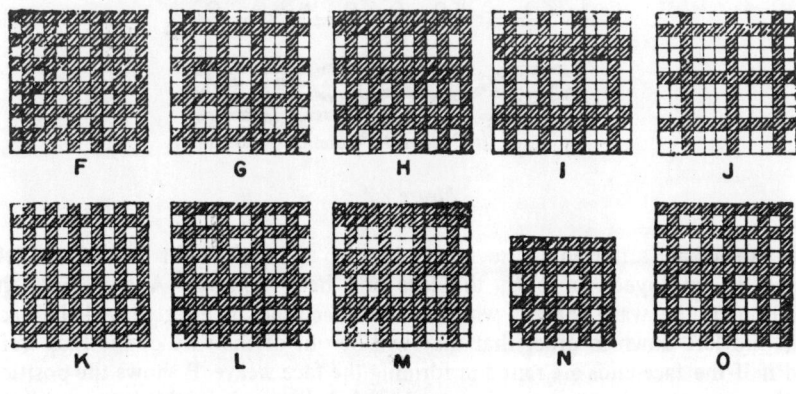

Figure 7.3

In deciding on the relative thicknesses of the face and back yarns, a good rule to follow is to have the relative counts about proportionate to the relative numbers of the threads per unit space. Thus, in a 1 face, 1 back double cloth the back yarn should be similar to, or not much thicker than the face yarn; the finest qualities of the structures being usually made with the same weave and counts for both fabrics. If arranged 2 face to 1 back, the back yarn may be proportionately thicker, or say, from two-thirds to one-half thicker than the face yarn; the back being made coarser than the face, particularly when worsted yarns are used for the latter, and woollen yarns for the former. The proportionate counts of the threads, however, depend upon the relative firmness of the face and back weaves, and the preceding proportions apply to the 2-and-1 arrangement when the back weave is firmer than the face weave, as described in the next paragraph. If the same weave is used on both sides of the cloth the back threads may be three or four times as heavy as the face threads in the 2-and-1 arrangement, especially when centre threads are employed for stitching.

Selection of the face and back weaves

When the threads are arranged in equal proportions the back weave is usually the same as the face weave, or contains about the same relative number of intersections, as, for instance, the 2-and-2 twill is suitable for backing the 3 up, 2 down, 1 up, 2 down twill. In other arrangements the backing weave is, as a rule, made with a relatively greater number of intersections than the face weave in order to compensate for the reduced number of threads. Thus, in the 2 face, 1 back arrangement, the plain weave is suitable for backing the 2-and-2 twill and the 2-and-2 hopsack; the 2-and-1 twill for backing the 3-and-3 twill; and the 2-and-2 twill for backing the 4-and-4 twill. However, in the making of cloths with a fine, smart face and soft back, the same weave may be used, in the 2-and-1 arrangement, for both the face and back textures; while for a similar type of cloth in a 1-and-1 arrangement of the threads, a looser back than face weave may be employed. The most regular effect is obtained by having the repeats of the face and back weaves equal, or one a multiple of the other. For example, the 1-and-3 twill is unsuitable for backing the 2-and-3 twill unless the threads are arranged irregularly in the proportion of 5 face to 4 back threads.

Tying or stitching

In double cloths the stitches joining the two fabrics together, if correctly placed, have no effect on the appearance of either the face or the underside of the cloth. When the method of stitching involves raising the back warp over the face picks then the back end can be used for tying only when it is away from the underside of the back cloth and the pick over which the tie is made must be away from the face of the top cloth. A stitch made in conformity with the above two conditions is invisible on either side of the double cloth as shown at D in *Figure 7.2*. Similarly, if the stitching is achieved by dropping a face end under a back pick both these elements must be away from their respective surfaces, as shown at E in *Figure 7.2*. The method of tying which is the more suitable is, in some cases, determined by the character of the face weave. If a warp satin, or a warp-faced twill weave is employed for the face fabric, tying by lifting the back warp only is suitable; while in the case of a weft sateen or a weft-faced twill weave, it is only advantageous to tie by dropping the face ends. When there is a choice of the two methods, other things being equal, the former method is usually preferable, as the back warp is less liable to show on the face than the back weft, which in the latter system is pulled upwards. This is because the back warp, as a rule, is a finer and smarter yarn than the back weft; during weaving the warp is under greater tension than the weft, and woollen and worsted cloths usually contract in finishing more in width than in length. In some cases both methods of tying are employed in combination, as previously explained, the object of double stitching being to obtain increased firmness of structure and also when other things are equal, so that both face and back warps will be at the same tension and only one warp beam will be necessary.

Construction of squared paper designs

Various factors, which require to be considered before a double weave is commenced, are fully dealt with in reference to subsequent examples. It is sufficient
at this stage to assume that in each example the face weave, the back weave, and
the ties, are placed in such positions relative to one another as will ensure that
the ties are covered on each side of the cloth as effectively as possible by the
adjacent floats.

In order to prevent confusion the different stages in working out a double
cloth design should be represented by different kinds of marks, as shown in
Figure 7.4, which illustrates, step by step, the construction of a double 4-and-4
twill structure in which the ends and picks are arranged 1 face, 1 back.

A and B represent the face and the back weave respectively which are marked
using the normal convention in which a mark equals warp up. Altogether, the
normal convention is preferred in designing for double cloths it being more positive
and easier to interpret. At C an area equal to one repeat of the double weave is
marked out with the order of arrangement of the face ends and picks and the
back ends and picks indicated clearly at the margins. In the example given, lines
around the repeat area indicate the back ends; in some of the subsequent
designs back ends are indicated by the shaded lines as already shown in
Figure 7.3. In practice it is easier to denote the order of arrangement by using
the letters f and b for the face and the back elements respectively. D shows the
first stage of actual double cloth construction which may be defined as: Insert
the face weave on the face ends and face picks only, according to the original
design. The second stage is similar except that it refers to the back weave:
Insert the back weave on the back ends and picks only, according to the original
design. E shows the appearance of the design after the completion of the second
stage. F shows the marks for the separating lifts which ensure that each series of
yarns weaves only with its own kind and this may be stated as: Lift all face ends
and back picks. Similarly, to complete the sequence, all back ends must be left
down on all face picks which means an absence of marks, i.e. all back ends down
on face picks. These lifts, in fact, determine which series becomes the face cloth
and which the back and by lifting the face ends out of the way when the back
picks are inserted separation of the two fabrics is achieved.

F represents a stage in which two separate fabrics are produced one above the
other. As there is no particular reason for producing two disconnected cloths in
this manner, it will be realised that this stage is the intermediate point in the
construction reached prior to the insertion of stitches or ties to bind the two
cloths together. Before the stitch marks are inserted it must be decided which
method of stitching is to be used and how frequently the cloths are to be
stitched. Assuming that it is required to stitch by lifting the back ends on the
face picks and that each back end is to stitch once in the repeat, the correct
positions of the ties are shown by the circles at G and by the black marks at K.
It will be observed that the rules outlined earlier for correct selection of stitch
points are obeyed, viz. (a) the back end is raised to make a stitch when it is
absent from the underside, i.e. the visible side of the back cloth; (b) it stitches
over the face weft when that weft is absent from the surface of the face cloth,
and (c) when it emerges near the surface of the face cloth it is covered by two
adjacent long floats of the face warp. In a construction in which both cloths are

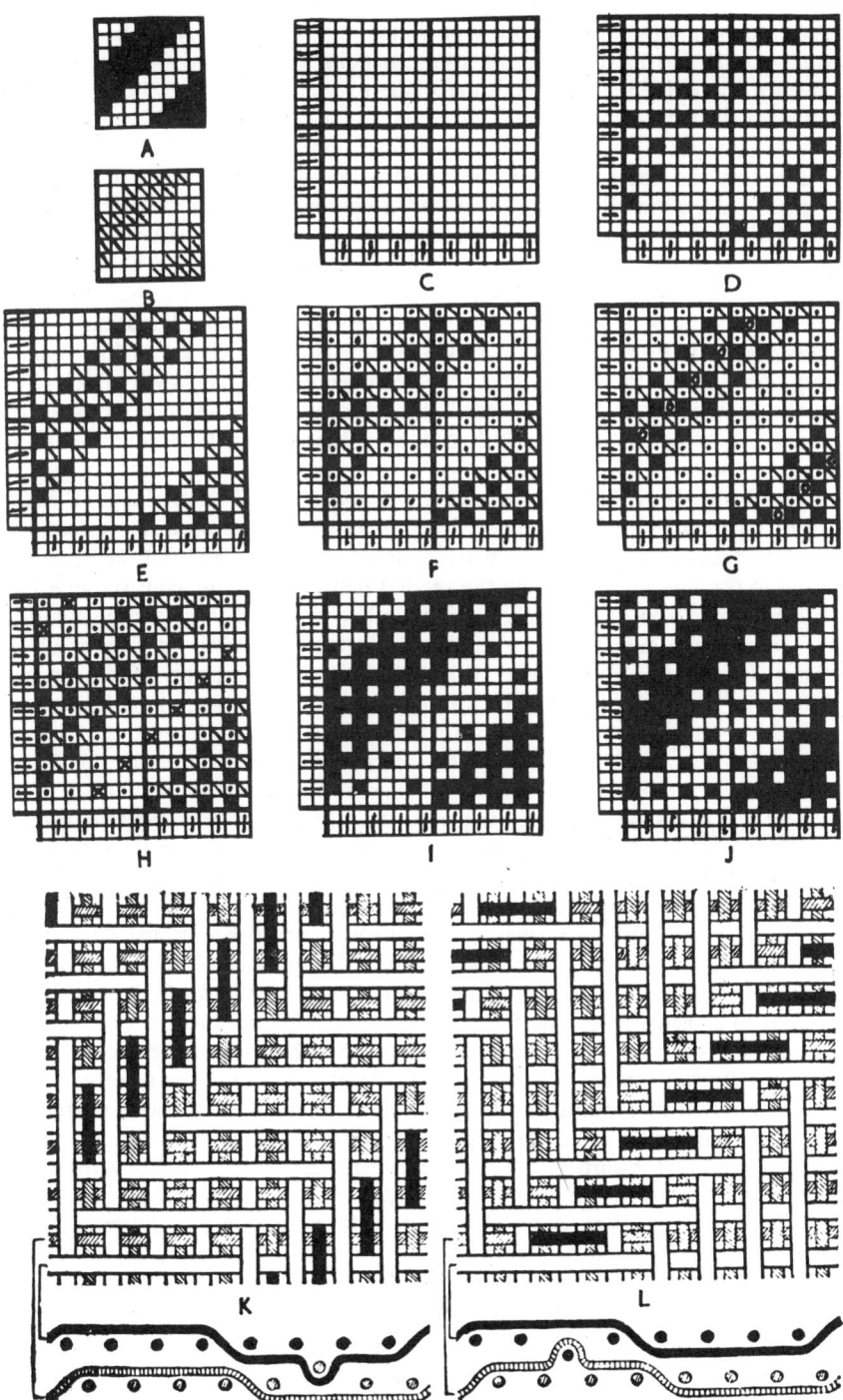

Figure 7.4

produced in a weave containing long floats it is usually easy to find suitable stitch points and in the example given an equally well-concealed tie could be achieved by lifting the back ends on a face pick either above or below the one actually selected. In some fabrics, however, the selection of suitable stitch points may be more restricted.

J shows the appearance of the construction G when only one colour of marks is used. Normally it is not necessary to prepare solid marked designs if it is stated along a design prepared in stages as at G how each type of mark is to be interpreted. In the case of G each mark means warp up. The design G corresponds with the interlacing diagram K in which the back threads are shown shaded. The positioning of the stitch points is emphasised by the black markings which show distinctly how the back warp stitching lifts are concealed by the adjacent long floats of the face warp. The warp cross-section given below K which refers to the interlacing of the first face and back pick also helps to visualise the manner in which a double cloth can be efficiently tied without disturbing either of the two visible surfaces.

The design H in *Figure 7.4* shows the second method of stitching, i.e. stitching by dropping the face ends on back picks. As in the previous system, one stitch per repeat is made, only in this case the face ends and not the back ends are used for the purpose. Using the second method stitching consists, in effect, of cancelling some of the separating lifts made in F and these are indicated by the crosses in design H. As a result, the instructions given in respect of the design H will be: Peg or cut all marks with the exception of the crosses. Otherwise, the design could be re-done by using only one type of mark, as shown at I, in which all marks indicate warp lifts. The design H corresponds with the interlacing diagram L and both indicate clearly the manner in which the ties are concealed, viz. the face end is dropped when it is absent from the surface of the face cloth, it stitches under a back pick which is absent from the underside of the back cloth, and the back pick which at the stitching point tends to be pulled up towards the surface is covered by two long adjacent floats of the face weft. This last point is particularly clearly shown by the diagram L in which the ties are shown by the black markings, and by the warp cross-section under L.

It will be noted from *Figure 7.4* that in both methods of stitching each end of the set used for the tying operates under or over each pick of the opposite set of weft in a regularly distributed order. The regularity of the distribution of ties should be attempted whenever possible as it aids the even distribution of crimp and yarn tension throughout the cloth and thus helps to produce a better fabric.

The method of constructing a double cloth in which the threads are arranged in the proportion of 2 face to 1 back in warp and weft is shown in *Figure 7.5*. This arrangement is suitable for fine face cloths woven with thicker yarns on the underside. The 10-thread fancy weave given at A in *Figure 7.5* is used for the face fabric, and the 5-thread sateen (with weft surface on the underside), given at B, for the back fabric. Both methods of tying are shown, the first being formed by raising the back ends in 5-sateen order over alternate face picks, as indicated at C, and the second by dropping the alternate face ends in a similar order under the back picks as represented at D.

In *Figure 7.5* the different stages of working are indicated separately in a similar manner to *Figure 7.4* but the back ends and picks are in this case denoted by the shaded squares. E shows the arrangement of the face and back threads;

the face weave is inserted at F; the circles in G indicate the positions of the ties using the first, and the crosses in H the positions of the ties using the second method of stitching; the back weave is inserted at I; while the complete design for each system of tying is given at J, and at K.

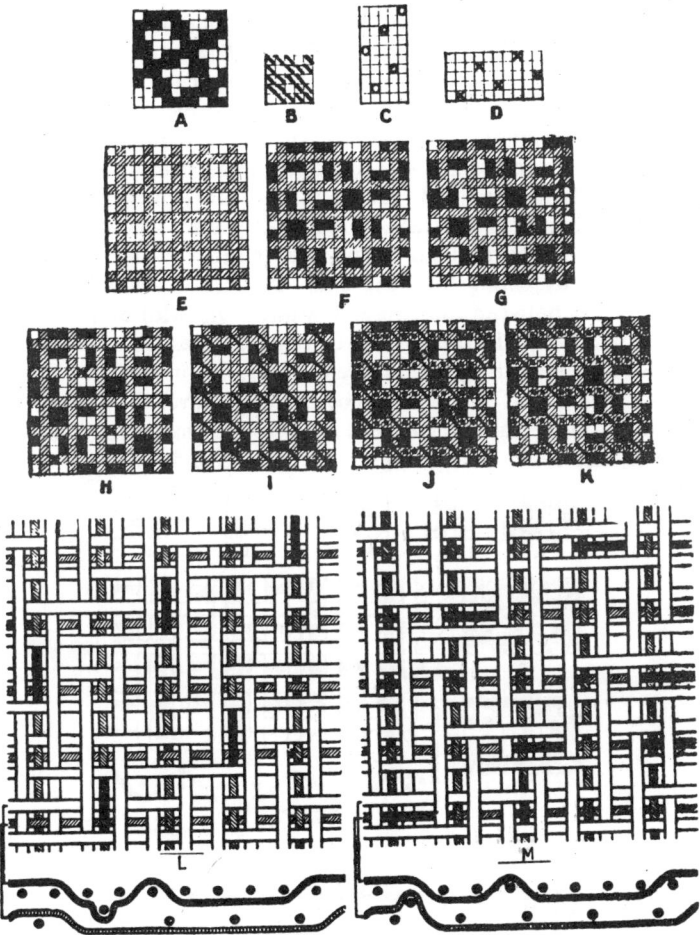

Figure 7.5

The interlacing diagram given at L in *Figure 7.5* corresponds with the weave shown at J, while the warp section below it shows the interlacing of the first back pick and the second face pick of L. Where the back warp enters the face cloth for tying it will be concealed by the face warp floats; also, as each back end (where raised for tying) is also raised on the backing pick which precedes and succeeds the tie, the face weft will be covered on the underside of the cloth by the back weft floats.

M in *Figure 7.5* shows the interlacing diagram corresponding with the weave given at K, with the appropriate warp section below. The ties will be concealed on the face of the cloth by the face picks which precede and succeed them, but on the back of the cloth the lowered face ends will be covered by the back warp on one side only. This cannot be avoided, because, on the underside there is only a warp float of one at a place in the backing weave. Thus in this construction stitching by lifting the back ends on the face picks is preferable.

In *Figure 7.6* the design is shown for a double cloth, in which the threads are arranged in a mixed order, the proportion being 2 face to 1 back in the warp, and 1 face to 1 back in the weft. This order of arrangement is specially applicable to face weaves which repeat on twice as many ends as picks. The broken 2-and-2 twill, given at A, is employed for the face fabric, ordinary 2-and-2 twill, B, for the back fabric, and the tying is effected in 4-thread twill order, given at C for one method of stitching and at D for the other. This is an example in which the back weave is looser than the face weave, taking into account that there are fewer ends on the underside than on the face. In order to enable positions to be selected for tying by lifting the back ends, the face weave is so placed that a backing end comes between the two ends which twill with each other, and not between two which cut. In this system of tying the ties are effectively covered on both sides of the cloth by the adjacent floats, but in tying by dropping the face ends to ensure that the ties will be perfectly covered on the back of the cloth as well as on the face, it has been necessary to change the positions of the back weave to that shown at M in *Figure 7.6*. E, F, G and H show the appearance

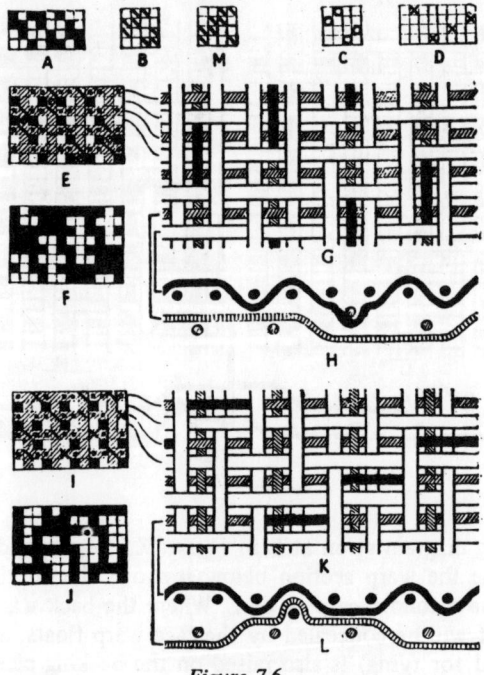

Figure 7.6

of the structure using the former method of tying whilst I, J, K and L show the corresponding views of the same structure stitched by the latter method. All marks indicate warp lifts with the exception of crosses in I.

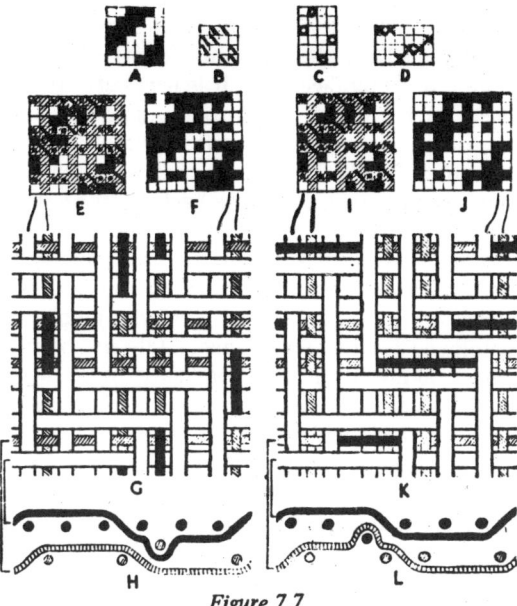

Figure 7.7

Figure 7.7 shows the design of a double cloth, in which the threads are arranged irregularly in the proportion of 6 face ends and picks to 4 back ends and picks. In this type of arrangement the face fabric may be made finer than the back fabric in almost any required proportion. The chief point to note is that suitable weaves are selected for the face and back fabrics respectively. Thus, if the threads are arranged 5 face to 4 back, a 5-shaft face weave should be combined with a 4-shaft back weave; if 4 face to 3 back, a 4-shaft weave with a 3-shaft weave, etc. In the example 3-and-3 twill is employed for the face fabric and 2-and-2 twill for the back fabric, and as the face fabric is finer than the back fabric in the proportion of 6 threads to 4, the 3-and-3 twill face will be similar in appearance to the 2-and-2 twill back. The cloth will therefore have the semblance of a double 2-and-2 twill, but its wearing property will be superior on account of the greater fineness of the face fabric. In the method of tying illustrated in *Figure 7.7*, no tie is placed on the second and fifth face pick, but in the method given at I a slight deviation from the ordinary system is illustrated. On the even back picks two face ends are dropped while on the odd backing picks only one is lowered. They are arranged in this way in order that all the face ends will be intersected by the backing picks, and to show one method of obviating the difficulty which frequently arises in weaving when only a portion of one series of ends is employed for tying. If, however, there is any possibility of the ties showing on the surface of the cloth, such a method should not be employed.

Construction of double cloth designs for looms with changing boxes at one end only

If the same kind of weft yarn is used for both the face and back fabrics, the method of constructing the design is not affected by the limitation in the boxing capacity of the loom, even though, as is sometimes the case, two or more shuttles are employed for the purpose of weft mixing. If, however, the back weft is different from the face weft, it is necessary for the face and back picks to be arranged on the design paper to alternate with each other in even numbers according to the relative proportions required. The arrangement of the face and back ends may be the same as in ordinary double cloths, and it is better that it should be the same, for if the back warp be employed for tying, the placing of the ties is then not influenced by the order in which the picks are inserted, so far as the face of the cloth is concerned. The covering of the corresponding face weft floats on the underside of the cloth is, however, not so easily effected when the picks are arranged in even numbers. The face warp should only be employed for tying when absolutely necessary, as the insertion of the picks in even numbers not only renders it more difficult for suitable tying positions to be selected in the majority of face weaves, but the interweaving of the backing weft with the face warp at intervals of 2 or 4 face picks, increases the tendency of the latter to group in 2's or 4's.

The system of construction is exactly the same as in the foregoing designs. In *Figure 7.8* the face and back threads are arranged in equal proportions, the

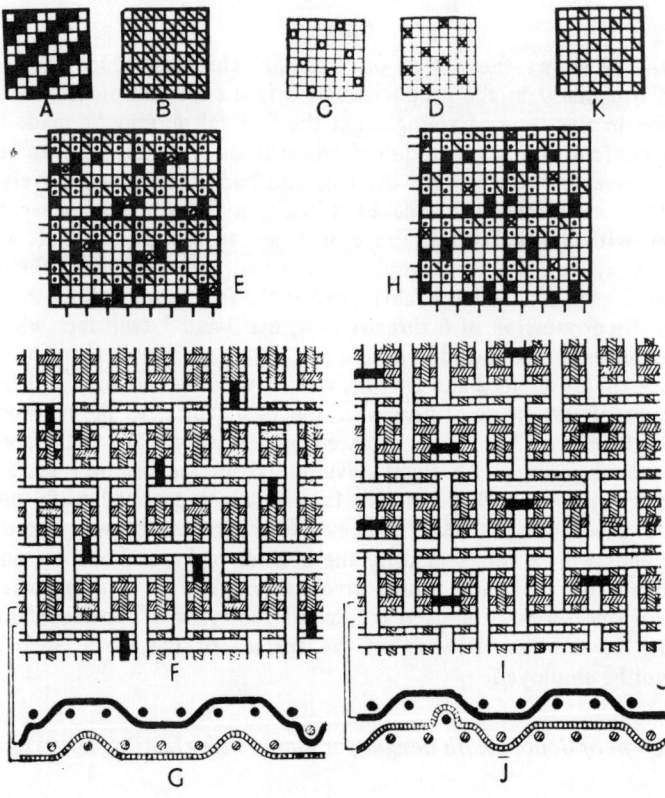

Figure 7.8

ends in the order of 1 face, 1 back and the picks in the order of 2 face, 2 back. The weave marks in *Figure 7.8* indicate warp up with the exception of the crosses which represent the face warp down for stitching. The 2-and-2 twill weave is employed for the face fabric, but for the back fabric the 3-and-1 warp twill, shown at B, is used when the back warp is raised for tying, and the 1-and-3 weft twill, shown at K, when the face warp is lowered for tying. Hence with the design E or F the underside of the cloth will have a weft surface, and with the design H or I a warp surface.

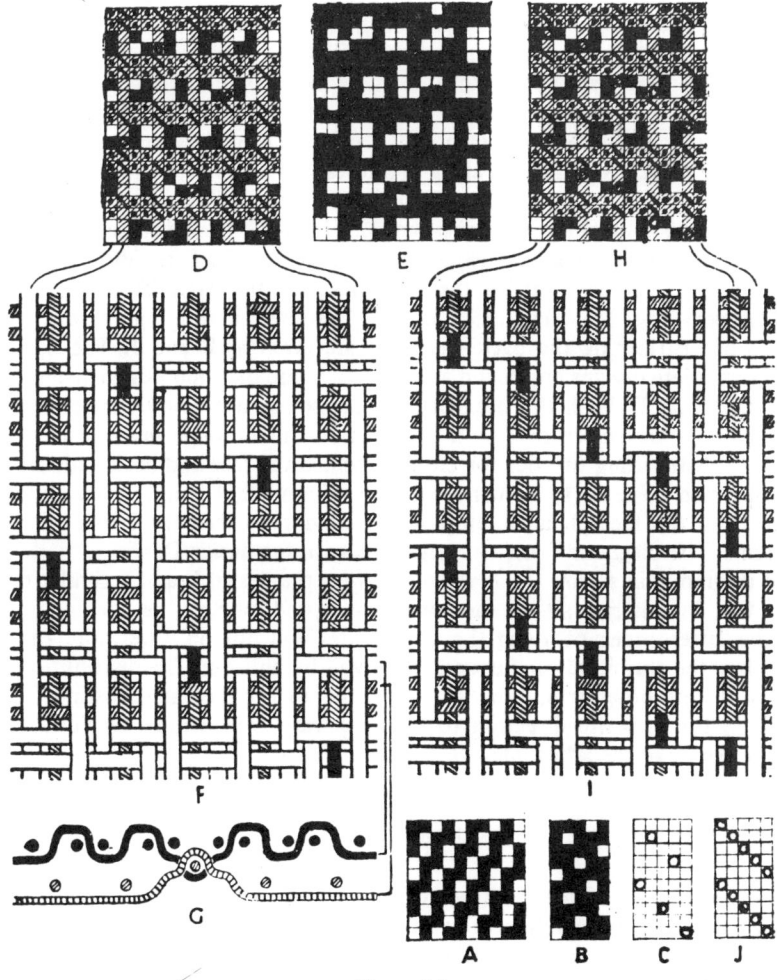

Figure 7.9

In the interlacing diagram given at F the warp ties are shown and arranged in 8-thread sateen order, as indicated at C, and it will be observed that on the face side of the cloth each tie is placed between two face warp floats. Also each back end is raised on the back picks which precede and succeed a tie, hence on the underside of the cloth each face weft tie occurs between two back weft floats. G in *Figure 7.8* represents the interlacing of the second face pick and the first back pick, from which it will be seen that the face ends and the back picks are quite separate and distinct from each other, while the face picks interweave at intervals with the back ends (in this case the eighth back end) for the purpose of uniting the two fabrics.

In the alternative stitching method illustrated at I in *Figure 7.8* no suitable positions for tying exist on which the even face ends could be dropped. The ties can therefore only be placed on the odd face ends, and though they are equally as well covered on both sides of the cloth a comparison with F will show that the distribution is less perfect in this system than when the back warp is employed for tying. Because of the difference in the arrangement of the ties the design E repeats upon twice as many picks as the design H, of which two repeats are shown. In the section given at J the face picks and the back ends are shown quite separate from each other, the union of the two fabrics being effected by the back weft interweaving at intervals with the face ends.

In *Figure 7.9* the threads are arranged in a mixed order, the proportions being 2 face to 1 back in the warp, and 2 face to 2 back in the weft. The 5-thread warp-face Venetian weave A is employed for the face fabric, and the 5-thread satin, with the weft on the underside, for the back fabric. Tying by dropping face ends on back picks is not illustrated in the figure, but two methods of distributing the back warp ties are given at C and J. The corresponding complete designs are shown at D and H, while the solid design given at E corresponds with D. With the ties distributed as shown at C and D, the alternate face picks only are passed over by the back ends, and although this is a standard method of distributing the ties for cloths in which the face and back ends are in the proportion of 2 to 1, in many cases it is found that a more even cloth is formed if all the face picks are employed for tying, as then the crimp of each is the same. The diagrams F and G correspond with the design D.

The diagram I in *Figure 7.9* corresponds with the design H and is similar to F except that in this case all the face picks are passed over by the backing ends. With the latter order of tying the two fabrics will not only be more firmly united, but the shrinkage of the weft picks will be uniform. The distribution of the ties in this order will, however, be liable to produce an indistinct twill running in the opposite direction to the face warp twill, which, by detracting from the clarity of the latter, may be a source of defect.

Reversible double weaves

The correct placing of the back weave in relation to the ties is of particular importance in the construction of reversible double weaves, such as are used for fine woollen and worsted overcoatings, in which the same effect is produced on both sides. Plans A to E in *Figure 7.10* illustrate the construction of a reversible 7-shaft whipcord, in which the threads are arranged in the order of 1 face, 1 back, in warp and weft. As the back of the cloth is warp surface, the same as the face, the back weave is exactly the opposite of the face weave. So far as regards the face of the cloth, the position of the back warp stitching lifts may be varied, as there are a large number of face floats to cover them. The positions of the back warp lifts must be selected carefully, and are indicated by the marks between the picks of the plans B, C, and D respectively, in which it will be noted that the position of the back weave is changed to accord with the position of the ties. Section F in *Figure 7.10* shows the interweaving of the first face and the first backing end with the weave and ties placed as at A and B;

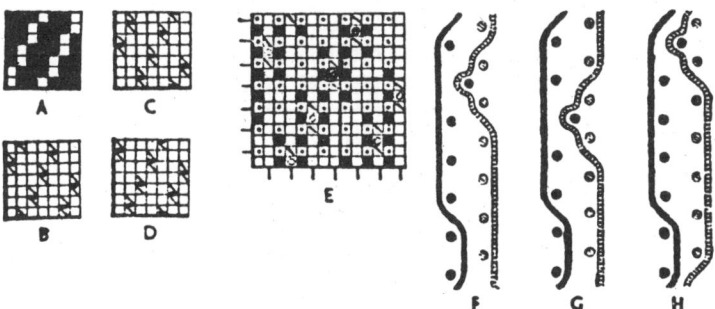

Figure 7.10

section G, as at A and C; and section H as at A and D. In each case the back of the cloth is as perfect as the face, the weaves being the same except that the twill runs in the reverse direction. E in *Figure 7.10* shows the complete double cloth design with A as the face weave and B the back weave.

Beaming and drafting of self-stitched double cloths

In double cloths the stitches put tension on the warp threads hence, other things being equal, the series used for tying requires to be longer than the unstitched series, and two warp beams are therefore necessary. By employing double-stitching, however, and using similar yarns and weaves of equal firmness for the two fabrics, a perfect double cloth can be woven with only one warp beam. Such a double weave is illustrated in *Figure 7.11*, in which the same weave—shown at A and B—is used for face and back, while both series of threads are used for tying, as indicated at C, in which the circles and crosses, respectively denote back warp up and face warp down, the cloth being double-stitched. The complete design is given at D, and the corresponding interlacing diagram at E, while F represents the interlacing of the first face and the first back end. It will be seen that the relative number of intersections is the same for each thread; hence, if the yarns in each fabric are similar, the contraction of the warp threads in weaving will be uniform, and the equal tension required for each series will be better obtained by using only one beam. If, however, the face yarns in a cloth are different from the back yarns, or, on the other hand, if the weaves are different as regards the relative number of intersections, it is better for two warp beams to be used, in order that the two series of threads may be separately tensioned. As a general rule, two beams are employed, and the proper tensioning of the two warps is then of great importance, because if the back warp is held tighter than the face warp the back warp stitches are liable to impair the softness of handle of the cloth, while if it is slacker the cloth is constructionally deficient. Normally, the two warps should be held at about the same tension.

With the exception that a set of healds is required for each fabric, the ordinary method of constructing a draft may be employed—i.e., the threads in each fabric which work alike may be drawn on the same heald. Therefore the

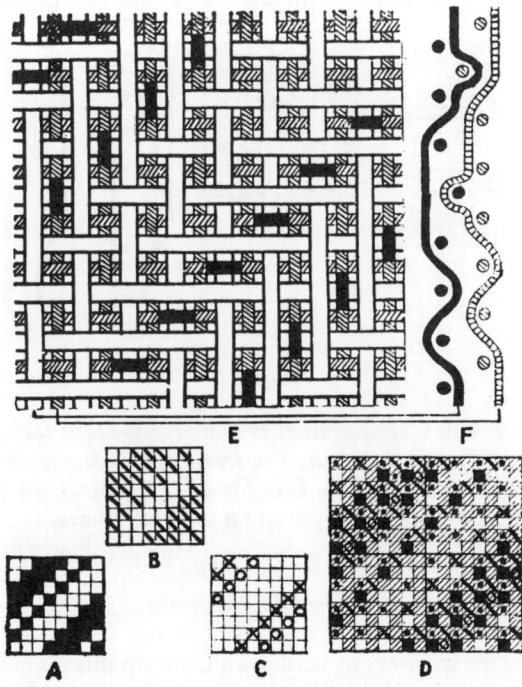

Figure 7.11

minimum number of healds in each set is decided by the number of threads in each fabric, which work different from each other. Thus, the design F in *Figure 7.12* requires eight back healds, although the back weave is on four threads because the back ends must be raised for tying independently of each other in the 8-thread sateen order given at C. Only four face healds are required, because the working of every fourth face end is the same. This will be understood from an examination of the interlacing diagram F and the section shown at G, which represents the interlacing of the first face and the first back end. The face ends work regularly in 2-and-2 order with the face picks, whereas the back ends, in addition to working in 2-and-2 order with the back picks, are raised for tying in 1-and-7 order with the face picks.

H, I, and J in *Figure 7.12*, in which different marks are used to distinguish the face ends from the back ends, show three different methods of drafting each suitable for the construction given at F. When the face and back healds are intermingled, as at H, it is convenient to employ as many healds as there are ends in the repeat of the design. Also, when a special draft such as I or J is used, the draft may be made upon the same number of face as back healds in order to give more scope in varying the weaves, and so that the healds will all carry an equal number of mails.

Sometimes it is convenient for the healds which have been used for the back fabric in a design to be subsequently employed as the face healds, and *vice versa*, in order that a change in the weave, or in the method of tying may be made without redrawing the warp. For example, if a face end be twisted to a back end, the drafts given at I and J in *Figure 7.12* may be employed for a double

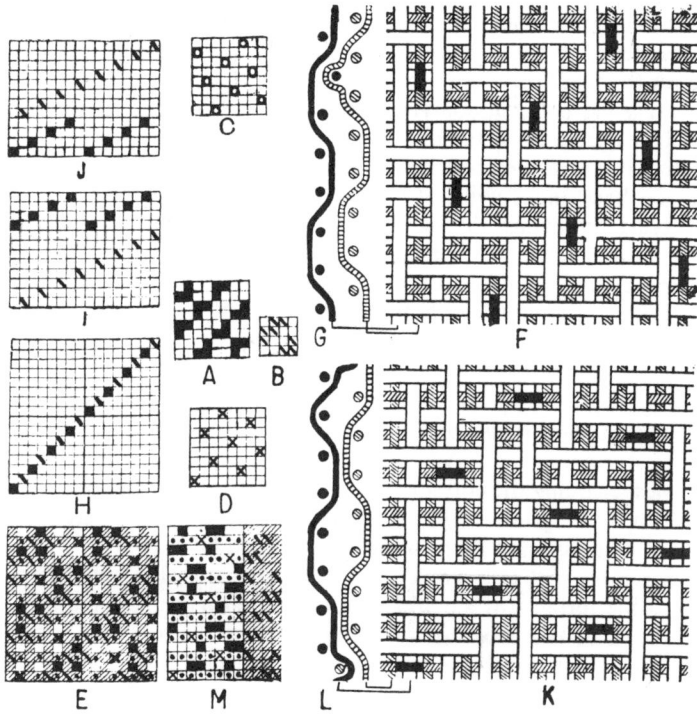

Figure 7.12

cloth in which the 8-shaft face weave shown at A is combined with the 4-shaft back weave given at B. As, however, there will then be only four back healds, it is impossible to effect the tying in 8-end order by means of the back warp lifts, but as there are eight face healds, the face ends may be depressed for tying independently of each other in the 8-end sateen order given at D in *Figure 7.12*. The complete double weave is shown at E, and M is the lifting plan for producing E on the draft given at I, the front eight healds of which are now used for the face ends, and the four rear healds for the back ends. The marks in M (with the exception of the crosses) represent healds raised. The interlacing diagram and a section showing the working of the first backing and the first face end are given at K and L in *Figure 7.12*. A comparison of F and G with K and L will show that when the first method of tying is employed, the back ends are affected, so that the number of back healds must be at least equal to the number of different tying positions in one repeat of the tying plan; while in the case of the second system of tying, the face ends are affected, hence the minimum number of face healds is determined in the same way.

From the above considerations it will be clear that the number of healds require in a self-stitched double cloth depends not only on the respective sizes of repeats in the face and the back cloths but also on the order of stitching. Thus, the construction given at F in *Figure 7.12* consisting of a double 2-and-2 twill would normally require only eight healds if it were stitched together by either method of tying in a 4-thread order. However, as the order of stitching consists of lifting the back ends on face picks in an 8-shaft sateen order, eight healds are needed for the back fabric alone and four more healds must be provided for the face fabric, thus increasing the total requirement to 12.

Considering the construction given at E and K in *Figure 7.12*, the face weave required eight healds by virtue of its repeat size and the back weave four. It will be clear that, if it is desired to stitch this cloth in an 8-sateen order, dropping the face ends on back picks will not increase the heald complement but the same order of tying by lifting the back ends on face picks will increase the total number of healds to 16. Thus, if it is necessary to economise on the number of healds employed, the method of stitching and the order of stitching must be considered in conjunction with the repeat size of each of the two weaves involved.

Selection of suitable stitching positions

General rules regarding the correct placing of the stitches or ties in a self-stitched double cloth were given in the opening sections of this chapter. To avoid confusion it was assumed that no difficulty would be experienced in placing a tie in such a manner that it would be adequately concealed by the structure in both the visible surfaces of a double cloth. It was also assumed that the order of the distribution of ties could always be arranged to the best advantage with the same number of ties placed on each end and each pick of the series which was employed for tying. Using the system of tying in which the back ends are raised for stitching over the face picks it may not be possible to realise the above assumption with some weave combinations and some face to back thread ratios because for perfect placement of the tie the following four conditions must coincide:

(1) The back end must be at that point away from the underside of the back cloth.
(2) It must 'surface' between two long warp floats of the face weave.
(3) The face pick over which the back end is raised must be absent from the surface of the face cloth.
(4) It must be only pulled down at a point at which its penetration into the back cloth level is covered by two adjacent weft floats on the underside of the back fabric.

Clearly, in some circumstances it will not be possible to achieve the simultaneous coincidence of all the four conditions.

Similarly, when the face ends are lowered for stitching under the back picks:
(1) The face end at that point must be absent from the surface of the face cloth.
(2) It must be lowered at a point at which two long back warp floats cover it on the underside of the back cloth.

(3) The back pick at the tie point must be away from the underside of the back cloth.

(4) It must penetrate towards the surface at a point at which it will be covered by two adjacent face weft floats on the surface of the face cloth.

Again, the simultaneous coincidence of the conditions may not, in some cases, be possible.

If it is conceded that the conditions (1) and (3) in each system of tying are absolutely compulsory then a certain degree of freedom must be accepted with regard to the conditions (2) and (4).

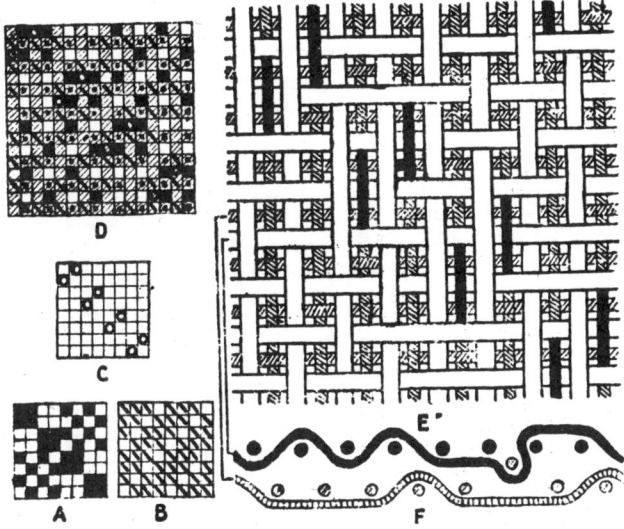

Figure 7.13

In the 1-face 1-back arrangement of the threads the ideal placement of stitches can usually be accomplished without much difficulty, except that all ties cannot be perfectly concealed in the case of such face weaves as the one shown at A in *Figure 7.13*. The complete cut which occurs in the weave between the second and third and the sixth and seventh ends and picks makes it impossible for the second and sixth back ends or picks to be concealed by the face picks or ends with a corresponding face float on both sides. In an example such as this, however, unless there is a considerable degree of contrast between the face and back yarns, it is better for a tie to fall on each thread of the series which is employed for tying rather than to select only those positions where the ties will be covered on both sides. C in *Figure 7.13* shows how the ties may be arranged for the back warp tying lifts, while at D the complete double weave is shown, the 4-thread twill given at B being used for the back fabric. The interlacing diagram of the structure is represented at E, and the interlacing of the fourth face and the fourth back pick at F. It will be seen that the ties on the fourth and eighth face picks are only covered on one side by the face warp. However, as there is the same number of ties on each back end and each face pick, the take up in weaving will be the same for each end, and the contraction in width the same for each pick. A more regular cloth will therefore be produced than would be the case if no ties were placed on the second and sixth backing ends.

When the threads are arranged in unequal proportions it is frequently impossible for the same number of ties to be placed on each face thread, although it is usually an easy matter to place the same number on each back thread. Thus, in the standard method of tying the 2-face to 1-back arrangement, usually only half the face picks are passed over when the back warp is used for stitching, and only half the face ends are lowered in the alternative system of tying. For instance, the warp ties for the Mayo weave, given at G in *Figure 7.14* are usually distributed as shown at I, the alternate face picks only being passed over by the backing ends. J shows the complete design with the 2-and-2 twill, given at H, as the back weave. The corresponding interlacing diagram of the structure is represented at K. However, by changing the position of the face weave to that shown at L, so that it is situated in relation to the back threads, as indicated at O

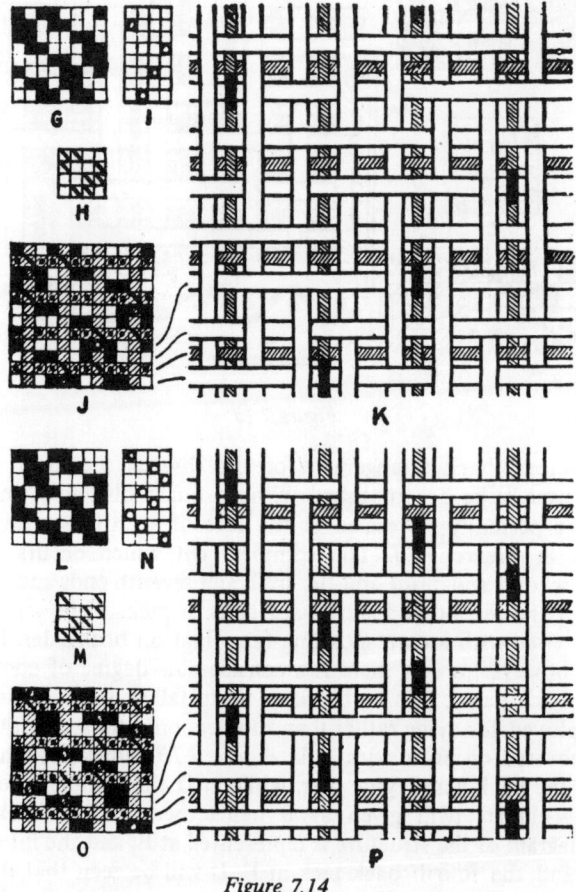

Figure 7.14

in *Figure 7.14*, it is possible for the ties to be distributed as represented at N. In this case, as shown in the design O (for which M is the back weave) and the corresponding interlacing diagram given at P, a tie is placed upon each face pick. In the repeat of the double weave, however, each back end is stitched twice to the face texture.

The interlacing diagram given at V in *Figure 7.15* shows the standard method of distributing the ties in the 2 face, 1 back arrangement when the face warp is dropped for tying. The corresponding face weave is given at Q, the backing weave at R, and the order of tying at S. When the even face ends only are depressed on the backing picks, they will, in ordinary weaves, be liable to take up in weaving more rapidly than the odd ends. This weave, however, is exceptional in the fact that the odd face ends interweave more frequently with the face picks than the even face ends, there being four intersections in eight picks in the former, compared with two intersections in the latter. This is shown in the section given at W in *Figure 7.15*, in which the dotted line shows the interweaving of the first face end, the solid black line of the second face end, and the

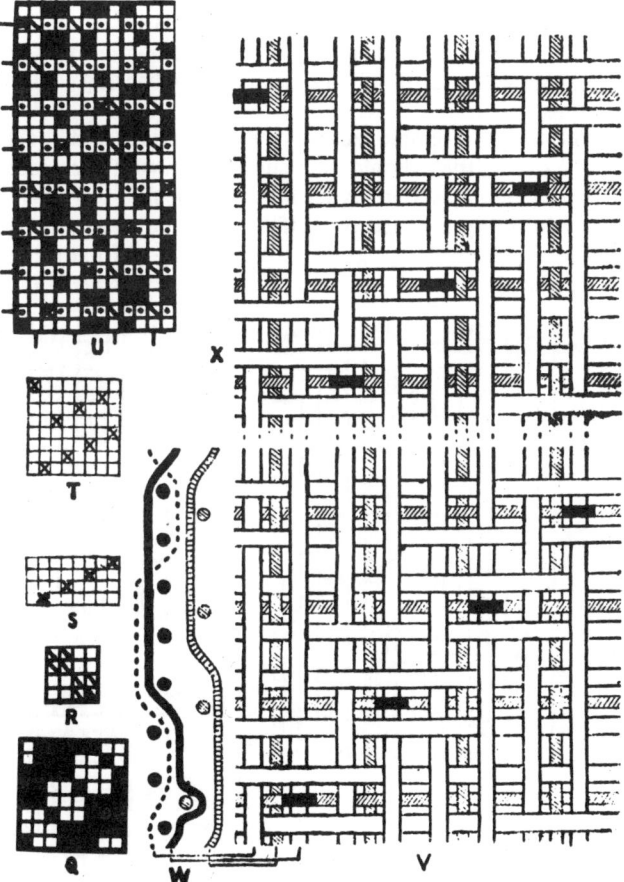

Figure 7.15

shaded line of the first back end. The placing of the ties on the looser woven ends will tend to neutralise the variation in the take-up caused by the difference in the number of intersections between the odd and even face ends. The distribution given at V in *Figure 7.15* will therefore yield the best results in weaves of this character. However, in order to illustrate a method of stitching on every face end, the interlacing diagram is extended to another repeat of the weave, as shown at X, the ties being placed on the odd face ends in the second repeat in order to balance those which are placed on the even ends in the first repeat. The plan of the ties for the two portions lettered V and X is given at T in *Figure 7.15,* while the complete double weave is shown at U. When the floats of the face weave permit it, this method of distributing the ties may be adopted with advantage for ordinary weaves. The difficulty which is frequently found in the 2 face, 1 back arrangement, of placing a tie on each face end, is one reason why tying by dropping the face ends is usually not so suitable as tying by lifting the back warp. It is better to have the ties unevenly distributed on the face picks than on the face ends, so far as the weaving of the cloth is concerned.

In order that a regular cloth may be obtained, the back weave should be suitable for the back fabric, and similar to the face weave. Thus, the loose face weave given at Q in *Figure 7.15* is backed with the 2-and-2 hopsack weave shown at R. This combination will permit the use of thick backing yarns and yield a soft under texture.

When a twill weave is employed for the face fabric and the ties are distributed in twill order, they should coincide equally with each face warp twill lines in the case of the back warp tying lifts or each face weft twill line in the case of tying by lowering the face ends. If they fall on alternate twills only, as shown in the double-cloth design given at A in *Figure 7.16* adjacent twill lines are liable to appear different from each other. The example is a double 2 up,

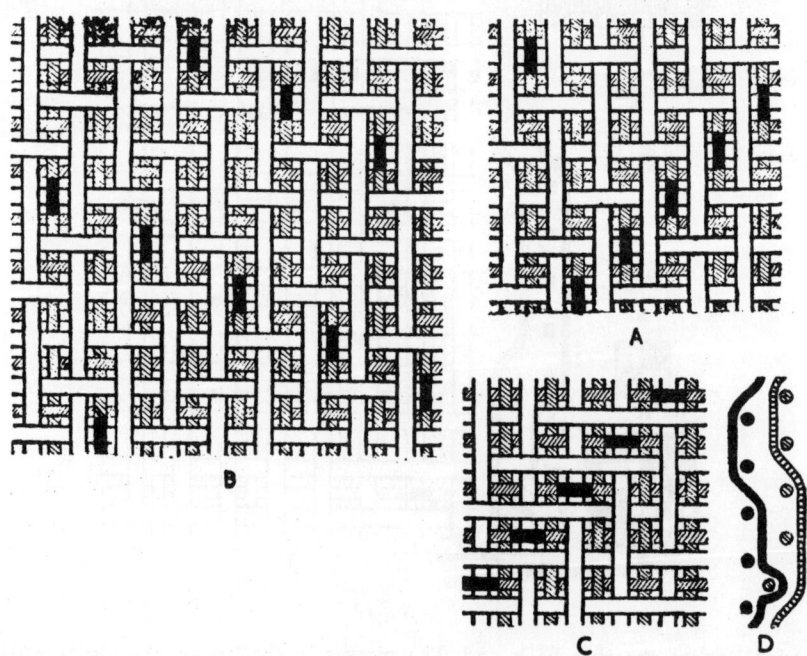

Figure 7.16

1 down twill structure, in which the threads are arranged in the order of 1 face, 1 back in warp and weft, the tying being effected by means of lifting of the back warp. By distributing the ties in 9-sateen order, as shown in the interlacing diagram given at B in *Figure 7.16* the ties will fall equally on each face warp twill line. In the construction B, however, the ties run somewhat distinctly in the opposite direction to the twill of the face fabric, and there is, therefore, a liability of a cross-twill showing in the cloth.

The interlacing diagram C in *Figure 7.16* illustrates that in face warp tying if there is a choice of two consecutive positions for a tie in the face weave, it is, as a rule, better to select that which will be covered by the greater number of following face picks. The section D, representing the interweaving of the first face and the first back end, shows this clearly. The example is a double 2 up, 3 down twill, and the threads are arranged 1 face, 1 back. It will be seen that the first face end may be lowered for stitching either between the first and second face picks, or the second and third. The former position is shown in the illustrations, and is preferable to the latter, because the beating up of two succeeding face covering picks gives a better opportunity of the tie being concealed.

The ties form the connection between the two fabrics in a double cloth, and are thus common to both weaves. Therefore, since it is first necessary for the ties to be placed according to the positions of suitable binding places in the face weave, the back weave should afterwards be suitably placed in accordance with the position of the ties.

There are three positions which a face warp or weft tie may occupy in relation to the floats on the underside of the cloth—viz., between two corresponding floats; with a corresponding float on one side, and an opposite float on the other; and between two opposite floats. Each position is illustrated in *Figures 7.17* and *7.18* which correspond with each other in every respect except that the back warp is raised for tying in *Figure 7.17*, while the face warp is employed for tying in *Figure 7.18*. The threads are arranged 1 face, 1 back in warp and weft, and the 2-and-2 twill weave is employed for both the face and the back of the cloth. In the face fabric the 2-and-2 twill is placed throughout as shown at A in *Figure 7.17*, while the ties are placed to suit the face weave as at B for *Figure 7.17* and as at C for *Figure 7.18*. The back weave, however, is placed in the three different positions given at D, E, and F in *Figure 7.17*. Thus N in *Figures 7.17* and *7.18* shows the flat view of the structure from the face side of the cloth, with the back weave placed as at D; R with the back weave placed as at E; and V, with the back weave placed as at F. Sections O, S, and W respectively show the interlacings of the first face and the first back pick of N, R, and V. The interlacing diagrams P, T, and X in each figure correspond with N, R, and V, and show the appearance of the underside when the cloth is turned over horizontally, as indicated by the numbers above the warp threads. Sections Q, U, and Y respectively show the interlacing with their respective picks of the first face, and the first back end (numbered 1 and 2) of P, T, and X.

It will be noted that in each arrangement the ties will be correctly covered on the face of the cloth by the corresponding face floats between which they are placed. By comparing the diagram N and the section I in each figure with the diagram P and the section Q, it will be seen that with the back weave placed as at D in *Figure 7.17*, the back of the cloth will be as perfect as the face, because all the four conditions of perfect placement of the ties mentioned earlier are satisfied simultaneously. A comparison of R and S with T and U, however, shows that with the back weave placed as at E, the back fabric will not be so

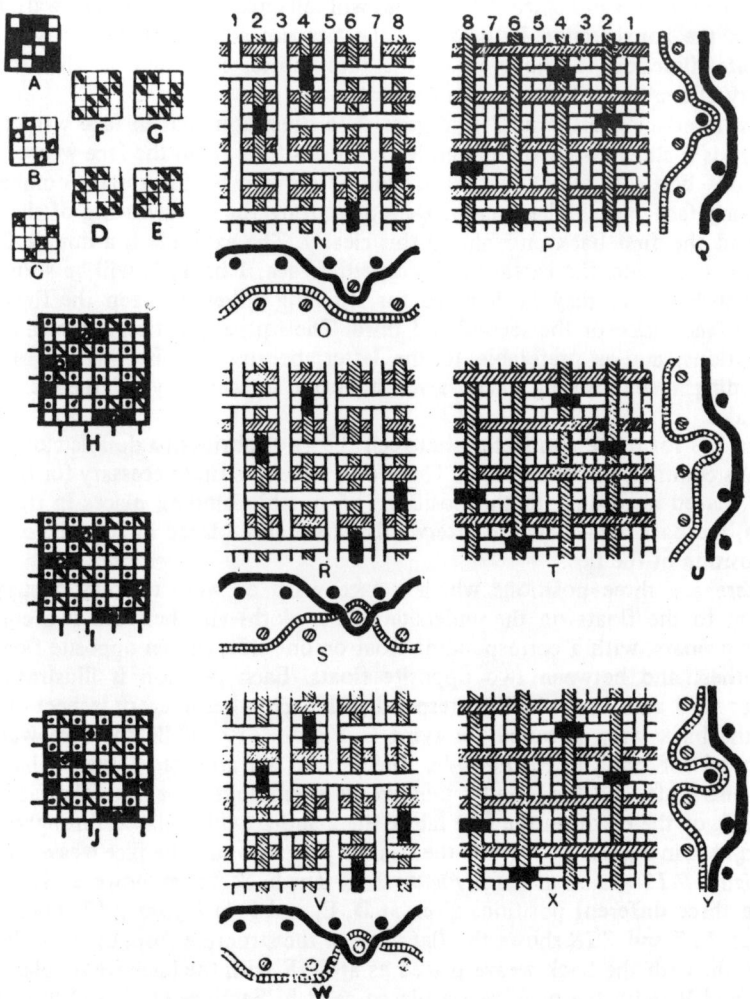

Figure 7.17

perfect as in the former case, because the position occupied by each tie on the underside is between a corresponding and an opposite back float. This kind of defect frequently cannot be avoided, as, for example, when a plain back weave is employed. Although in well set cloths the defect is practically invisible, it should only be allowed to occur when absolutely necessary. A comparison of V and W with X and Y in which the backing weave is placed as at F, shows the most serious defect which can occur in the back fabric. In this case the position occupied by each tie on the underside is between two opposite back floats. When the back warp is employed for tying, this causes the warp floats in the underside to be broken by the face weft, as shown at Y in *Figure 7.17,* while when the face warp is employed for tying the back weft floats on the underside are broken by the face warp as shown at W in *Figure 7.18.* This not only results in the ties showing prominently on the underside, but as the intersections of the back threads are correspondingly increased, the back fabric is made firmer and harder than it should be.

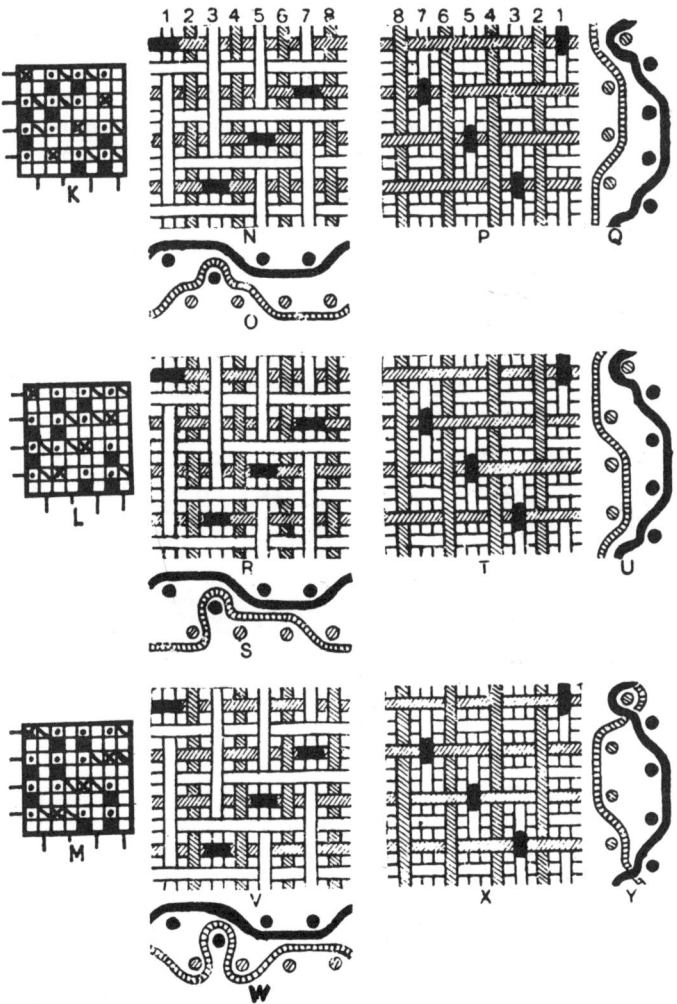

Figure 7.18

The complete plans for the interlacing diagrams given in *Figure 7.17* are shown respectively at H, I, and J. A comparative examination shows that in tying by lifting the back warp the most perfect back fabric is obtained when each back end is raised on the back picks which precede and succeed the face pick on which the tie is placed. In the same way a comparison of the plans given at K, L, and M with the corresponding diagrams in *Figure 7.18* shows that in tying by lowering the face ends the most regular back fabric results when back ends float under the back picks, one on each side of the face end which makes the stitch (crosses in K, L, and M indicate warp down). The back weave may also be placed in respect of the face weave as shown at G in *Figure 7.17*, but this will produce a similar defect to that produced by placing it as at E.

For some of the simpler standard double cloths in which the ties are dis‑ tributed in regular order, the position of the back weave in relation to the ties can be reasoned out without difficulty. Thus, in the foregoing example the best result is obtained with the face and back weaves occupying corresponding positions, as shown at A and D in *Figure 7.17*. A weft or warp float on the surface of the upper fabric should be above a similar float on the top side of the under fabric, so that where the two fabrics are in contact, a warp float of one is against a weft float of the other. The best conditions are thereby obtained for the interweaving of the warp threads of one cloth with the weft threads of the other cloth. For example, in a 1 face, 1 back arrangement of the threads, the 4-and-4 twill, shown at A in *Figure 7.19*, may be backed with the same weave in a similar position, as shown at B; while in a 2 face, 1 back arrangement it may be backed with the 2-and-2 twill in the position shown at C. If the threads are arranged in the proportion of 7 face to 5 back, the 3 up, 4 down twill in the position shown at D may be backed with the 2 up, 3 down twill in the position shown at E.

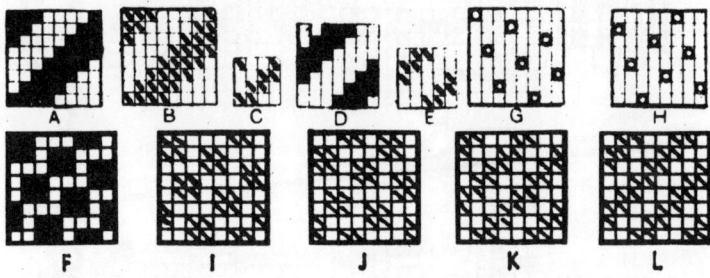

Figure 7.19

In only the simplest cases, however, it may be safely assumed, without experiment, that the best relative position of the back weave has been ob‑ tained. Thus, taking F in *Figure 7.19* as the weave for both the face and the back of a 1 face, 1 back double cloth the back warp stitching lifts may be either placed as shown in the plan G, or with the same result, so far as the face of the cloth is concerned, they may be placed one pick lower, as shown at H, the marks indicating back ends raised over face picks. With the ties placed as at G, the most perfect under-fabric will be obtained by commencing the back weave exactly like the face weave, as shown at I, but with the ties placed as at H, it will be necessary for the back weave to be changed to the position shown at J in order to secure the best results. In the same way assuming that the weave F is required to be backed with a 2-and-2 twill, K shows the best position of the weave for tying as at G, while L is the best for tying as at H. It is evident, therefore, that the positions of the face weave, the ties, and the back weave, cannot be decided upon haphazardly, but that they should bear a definite relationship to each other.

WADDED DOUBLE CLOTHS

A wadded double cloth consists of a face and a back fabric, tied together by floating back ends on face picks, or face ends under back picks as in ordinary self-stitched double cloths, with the addition of a special series of weft or warp threads introduced independently of the face and back yarns. The weft-wadded cloths thus consist of three series of weft and two series of warp threads, while in the warp-wadded cloths there are three series of warp and two series of weft threads. The wadding threads lie between the two fabrics, and are visible neither on the face nor back; hence a thicker and cheaper yarn than that used for the face and back may be employed for wadding without the appearance of the cloth being affected. The type of construction is therefore useful in cases where increased weight and substance are required to be economically obtained in conjunction with a fine face texture. The wadding threads may be introduced into any arrangement of the face and back threads, but the common proportions are 1 wadding to 1 face and 1 back, 2 face and 2 back, or 2 face and 1 back. The first arrangement is suitable when the wadding yarn is not so much thicker than the face yarn, and the second and third when very thick wadding is used.

Weft-wadded double cloths

The construction of designs for these cloths is illustrated by the examples given in *Figure 7.20* in which A is the plan of the face weave, and B of the back weave. Since the wadding yarn simply lies between the two fabrics without interweaving with either, the same conditions are necessary, so far as regards the face weave, the ties and the back weave, as in the construction of ordinary double cloths.

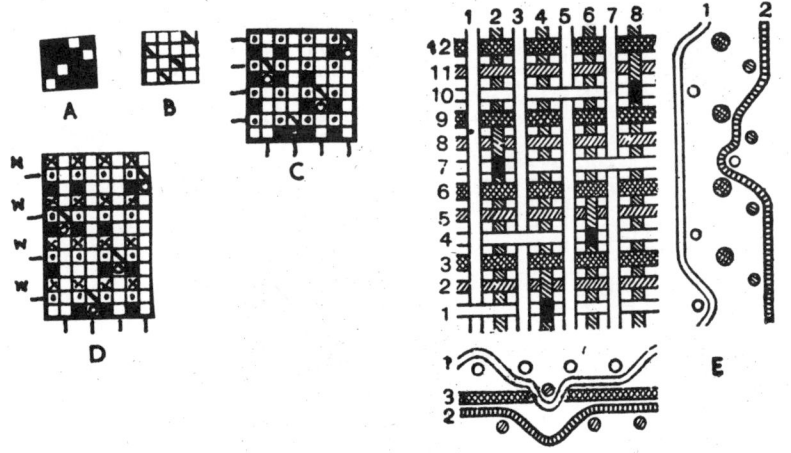

Figure 7.20

The wadded design is therefore exactly the same as the ordinary double design except for the inclusion of the wadding threads; and in order that comparisons may be made, the double weave without the wadding is given at C. In the complete design, given at D the solid squares indicate the face weave, the circles the ties (back warp up on face picks), the diagonal marks the back weave, the dots the face ends up on the back picks, while crosses are inserted to show the lifts of the wadding threads. It will be noted that in weft-wadded structures all face ends are up, and all back ends are down, on wadding picks.

In the interlacing diagram given at E in *Figure 7.20*, which corresponds with the complete design shown at D, the back and the wadding threads are shaded in different ways in order that they may be readily distinguished. The wadding threads are also represented as being of a larger diameter than the face and back threads. In the interlacing diagrams the threads, for convenience, are

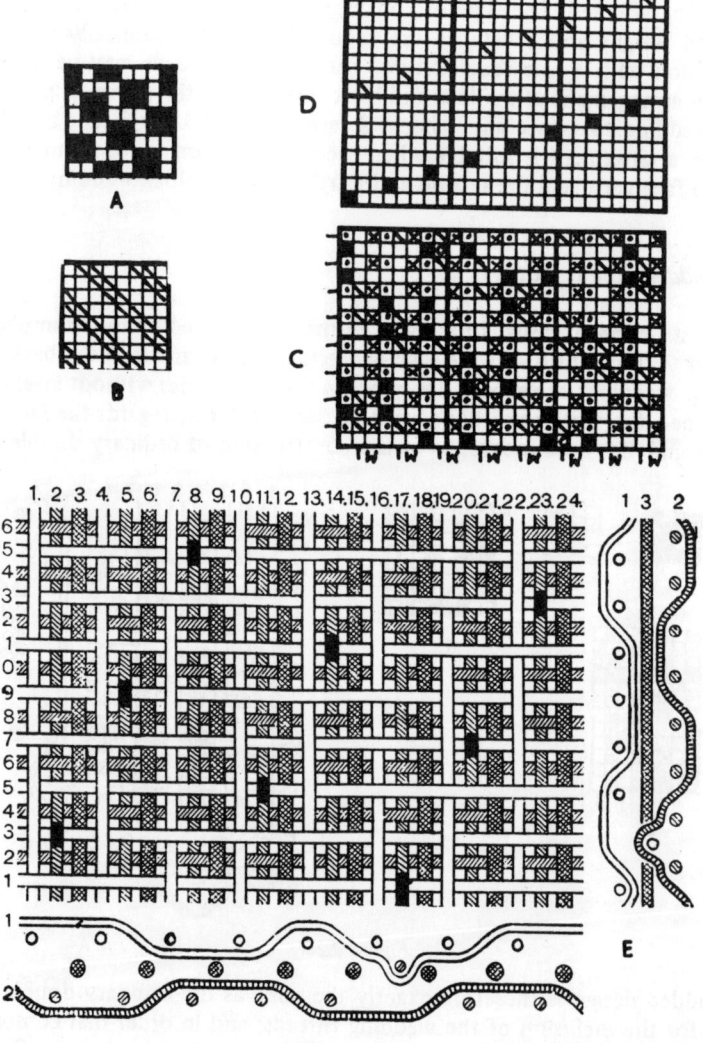

Figure 7.21

placed alongside each other at approximately uniform distances apart in the same order as in the designs. The positions of the ties are indicated by the solid marks.

In the example given in *Figure 7.20* the picks are arranged in the order of 1 face, 1 back, 1 wadding; and the ends 1 face, 1 back. The 4-thread satinette weave, warp surface on both sides of the cloth, is employed, the tying being effected by raising the backing ends in a similar order over the face picks. In the corresponding sectional views, the section on the right of the interlacing diagram shows the interweaving of the ends 1 and 2, and that below of the picks 1, 2, and 3.

Warp-wadded double cloths

The wadding yarn is more economically and conveniently introduced in the warp than in the weft but the greater strain put on the warp threads in weaving necessitates the use of a better quality of wadding material. The construction of the designs is illustrated in *Figure 7.21* in which the face and back weaves are given at A and B respectively, while the complete design is given at C and the draft at D.

In the corresponding interlacing diagrams at E the ends are arranged in the order of 1 face, 1 back, 1 wadding, and the picks 1 face, 1 back. The face weave is the 8-thread twilled hopsack, the back weave is 2-and-2 twill, and a sateen order for back warp tying lifts is used. In the warp-wadded structures the wadding ends must be raised on all back picks and left down on all face picks.

The draft for the design C in *Figure 7.21* is given at D. The wadding ends require only one heald, but in fine setts, to avoid crowding, they may be drawn on two or more healds which are then operated as one.

The introduction of wadding threads increases the strength of a double cloth in the direction of the wadding yarn; and sometimes, for the purpose of obtaining increased firmness the wadding threads are stitched to the double cloth, these stitches being placed next to the ordinary stitches in order to minimise their effect. Thus, in stitching the wadding weft in *Figure 7.20*, each back end would pass over the wadding pick which precedes the normal stitch. In *Figure 7.21* wadding ends would also lift over the face picks on the right of each backing warp stitch.

CENTRE-STITCHED DOUBLE CLOTHS

In wadding a double cloth the chief object is to get a heavy structure by introducing a yarn which is usually thicker and cheaper than the face and back yarns. In centre stitching, however, although the threads may be introduced in the same order as in wadding, and additional weight thereby be obtained, the specific purpose is to bind the two fabrics together with the centre threads, which as a rule are finer than either the face or backing threads. In this system the threads of one fabric do not interweave with those of the other fabric; the centre threads oscillate between one and the other, and lie between them when not employed for tying. The two fabrics are less firmly united than with the self-stitching, and the cloth has a softer and fuller handle. It is a useful method for cloths in which there is a great difference either in the thickness or the colours of the face and back yarns, such as overcoatings in which a check lining is woven with the face fabric, and heavy cloaking and mantle cloths which are made with coloured checks on one side and solid shades on the other. In such cloths the ordinary method of tying is not suitable, as the contrast in colour and the difference in thickness between the face and back yarns make the ties liable to show.

In the accompanying interlacing diagrams the back and centre threads are shaded in different ways, and the latter are also represented as being of smaller diameter than the face and back threads. The face and back weaves are given separately for each example. In the complete designs the solid squares indicate the face weave, the diagonal marks inclined to the left the back weave and the dots the face ends up on the back picks. The lines at the side and below the designs indicate the positions of the back threads and the centre-stitching threads are denoted by an 'S'.

Centre-warp stitching

In centre-warp stitching the following procedure needs to be observed: (a) Where no ties occur the centre warp lies between the face and the back fabric and, therefore, must be lowered on the face picks and raised on the back picks. These separating lifts are indicated in the designs by the dots. (b) In tying to the face cloth the centre ends are raised over the face picks where these are absent from the face, i.e., where they are covered by two adjacent floats of the face warp. These tying lifts are indicated by the circles in the designs. (c) In tying to the back cloths the centre ends are lowered on the back picks where these are absent from the underside, i.e., where they are covered on the under-side of the back cloth by two adjacent floats of the back warp. These tying positions are represented by the crosses in the designs and indicate centre warp down.

Thus, the instructions in respect of all the designs are: Peg or cut all marks except the crosses.

The plans in *Figure 7.22* are illustrative of the construction of double cloths arranged 1 face, 1 back, in which the two fabrics are stitched together by means of centre warp. The design D is a double 2-and-2 twill, the face weave being as at A, and the back weave as at B, while the ends are arranged in the proportion of 4 face and 4 back to 1 stitching, as indicated at D. The interlacing diagram C corresponds with the design D, the section on the right of the flat view showing the interweaving of the ends 7, 8, and 9, and that below of the picks 2 and 3.

As each repeat of the double weave given at D contains only one stitching end, the ties always occur in the same line, both on the face and back of the cloth. A better arrangement is given in the design E, in *Figure 7.22*, in which the ends are in the proportion of 2 face and 2 back to 1 stitching. The face weave and the back weave are the same as in the design D. In this case there are two stitching ends in one repeat of the double weave, which not only causes the fabrics to be more firmly united, but enables an alternate distribution of the ties to be made. This is clearly shown in the corresponding interlacing diagram given at F. The section on the right of the flat view shows the interweaving of the ends 3, 4 and 5, and that below of the picks 2 and 3. The draft is given at G, and the lifting plan at H.

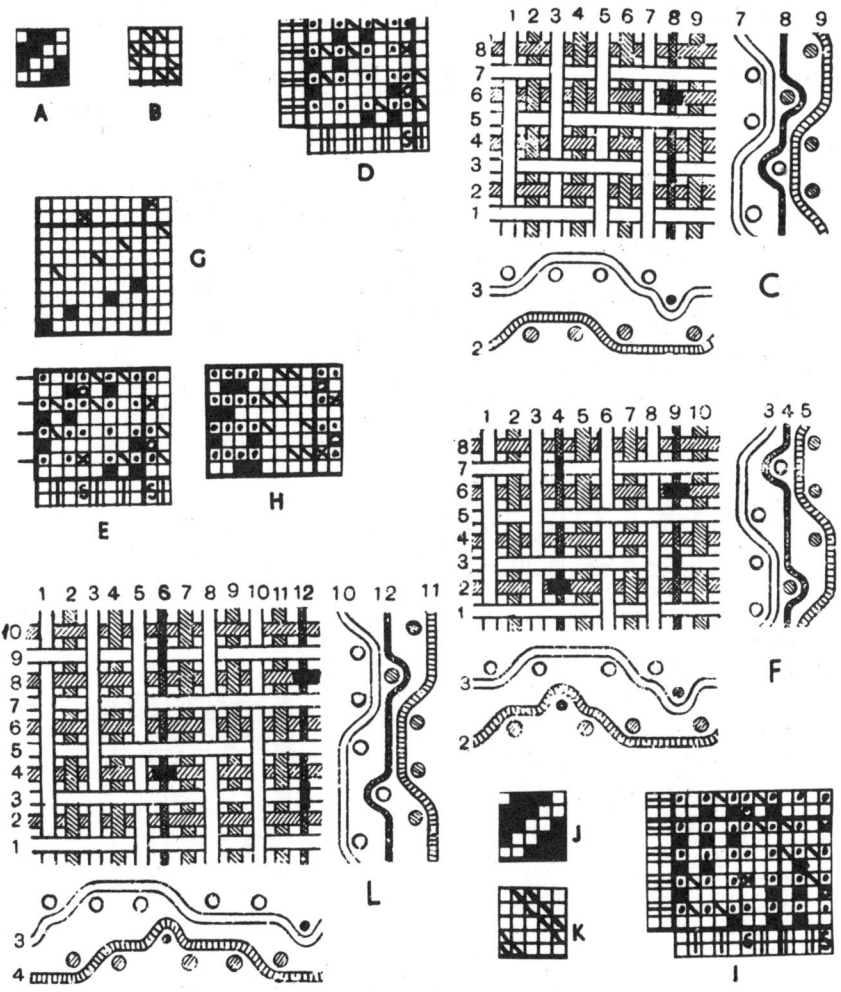

Figure 7.22

The design I in *Figure 7.22* is a 3 up, 2 down twill weave backed by a 2 up, 3 down twill with two stitching ends in each repeat. In this example the face and back weaves (given at J and K respectively) are so arranged that the direction of the twill line when the piece is turned over is the same as on the face side, the cloth being thus perfectly reversible. The corresponding interlacing diagrams are given at L, the section on the right of the flat view showing the interweaving of the ends 10, 11, and 12, and that below of the picks 3 and 4. As shown here, in centre-stitched cloths the back weave requires to be placed in such a position in relation to the face weave that the ties on each stitching thread will be about half the repeat distant from each other. Thus the second stitching end (end number 12) at L is raised for tying on the third pick, and depressed half the repeat distant on the eighth pick.

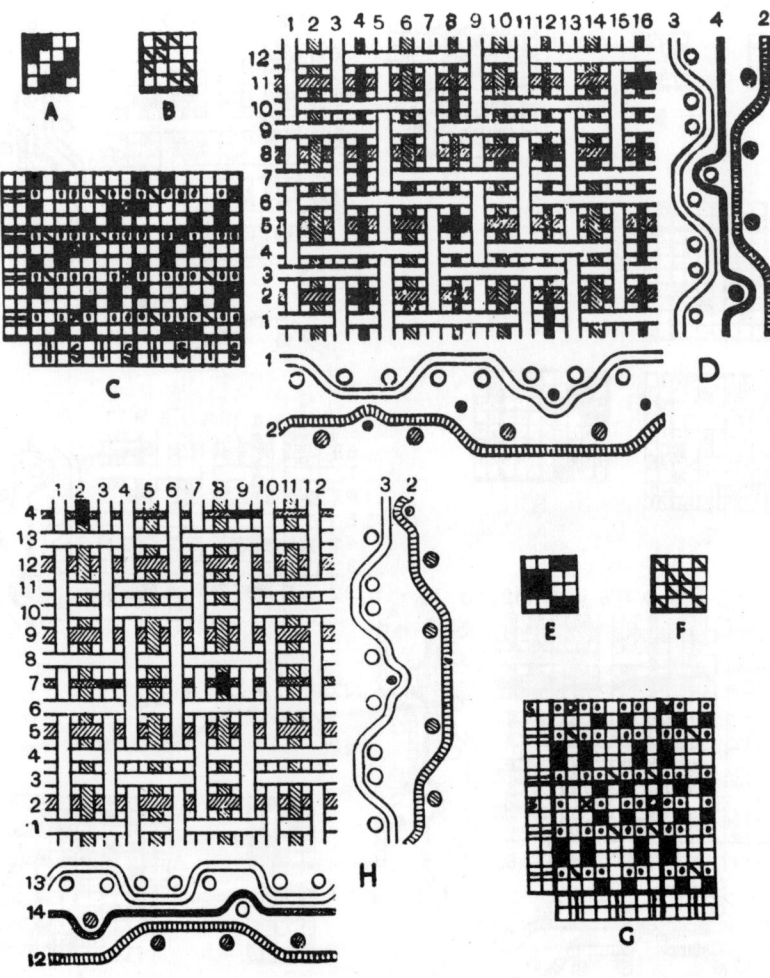

Figure 7.23

An example is illustrated at A to C in *Figure 7.23,* in which the proportion of face threads to back threads is 2 to 1, an arrangement which permits the use of very thick yarns in the under-fabric. C is a double 2-and-2 twill, with the face weave as at A, and the back weave as at B. The tying is effected by means of four centre ends in the repeat, the complete order of warping being 1 face, 1 back, 1 face, 1 centre, as shown at C. The corresponding interlacing diagrams are given at D in *Figure 7.23,* the interweaving of the ends 2, 3, and 4 being shown on the right, and of the picks 1 and 2 below the flat view. The example shows how a tartan-lined overcoating cloth is constructed. The tartan-check side is composed of the finer yarns, and is taken as the face in weaving, although in the made-up garment it forms the back; while the solid side consists of the coarser fabric which forms the back in weaving and the face when made up.

Centre-weft stitching

This form of stitching is not very often used as it reduces the rate of cloth production. This is due to the fact that when the centre weft picks are introduced the take-up must be rendered inoperative and thus the picks do not contribute to the length of cloth being produced. In constructions in which the use of centre stitching threads is essential it is, therefore, preferable to use the centre warp stitchers. However, there are some situations which make it necessary to use the centre weft and one reason for the use of this method occurs when all the existing jacks in a dobby are required to operate the face and the back healds and none are left to control the centre warp ends. Occasionally the centre weft is also used if the mounting of an extra beam required by the centre warp threads presents a particular difficulty in respect of the control or access to the warp yarns.

In using centre-weft stitching the following procedure needs to be observed: (a) Where no ties occur the centre weft lies between the face and the back cloth. To achieve this on centre weft picks the face ends are raised (as indicated by the dots in design G, *Figure 7.23*) and the back ends are lowered. (b) To achieve a face fabric stitch a face end must be dropped on a centre pick at a point at which it is absent from the surface, i.e. when it is covered by two adjacent floats of the face weft. This is indicated in the design G by the crosses. (c) To achieve a back fabric stitch a back end is raised on a centre pick at a point at which it is absent from the underside of the cloth, being covered by two adjacent floats of the back weft. This is indicated by the circles in the design G. Thus, the lifting instructions for the design G are: Peg or cut all marks with the exception of the crosses.

The plans E to G in *Figure 7.23* illustrate the principle of stitching by means of centre weft. The double 2-and-2 hopsack weave is employed, the face weave being given at E, and the back weave at F. The picks are in the proportion of 4 face and 2 back to 1 stitching as indicated at G, one repeat of the double weave thus containing two centre picks. The complete design is given at G, and in the corresponding diagrams represented at H, the interweaving of the ends 2 and 3 is shown alongside, and of the picks 12, 13, and 14 below the flat view.

8

Interchanging Double Cloths

In the introduction to the previous chapter it has been stated that double cloths could be joined together by interchanging fabric layers. Each interchange represents in effect a stitch and when such interchanges are frequent the cloths are very firmly united but if the intervals between the changes are considerable then the two layers remain separate over a large area which may give rise to puckering or other forms of cloth distortion. It may also lead to undue wear as the separate cloth 'pockets' between the interchange points would be free to 'ride' one upon another thus creating interlayer rubbing. In such circumstances it may be advisable to stitch such areas together additionally by one or the other method of self-stitching obeying the normal and well established rules explained previously.

The interchanging of the threads means that the series which actually alternates between the face and back of the cloth can no longer be designated as the face or the back yarn because in effect it will occasionally be the one and occasionally the other and should be best identified by the colour or by the position which it occupies, i.e., say, odd or even. It will be appreciated from the principles of double-cloth construction established previously that the face cloth is brought to the top by means of the separating lifts. Thus in a cloth in which the ends and the picks are both arranged in an alternate 1 black, 1 white order the black cloth will form the face when, irrespective of its weave, all the black ends are raised on all the white picks, and, by converse, a white cloth will occupy the top position when all the white ends are raised on all the black picks. The separating lifts do not prejudice the interweaving of the threads within their own cloths but merely ensure that given layers occupy either the top or the bottom positions in the composite structure.

INTERCHANGING DOUBLE PLAIN CLOTHS

These fabrics are produced in a variety of materials for suitings, over-coatings and furnishings in designs consisting of stripes, checks, spots and more elaborate figured effects. In the simple design ranges the effects can be achieved by two different methods: (a) By retaining a constant and 1-and-1 colour arrangement in the warp and in the weft whilst changing the position of the separating lifts; (b) by retaining the same weave and changing the colour pattern. The former method is more common because it is capable of greater versatility of effect, the latter being mainly confined to the production of stripe patterns. Whichever method is employed it is preferable to commence the design by constructing first a simplified and condensed motif to indicate which colours of the threads are going to appear on the face and which on the back at any given area of the design. The degree of condensation in the motif is by two in each direction so that one square of the motif represents two ends, 1 dark and 1 light, and two picks. Each fully worked-out design in the examples which follow is preceded by such a motif which shows clearly the intended disposition of the two differently coloured layers within a repeat.

Effects due to changes in the position of separating lifts with continuous 1-and-1 colour arrangements.

The motif A and B in *Figure 8.1* indicates a stripe design in which at A four dark ends together with the dark picks work on the surface whilst at the same time the light ends and picks form the back cloth, and at B the reverse takes place with the next four light ends forming the face and the corresponding four dark ends producing the back cloth. The fully worked-out designs and interlacing diagrams are given separately at C to correspond with the portion A and at D for the portion B of the motif. The colour arrangement of 1 dark, denoted by the solid marks, and 1 light, denoted by the diagonal marks, is shown on the margins of the fully worked-out designs and this notation is retained throughout this chapter. In the actual fully worked-out designs the solid squares indicate the weave of the dark threads, the diagonal marks the weave of the light threads, and the dots the separating lifts using the normal convention in which a mark equals warp up. Comparing the design C with D it will be noted that the weave marks in both are identical, the only difference being in the placement of the separating lifts. In the design C these lifts cause all dark ends to be raised on all light picks and in the design D result in all light ends being lifted on all dark picks thus achieving the exactly opposite disposition of the two cloths clearly indicated by the interlacing diagrams C and D and by the sections underneath them which show the interlacing of the first dark and the first light pick in each case.

When the two weaves are combined in stripe formation a compact structure results due to the interchange of the dark and the light weft between the face and the back. If however, a broader stripe is required the interchange which in effect stitches the two layers together may not be sufficiently frequent to provide the necessary degree of interlayer cohesion and in such circumstances it may be necessary to self-stitch the two cloths which is usually done by lifting some back ends on face picks at infrequent intervals as shown by the design E in *Figure 8.1* in which the circles denote the stitching lifts. In the case of design E the light threads form the back cloth and, therefore, the light ends are raised for tying on the dark picks when, however, the layer positions are reversed and the dark threads form the back cloth then the dark ends would be raised for stitching on the light picks. As in a plain weave where adjacent floats do not exist it is difficult in some constructions to provide perfect concealment for the stitch points. For this reason the interchanging plain cloths are self-stitched only when this is deemed absolutely essential.

It will be appreciated that the cloth layers can be combined in a variety of ways to produce coloured stripes of different widths. Smart single, double, or

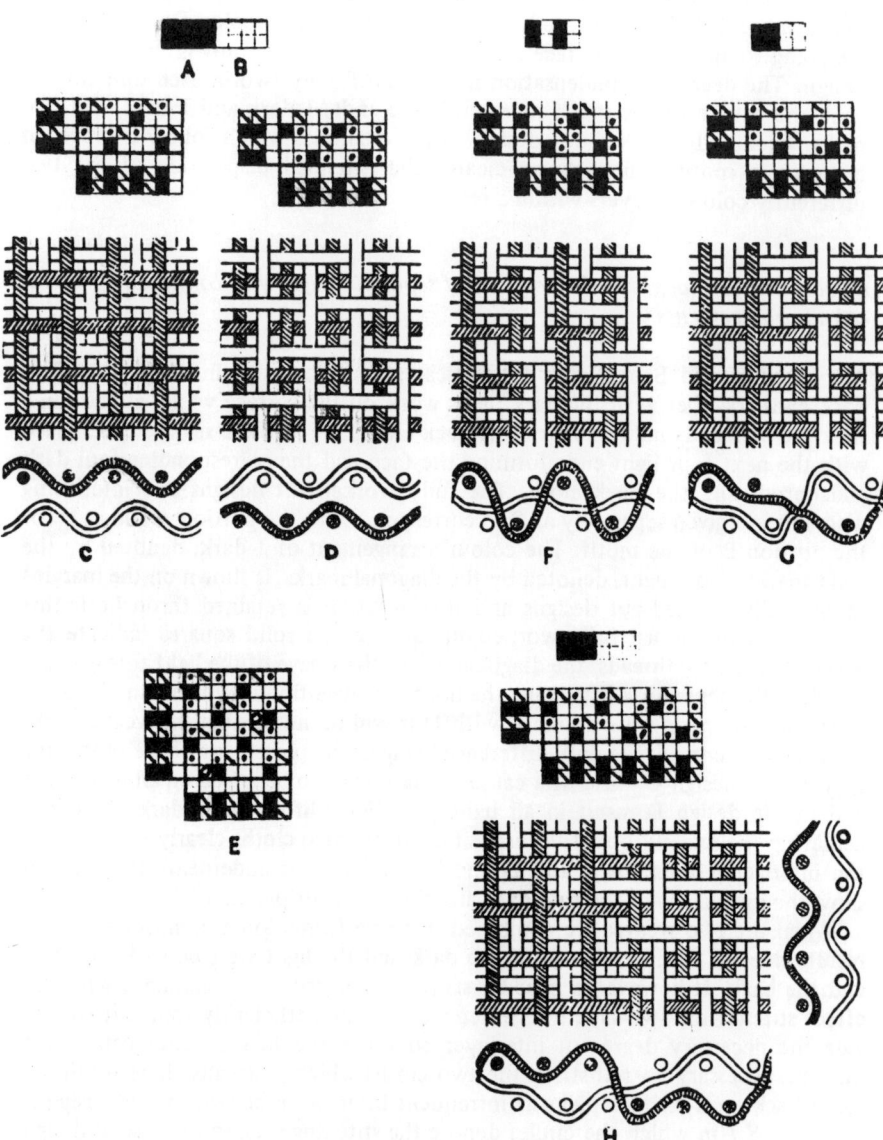

Figure 8.1

treble hairlines are most favoured but wider stripe combinations are also produced. The motifs, designs and interlacing diagrams given at F, G, and H in *Figure 8.1* illustrate respectively the construction of the single, double, and treble hairline effects. It will be noted that in each example the thread colours is 1 dark,

1 light in the warp and in the weft and the occurrence of the alternate stripes is due to the appropriate placement of the separating lifts. The warp cross-sections underneath the interlacing diagrams at F, G and H show that the stripe effect is due to the order of interchange of the weft yarns while the warp ends in each stripe remain continuously either in the top layer or in the bottom as decreed by the separating lifts. This is shown by the weft cross-section to the right of H which illustrates the interlacing of the first dark and the first light end. The appearance of a cloth in which stripes of varying width are combined together is illustrated by the sample A in *Figure 8.2*.

In order to ensure that correct junctions are obtained at each interchange it is necessary to retain exactly the same weave relationship between the dark and the light cloth. It will be clear from the examination of the designs and inter-lacing diagrams A to H in *Figure 8.1* that such a relationship has been main-tained. Thus, on the odd picks of each cloth the even ends are raised and on the

Figure 8.2

even picks the odd ends of each layer form the top shed. The disturbance of this relationship is liable to result in unwanted cut marks and floats which dis-figure the clean appearance of these constructions.

In addition to vertical stripe designs the technique of the displacement of separating lifts can be used to produce horizontal hairlines as shown at I in *Figure 8.3*. Normally, however, the horizontal hairlines are combined with vertical stripes to form checks or designs similar to the one illustrated in the motif J and the corresponding fully worked-out structure in *Figure 8.3*. The same type of design is also shown in the cloth sample B, *Figure 8.2*, incorporated within a more elaborate stripe effect.

Another method of adding variety of effect is to produce a mixed colour stripe in which the dark warp is made to weave plain with the light weft in the top cloth whilst the light warp weaves plain with the dark weft in the bottom cloth. In this way an intermingled stripe is obtained in which the resultant hue, due to the close juxtaposition of the two colours, is midway between the one and the other. Such a stripe is illustrated in the sample B in *Figure 8.2* and is also shown in the motif K in *Figure 8.3* and in the accompanying fully worked-out design and diagrams. In the motif the mixed colour stripe is indicated by the shaded squares whilst in the full design the plain weave lifts of the dark ends are indicated by the crosses and similar lifts of the light ends by the circles to make identification of this portion of the design easier.

In the foregoing solid stripe designs it will be noted that 2 face or 2 back ends are invariably brought together where the change from one colour to the other is made. This tends to create a crack in the surface and it is sometimes possible, with the same order of colour, to avoid it by arranging the weaves in a special manner as indicated at L in *Figure 8.3*. In this case each stripe section contains an odd number of ends, so arranged that the two ends which are brought together where the interchange takes place are face ends. This, unavoidably, causes more ends to appear on the face than on the back, although the picks are equal on both sides. The interlacing diagram at L and the section below it show that with this form of arrangement the cloth is no longer reversible, the pattern formed on the surface of the cloth being 3 dark, 3 light, and on the back 2 light, 2 dark. Thus, in this form of arrangement each stripe section contains one back end less than the number of face ends and the method could be extended to achieve similar effects with a pattern of, say, 2 dark, 2 light on the face, and 1 light, 1 dark on the back, or, 4 dark, 4 light on the face, and 3 light, 3 dark on the back, etc. One necessary precaution which must be taken when this system is used is to ensure that the same number of face ends is sleyed through each split of the reed. It will be noted that in the case of this structure no motif has been given—the omission is deliberate because the motif is applicable to reversible effects only in which it is assumed that the reverse of the effect shown on the face is equal in size but opposite in colouring. In the structure given at L the size of the reverse side effect is different from that achieved on the face.

So far, the interchanging double plain weave has been shown capable of producing attractive stripe effects in solid and intermingled colourings. It is, however, equally suitable for the production of check, spot and figured effects. The elaborately figured constructions are considered in Chapter 10; at this point two simple examples are shown to indicate the manner in which both the thread elements, i.e., the warp and the weft can be made to interchange. M in *Figure 8.4* represents the motif and the fully worked-out design for a popular counterchange effect. It will be noted from the design that the same weave relationship is retained throughout the repeat and the effect is due, as before,

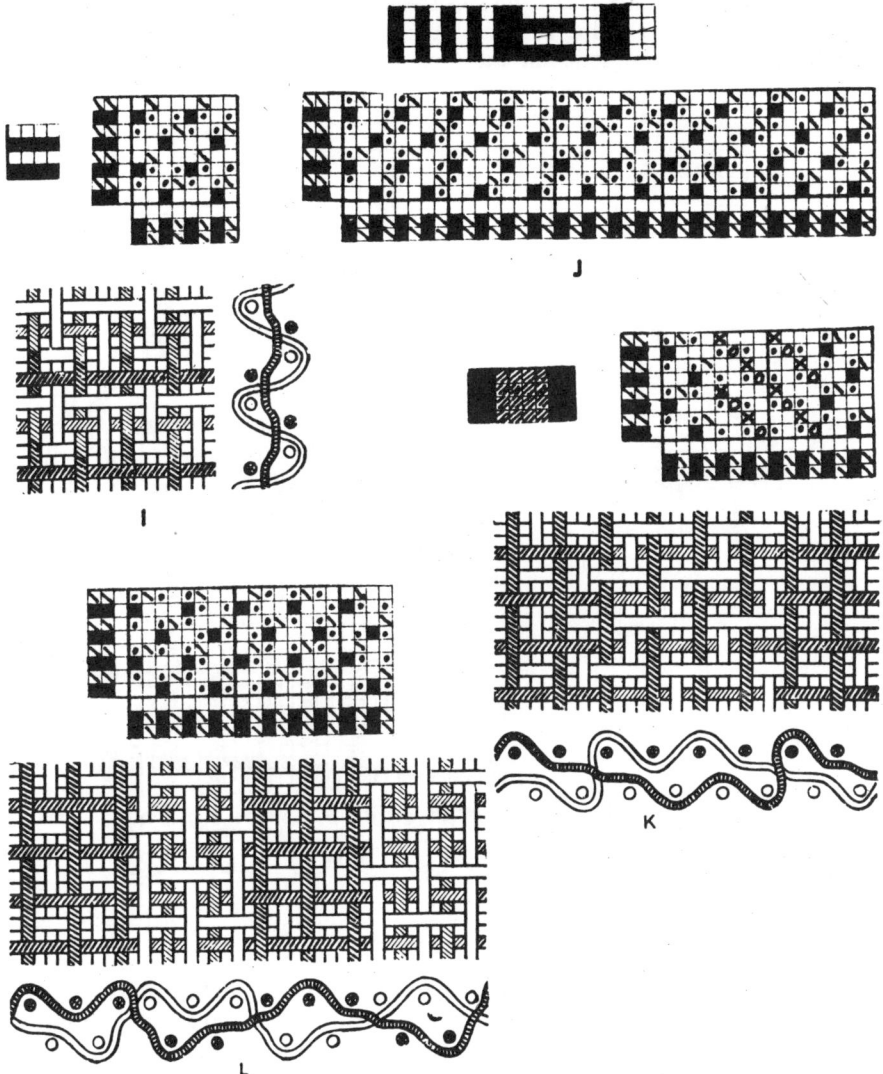

Figure 8.3

to the positioning of the separating lifts. Thus, where the dark cloth is required on the surface the dark ends are raised on the light picks and where the light coloured cloth is displayed on the face the light ends are raised on the dark picks. The cloth is perfectly reversible and the effect is illustrated in the sample C in *Figure 8.2*.

A light coloured spot on a dark ground given at N in *Figure 8.4* is constructed in a different way. The effect is slightly dissimilar on each side because due to the special arrangement of the separating lifts the light spot on the face is produced by three light ends and picks but the corresponding dark spot on the back comprises only two dark ends and picks the method being similar to the one adopted in producing the stripe design L in *Figure 8.3*. The spot produced in this manner appears more solid on the face and is not separated from the dark

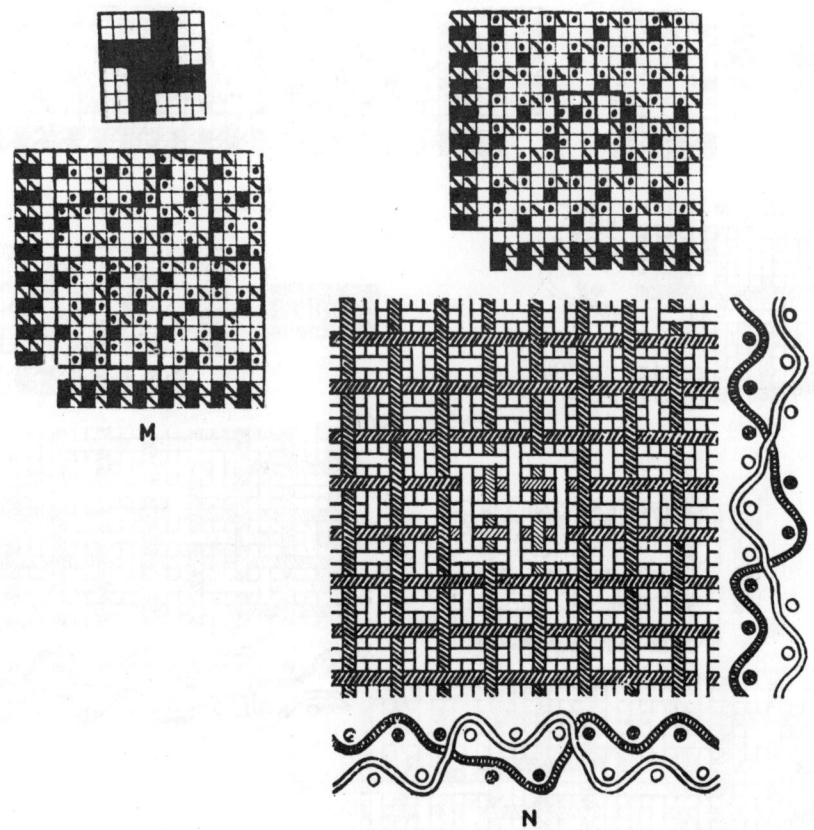

M

N

Figure 8.4

ground by the distinct cuts or cracks which are liable to develop at each interchange when the method illustrated at M in *Figure 8.4* is employed. The interlacing diagram and the design N in *Figure 8.4* show clearly this method of construction and the warp and weft sections below and alongside the interlacing diagram which respectively indicate the interweaving of the picks 6 and 7, and the ends 6 and 7, illustrate the manner in which both the warp and weft threads change from one cloth layer to the other.

Effects due to changes in the colour arrangement

In this method of producing stripe and check effects the position of the separating lifts is constant and the different colours are brought to the surface by changing the relationship or the order of the dark and light threads. Thus, if it is the odd ends that are raised on the back picks and the odd ends in one stripe happen to be dark then a dark stripe will result; if, on the other hand, they happen to be light then a light coloured stripe will be produced. It follows, therefore, that if the odd ends be designated the face ends and the even ends the back ends then to obtain the necessary colour change in the stripe in one section the ends will be arranged 1 dark, 1 light, whilst in the next section the arrangement will have to be 1 light, 1 dark. The picks are usually arranged in a continuous 1 dark, 1 light order as, apart from the check effects which will be dealt with later, there is no benefit to be derived from changing the wefting order.

The method of construction is demonstrated by the designs, interlacing diagrams and sections at A, B, C and D in *Figure 8.5*. The simplified motifs above the designs show that at A and C a dark coloured cloth is on the face whilst in B and D the light coloured threads form the face fabric. Previously used notation is employed, i.e. the weave lifts of the dark ends are shown solid, those of the light ends by the diagonal marks and the separating lifts by the dots. The colour arrangement is denoted by the solid squares for the dark ends and picks and the diagonal marks for the light ends and picks around the margins of each fully worked-out design as before.

It willl be noted that identical face effects are formed in A and C, and also in B and D. The only difference between the two pairs of similar effects is the relationship of the back weave to the face weave, a different tabby of the plain weave being used between A and C, and again between B and D. The arrangement A or C may be equally well combined with either B or D so that greater latitude is available in joining the weaves in this than in the previous system. Other advantages of this system are that a more even cloth results because the bringing of two face or two back ends together at the point of interchange is avoided,

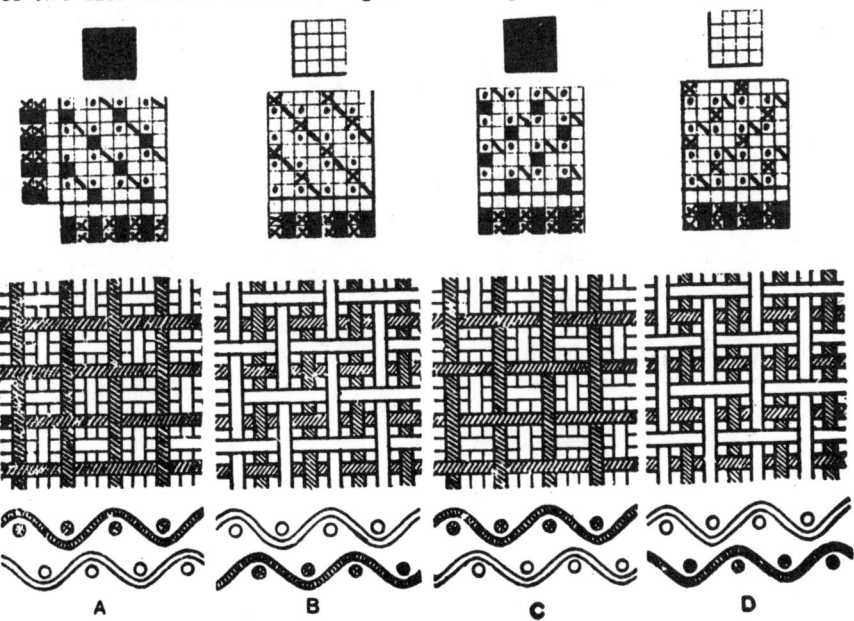

A   B   C   D

Figure 8.5

and if required, the weaves at the interchange can be made to cut. For worsted cloths cutting of the weave is frequently preferred as the pattern in clear finished cloths is thus brought out smartly and definitely. However, cutting in small sections and several times in succession should be avoided as it tends to produce a ribbed cloth with an unnecessarily harsh handle.

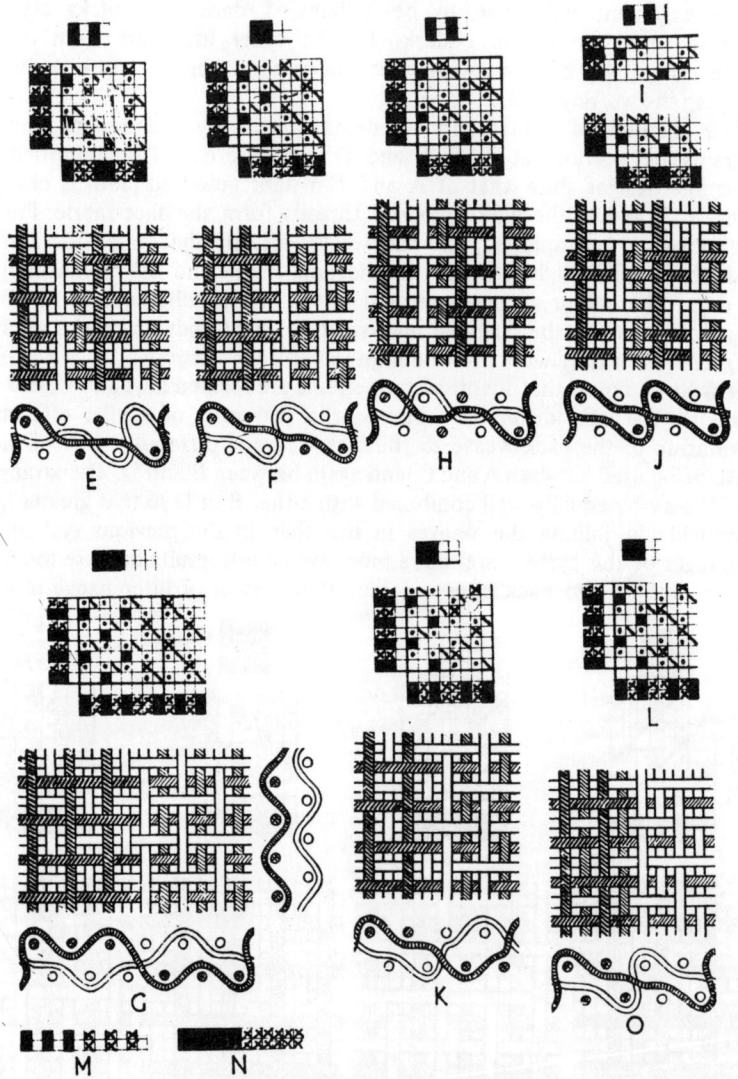

Figure 8.6

In E, F, and G in *Figure 8.6,* the weaves are arranged to cut at each change of the pattern, as is shown in the designs. In each case, however, the weaves may be arranged without the cutting, or to cut at one change and not at another. Thus, H shows how the vertical hairline may be obtained with the weaves cutting every 4 ends, instead of every 2, as in the design E; while I shows how the same effect may be formed without the weaves cutting. The most common method of producing the single-thread vertical hairline, with the weaves cutting every four ends, is illustrated by the design J in *Figure 8.6,* and the corresponding diagrams. In this case, while the arrangement of the warp colouring is the same as that for the designs E, H, and I, the threads interlace in the same order as in the design F. The wefting, however, is changed to 2 dark, 2 light, hence the pattern may be produced in looms with changing boxes at one side only. Design K in *Figure 8.6* shows how the 2-thread stripe pattern may be formed with the weaves cutting every eight ends and L without the weaves cutting, the colour effect in each case being the same as that produced by the design F.

If the plans E to L in *Figure 8.6* are carefully analysed and compared with the corresponding interlacing diagrams, it will be seen that each colour of weft interweaves only with its own colour of warp, and solid lines of colour are formed on both sides of the cloth. It will also be seen that with the warp colours arranged on the principle indicated below the designs E to L, the pattern on the underside of the cloth is exactly the reverse of that on the face, which in many cases is a distinct advantage. If, however, the colour pattern on the underside is of little or no importance, the back ends may be in either dark or light, or practically any colour, since these ends remain on the back all the time, while the face ends are continuously on the face, as shown in the section alongside the interlacing diagram given at G in *Figure 8.6.* In such a case, therefore, so long as the face ends are arranged as to colour in accordance with the form of pattern which is required on the surface, the order of warping may be materially simplified by colouring the back warp in sections to conform with the face order of colouring. For example, the design given at G in *Figure 8.6* will produce the pattern 3 dark, 3 light on the surface, if the face ends are arranged in the order of 3 dark, 3 light, as shown at M. So far as regards the face of the cloth, the complete warping plan may therefore be 6 dark, 6 light, as shown at N, which is a simpler arrangement than the warping order which is necessary in order that the face and back will be alike. In the same manner, of course, the warping plan for the 2-dark, 2-light stripe produced by the design F may be 4 dark, 4 light. The back warp may also be in a different colour from either of the face warp colours, and it may be different in thickness. The weft, however, should be in the same colours as the face warp, and similar in thickness, and it is usually an advantage to insert more picks than ends per cm, as greater solidity of colouring is thereby obtained. The interlacing diagram of the design F, with the warp colours arranged 4 dark, 4 light, is given at O in *Figure 8.6,* while the section shows the interweaving of the picks 1 and 2. If the diagram given at F is compared with O, it will be noted that on the face the patterns are respectively the same, the change in the warping plan simply affecting the underside, where an intermingled colour effect is produced.

The method of constructing a specific stripe design is illustrated at P to T in *Figure 8.7.* It is assumed that the ends are arranged 1 face, 1 back, the picks 1 dark, 1 light, and that the pattern to be formed on the surface of the cloth is 3 dark, 2 light, 3 dark, 1 light, 2 dark, 1 light (shown in the motif above P). The position of the back ends, the order of wefting at the side, and the face warping plan below the ends, are first indicated, as shown at P. The complete order of colouring the ends may afterwards be arranged, as described with reference to

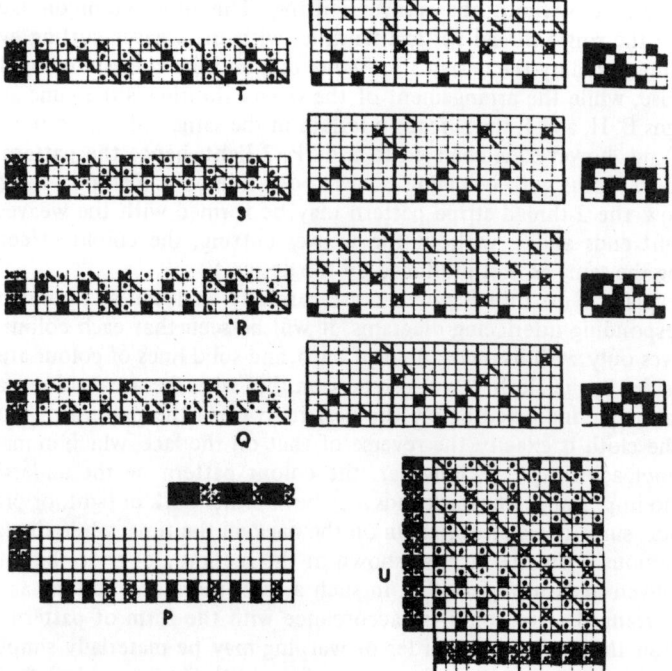

Figure 8.7

the weave G, *Figure 8.6*, according to the effect which is required on the underside. Each colour of weft is arranged to interweave on the face only with its own colour of warp; and if the weave A or C, *Figure 8.5*, is inserted in the dark sections, and the weave B or D on the light, the required colour pattern will be formed on the surface of the cloth, although, as has already been shown, the weaves may be combined in many different ways. Q shows the weaves arranged to cut at each change of the pattern, while in R a cut is made only at each side of the 3's of dark colour. S shows another arrangement with weaves cutting at each change, and T with the cutting as at R. Other combinations may be made which will produce the same colour pattern on the surface, but one of the chief objects to note in arranging the weaves is the simplification of the drafting.

The respective drafts are given alongside the design Q, R, S, and T, and the lifting plans on the right of the drafts. Two sets of four healds each, are required in each case but while in the drafts for Q and S the front four healds produce the dark sections of the pattern, and the back four healds the light sections, the drafts for R and T are arranged with the face threads drawn on the front four healds and the back threads on the back four healds. The draft for Q is the simplest arrangement, because not only are the mails per unit space of the healds in each set the same, but the order in which the threads are drawn in can be readily followed. The order of drafting is 1, 2, 3, 4, throughout, a change from one set of healds to the other being made at each change of colour. In the draft for the design R a definite system is also employed for the simplification of the drawing-in. Thus, the dark face ends are on the odd healds, and the light face

ends on the even healds of the front set. The first heald of the front set is followed by the first heald of the back set, the second by the second, and so on.
Further, in the case of tappet shedding, the design and the draft may, with care,
be made to conform with any given lifting plan. For example, if the threads 19
to 22 of the draft for the design Q are drafted 1, 2, 3, 4, instead of 3, 4, 1, 2,
the lifting plan for Q will produce the design R. The design S may be taken to
illustrate a defective combination of the weaves to fit a given lifting plan. In this
case with the lifting plan the same as for Q the draft is unsatisfactory, because
not only is it difficult to follow, but there is an extreme variation in the setts of
the healds.

The method of arranging the colours and the weaves for producing a *check*
pattern in sections of 4 dark, 4 light colouring on both sides of the cloth is
illustrated by the design U in *Figure 8.7*. The order of colouring in the weft is
the same as that in the warp—viz., 1 dark, 1 light for four times, and 1 light,
1 dark for four times. The pattern is obtained by the interchanging of both the
warp and the weft threads.

Effects in three and four colours

As there are only four picks in the repeat of the double plain weave, the limit
as to the number of colours which can be introduced is four, if each line on the
surface is required to be solid in colour. If one colour is brought to the surface
for two consecutive threads, in order to form a plain weave there must be at
least two picks of that colour out of the four in the repeat of the wefting plan.
Hence, in such a case, the limit as to the number of colours is three, of which
two must form single lines of colour.

In the designs given in *Figures 8.8* and *8.9* the differently coloured threads
are represented by different marks along the bottom and at the side of each
plan. The marks which indicate where the face warp lifts over the face weft
correspond with the marks which are used to represent the colours; the diagonal
marks indicate the back weave, and the dots the face ends up on the back picks.
The motif above each design indicates the surface colour of each section of the
pattern.

Interlacing diagrams and sections showing the interweaving of the threads
are given of the majority of the designs; and in order that comparisons between
them may be readily made, the threads in the diagrams are shaded in different
ways to represent the colours. The colours are also indicated by numbers—shade
1 corresponding with the full squares on the design paper, shade 2 with the
circles, shade 3 with the crosses and shade 4 with the vertical lines.

The designs A and B in *Figure 8.8* and the corresponding diagrams illustrate
the method of producing patterns in four colours in the system of arrangement
in which two face or two backing threads are brought together where the weaves
interchange. The shades in the warp are arranged in the order of 1, 2, 3, and 4
throughout, and in A the wefting is in the same order. The weave in section 1 of
A brings the odd ends and picks in shades 1 and 3 to the surface, while the even
ends and picks in shades 2 and 4 pass to the underside. Each face end floats
under its own colour of weft and over the other colours, vertical lines in shades
1 and 3 being formed on the face. The weave in section 2 of A brings the even

ends and picks in shades 2 and 4 to the surface, while the odd ends and picks in shades 1 and 3 pass to the underside. Again each face end floats under its own colour of weft and over the other colours, with the result that vertical lines in shades 2 and 4 are formed on the face. The complete pattern produced on the surface by the design A is a single thread stripe with the shades arranged in the order of 1, 3, 1, 3, 2, 4, 2, 4. By combining four threads of section 1 with four threads of section 2, the single thread vertical hairline in four colours is formed.

The design B in *Figure 8.8* is exactly the same as the design A, except that in the weft the shades 2 and 4 are reversed in order that they will occupy different positions in relation to the corresponding shades in the warp. An examination and comparison with the diagrams given at B will show that in section 3 each face pick passes under its own colour of warp and over the other colours, with the result that horizontal lines in shades 2 and 4 are formed on the surface. In the same manner, if the shades 1 and 3 are reversed, section 1 will produce horizontal lines. The complete effect produced on the surface by the design B is a stripe arranged—one of shade 1, one of shade 3, one of shade 1, one of shade 3, and four of the horizontal hairline in shades 2 and 4.

With regard to the underside of the cloth, an examination will show that in A, section 1 produces a horizontal hairline in shades 2 and 4, and section 2 a similar effect in shades 1 and 3; while in B, section 1 produces a vertical hairline in shades 2 and 4, and section 3 a horizontal hairline in shades 1 and 3.

It is evident from the foregoing that with the warp arranged 1-and-1 throughout in four shades there is considerable scope for producing variety of pattern in stripe form by varying the spaces occupied by the weaves and by changing the order of wefting. The weaves may also be arranged and combined to form figured styles in which the pattern is due not only to contrast in colour, but the direction of the lines of colour may be varied as desired. Three-colour effects may be obtained by employing the same shade for all the odd or for all the even threads.

Stripe patterns in three and four colours are usually produced in the system in which the ends are arranged 1 face, 1 back throughout. In the warping plan the chief point to note is that the face ends are coloured according to the form of stripe which is required on the surface, and in the following examples the colouring of these ends only is indicated below the designs.

Examples C, D, E, and F in *Figure 8.8* show different ways of constructing a vertical hairline in three colours. It will be noted that the arrangement of the face warp colours is the same in each case, but in C and D the shades in the weft are in the order of 1, 2, 1, 3, and in E and F in the order of 1, 1, 2, 3. In C and E the weaves do not cut; in D one cut, and in F two cuts are made. The interlacing diagrams and the sections of the weaves C and F represent two repeats of each construction. In the diagrams the back ends are not shaded, because, so far as regards the face of the cloth, they may be in any colour. Thus, the complete warping plan may be two of shade 1, two of shade 2, and two of shade 3, etc. If, however, solid lines of colour are required on the underside, it is necessary for each back end to be in the same colour as the pick over which it is raised. For example, in the design C, the first back end is raised on the pick in shade 3, and the second and third on the picks in shade 1. If, therefore the back ends are coloured to correspond, when the cloth is turned over the shades on the underside will be in the order of 1, 1, 3. The arrangement of the back ends as to colour,

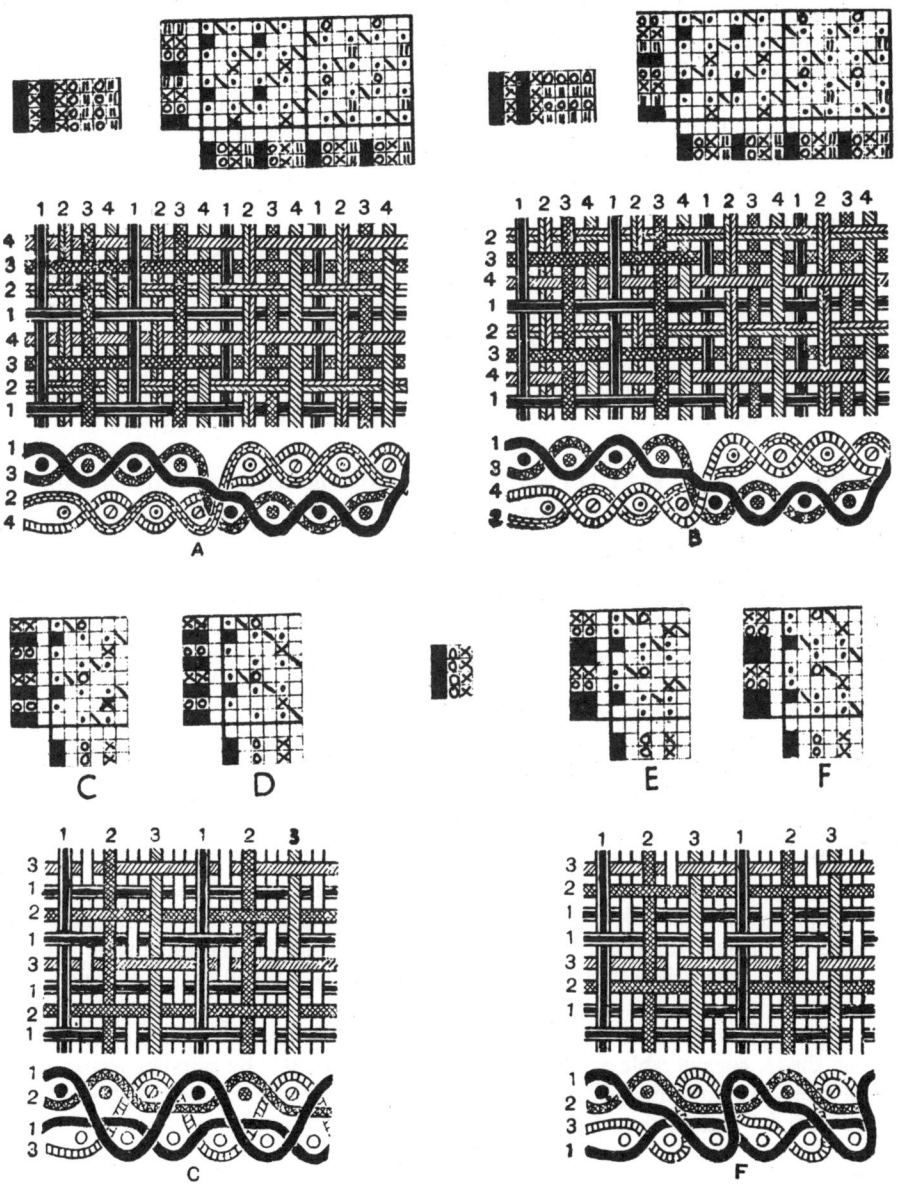

Figure 8.8

which will produce solid lines on the underside, is indicated above each plan. It will be noted that the pattern is not the same as that on the face, except in the design E, for which the shades in the complete warping plan will be in the order of 1, 3, 2, 1, 3, 2.

Two standard three-colour hairline arrangements are given at G and H in *Figure 8.9*. In the pattern produced by G the shades on the surface are in the order of 1, 2, 1, 3, and by H in the order of 1, 1, 2, 3, as indicated by the numbers above the warp threads in the interlacing diagrams. The complete warping plan for G may be two of shade 1, two of shade 2, two of shade 1, and two of shade 3; and for H, four of shade 1, two of shade 2, and two of shade 3; but if the back ends are coloured in the order indicated above the designs, the pattern on the underside will in each be exactly the same as that on the face.

So long as two of the colours are used only to form single-thread stripings, a great variety of stripe effects can be produced by varying the spaces occupied by the third or ground colour. An example is given at 1 in *Figure 8.9*, which will produce the following pattern on the surface:

```
Shade 1 — 1  .  2  .  3  .  2  .  1  .  1  .
Shade 2 — .  1  .  1  .  1  .  1  .  .  .  .
Shade 3 — .  .  .  .  .  .  .  .  .  1  .  1
```

The designs J, K, and L in *Figure 8.9* show different methods of arranging the weaves and the colour for producing the vertical hairline in four shades, with the ends arranged in the order of 1 face, 1 back. It will be noted that in J the weft colours are in the same order as the warp colours, but in K and L the second pick is in the same colour as the third face end, and the third pick as the second face end. For the reason that each colour of weft must pass over its own colour of warp in producing the vertical hairline, the arrangement shown at J is somewhat defective, because the intersections of the face ends form a twill line. With the double plain weaves combined as in J, the warp-backed 3-and-1 warp twill is really formed. In the same manner K is the warp-backed 4-thread warp satinette weave. In the design L, however, the weaves are purely double plain, arranged to cut every four ends. The interlacing diagrams (showing two repeats) and the sections given at J and L will show that in each case vertical lines of colour in the order of 1, 2, 3, and 4 are formed on the surface.

The same remarks apply with reference to the colouring of the back ends, as in the case of three-colour effects. If they are arranged in the order indicated above the plans, solid lines of colour will be formed on the underside. The shades in the complete warping plan for L, in *Figure 8.9* will then be arranged in the order of 1, 4, 2, 3, 3, 2, 4, 1.

Although in the four-colour effects the pattern is limited to single lines of each shade, considerable variety of effect can be obtained by suitably arranging the face warp colours and the weaves to correspond. For example, the design M in *Figure 8.9* will produce the following single-thread stripe:

```
Shade 1 — 1  .  1  .  .  1  .  1  .  .  .  .  .  .  .  .  .
Shade 2 — .  1  .  1  .  .  1  .  1  .  .  .  1  .  .  .  .
Shade 3 — .  .  .  .  1  .  .  .  .  1  .  1  .  .  1  .  1  .
Shade 4 — .  .  .  .  .  .  .  .  .  1  .  1  .  .  1  .  1
```

In constructing such a style, however, it is necessary to remember that when more than two colours are used, a change of colour may not cause an interchange of the picks to take place. Thus in M no interchange takes place between the shades 1 and 2 and the shades 3 and 4. Care must therefore be taken to group

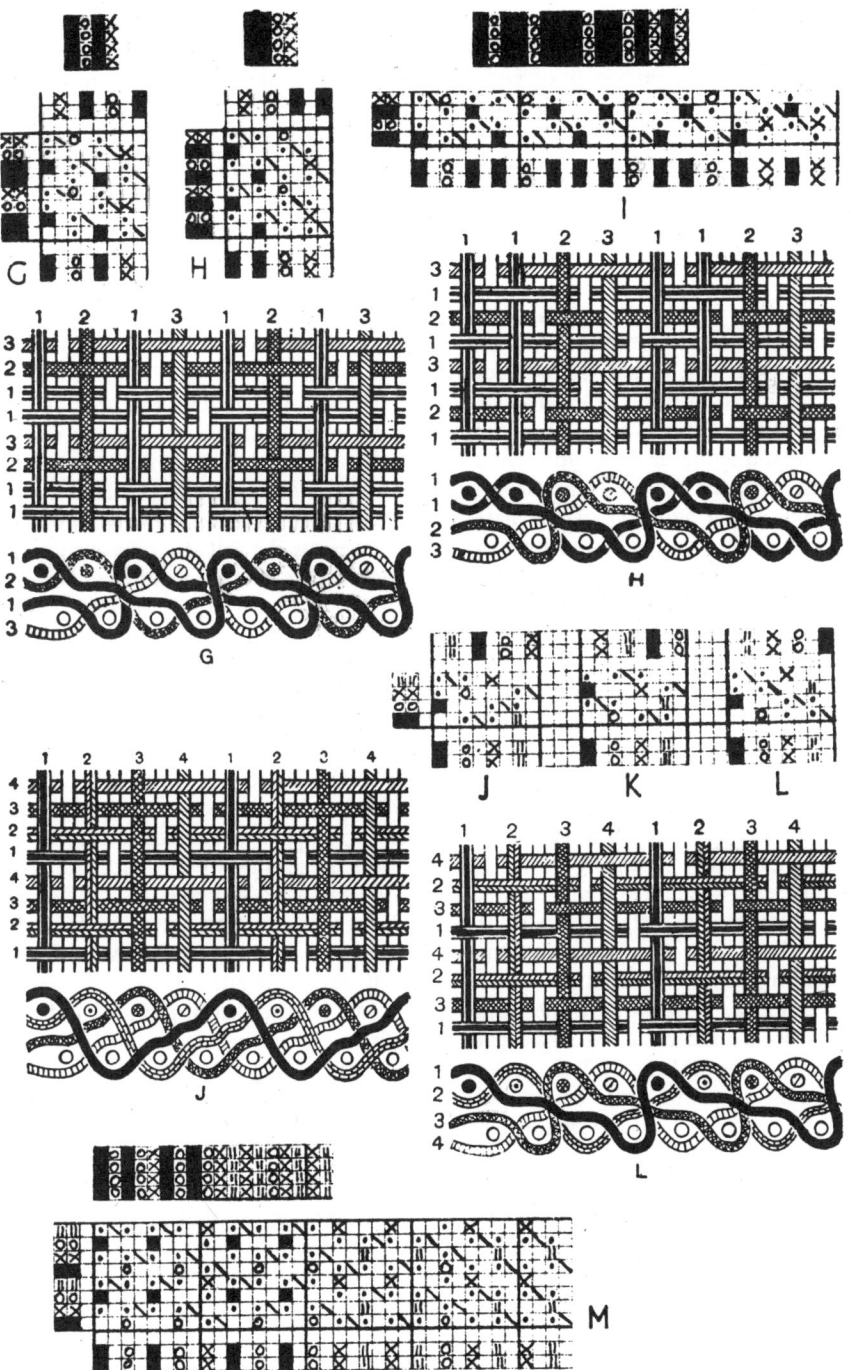

Figure 8.9

the shades in such a manner that the weave in any one section does not occupy too large a space, or the cloth will be liable to cockle. In making a design the weave marks for each pick of weft require first to be related to its own colour of face warp, the marks on the back ends being then added to the best advantage. Experiments may be made in changing the order of wefting and altering the weaves to correspond to find the most satisfactory arrangement.

INTERCHANGING DOUBLE TWILL AND SATEEN STRIPE DESIGNS

While the double plain, owing to the neat appearance, firmness, and good wearing quality of the cloth, is the standard weave used in the production of patterns in which each section requires to be in solid colour, similar effects may be obtained by using other double weaves. These also enable a larger number of colours to be introduced. Thus the double 3-thread twill permits the use of six colours, the double 4-thread twill and the 4-thread sateen the use of eight colours; while with the double 5-thread sateen any number of colours up to 10 may be employed. In addition, if the pattern is required in large sections, firmness of structure may be obtained by tying the weaves on the self-stitching principle. In small patterns, however, the interchanging of the threads where the weaves are combined gives sufficient firmness.

Numerous examples might be given to illustrate the various ways in which patterns may be formed; but as the principles involved are the same as in the construction of the double-plain effects, the examples N to Z in *Figure 8.10* will be sufficient for the purpose. N and O are the opposite double 3-thread warp-faced twill weaves, arranged on the system in which two face or two backing ends are brought together where the weaves are combined. In N the odd threads are on the surface and even threads on the back, while in O the even threads are on the surface, and the odd threads on the back. Various schemes of colouring, each necessarily repeating on six threads, are indicated above and alongside the plans, each shade being represented by a different kind of mark. At P the order of colouring in warp and weft is 1-and-1 throughout; and assuming that six threads of each weave are combined in stripe form, as shown in *Figure 8.10,* the pattern formed on the surface will be three threads of shade 1 and three threads of shade 2, as shown by the motif to the left of the warping plan P. Q will produce a solid coloured stripe pattern in three shades, arranged on the surface in the order of 1, 1, 2, 2, 3, 3. R is a four-colour arrangement, the shades being brought up in the order of 1, 1, 3, 2, 2, 4. S is in five shades, the order on the face being 1, 1, 2, 3, 4, 5; while T produces an effect in six shades in the order of 1, 2, 3, 4, 5, 6. The surface arrangement of the threads is clearly indicated by the motifs on the left of each warping plan. It will be noted in the designs that each face end passes under its own colour of weft and over the other colours, while each back end is raised over its own colour and passes below the other colours. Solid vertical lines of colour are thus formed on both sides of the cloth. The double 3-thread weft-faced twill weave may be arranged in the same manner to form horizontal lines.

Although the list is by no means complete, the foregoing examples illustrate the diversity of effect which can be obtained in one design by varying the arrangement of the threads as to colour. In any of the schemes of colouring, however, still further diversity can be produced by varying the spaces occupied by the weaves. In addition, the weaves may be combined as in the interchanging double plains to form check and figured patterns in two or more colours.

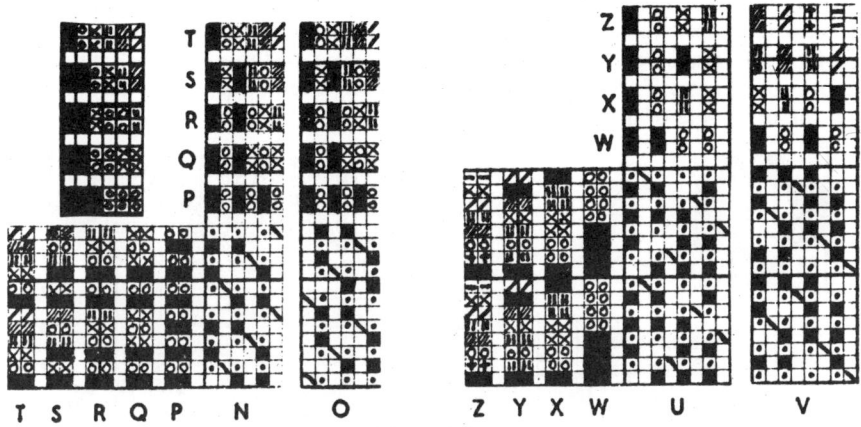

Figure 8.10

The plans given at U and V in *Figure 8.10* are the opposite double 4-thread warp satinette weaves constructed on the system in which the ends are arranged 1 face, 1 back throughout. Four different colour plans for the face ends are given above the designs at W, X, Y, and Z, and the corresponding weft colour plans, similarly lettered, are shown alongside. It is assumed that eight threads of each weave are combined in stripe form, although, as will be understood, the space occupied by each may be varied as desired. The face warping plans indicate the colour patterns which will be formed on the surface, while the chief point to note in arranging the weft colours is that each pick passes over its own colour of warp. W is a two-colour pattern, the shades on the surface being in the order of 1, 1, 2, 2, 1, 2, 1, 2. X is in four shades in the order of 1, 2, 4, 3, 3, 4, 2, 1; Y in six shades in the order of 1, 2, 1, 3, 4, 5, 4, 6; while Z shows how a single-thread stripe in eight shades may be arranged. The colour of the back ends is not indicated, but if solid lines are required on the underside, each must be in the same colour as the pick over which it is raised.

CUT EFFECTS IN INTERCHANGING DOUBLE CLOTHS

It has been demonstrated in the preceding sections of this chapter that distinct cut marks, similar to those obtained in ordinary herringbone and diaper constructions (see *Watson's Textile Design and Colour*), could be obtained in double plain cloths by selective use of the cloth interchange principle. The same techniques could be extended to double twill and other weaves if required.

In the following, several other methods are shown which further emphasise the structural versatility of the interchanging double cloths.

Cut effects produced by interchanging the threads

The effect indicated by the motif A, the fully worked-out design B and the interlacing diagram and sections given at C in *Figure 8.11*, represents a simple structure which is used to produce a reversible cloth with a soft handle due to loosely interlaced threads which are displayed on the face and on the back. The surface appearance of this structure is similar to a 10-thread hopsack weave as shown by the motif. This is achieved by floating, alternately, on the surface and on the underside of a plain ground cloth extra warp and weft threads which are

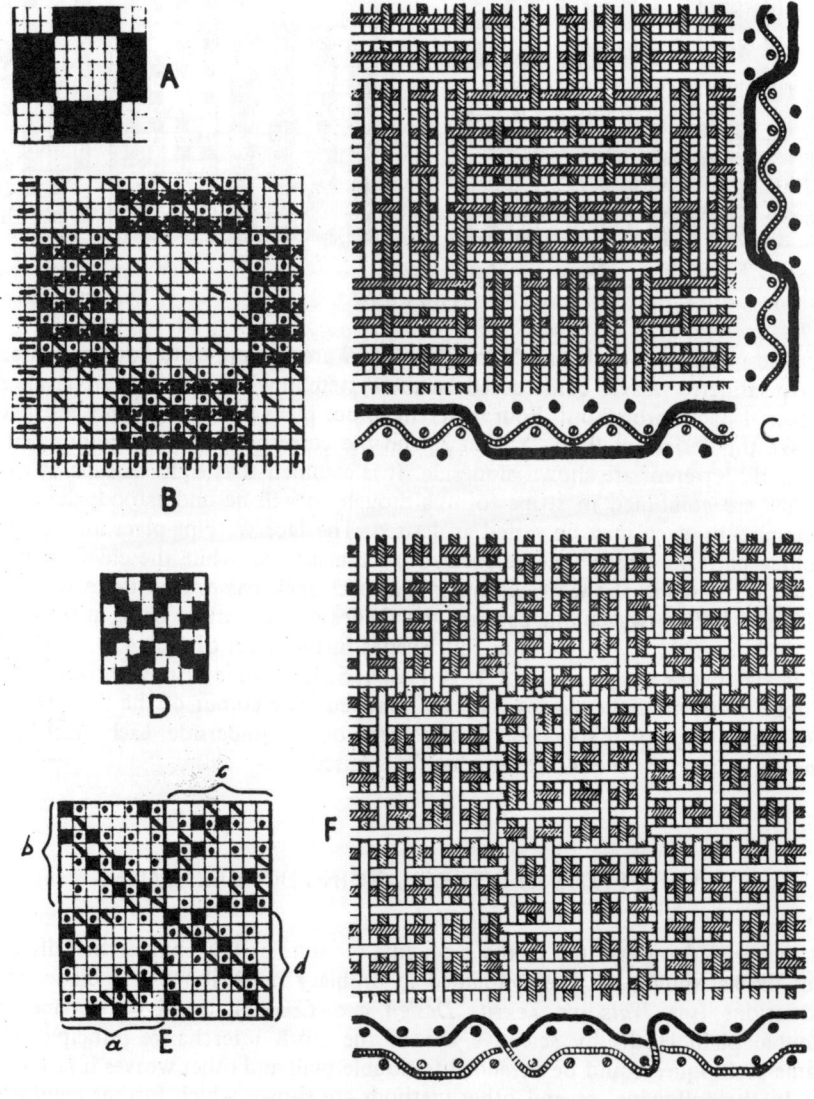

Figure 8.11

considerably thicker than those used for the ground fabric and which, therefore, cover it completely. The fine, even numbered ends and picks weave plain, as indicated at B by the diagonal marks, and form the firm centre structure through which the odd yarn elements interchange.

Where the odd ends are on the face, which corresponds to the solid portions of the motif A, the odd picks are on the underside of the ground cloth. This is obtained by lifting the odd ends on all the picks, as represented by the solid marks and the dots at B and by lifting the even ends on odd picks, as indicated by the crosses at B.

When the odd picks are displayed on the surface, which corresponds to the blank portions of the motif, the odd ends are on the back of the plain ground cloth. This is achieved by dropping all the ends on odd picks and the odd ends also on the even picks.

The interchanges are shown clearly by the interlacing diagram and the sections at C in *Figure 8.11* from which it will be also apparent that structurally this fabric may be classified as a continuous extra warp and extra weft cloth. It is included here because it defines the principle of the thread interchange particularly well. The cut lines in this cloth run both vertically and horizontally, and occur at each point at which the warp float is opposed to the weft float; they are particularly distinct because each cut mark also represents the point at which the floating threads alternate between the face and the back of the plain ground structure.

D, E and F in *Figure 8.11* indicate respectively the motif, the fully worked-out design and the interlacing diagram of a different construction in which distinct cuts are produced by the thread interchange. The motif shows a typical diaper arrangement in which cut marks would be produced even in a single cloth structure due to the opposition of the warp float by the weft float at the boundary of each section. In this construction the effect is more clearly defined because at the boundaries of the sections certain ends or picks are also made to interchange between the face and the back cloth. Thus, in section *a*, odd ends and picks form the face cloth, in section *b*, the face is formed by even ends and odd picks, in section *c*, by even ends and even picks, and in section *d*, by odd ends and even picks, whilst the opposite sets of threads in each respective section produce the back cloth. In the fully worked-out design E the weave lifts of the odd ends are represented by the solid marks and those of the even ends by the diagonal marks. The dots represent the separating lifts which in an interchanging cloth indicate the lifts of whichever ends happen to be, at any given part of the construction, the face ends over whichever picks happen to be at the same point, the back picks. The interlacing diagram, which consists of 1½ repeats in each direction, shows the interchanges clearly as the threads are shaded when they pass from the face to the back of the cloth. The warp section at C represents the interlacing of the first and second pick of the construction.

As the threads in this structure alternate between the face and the back it will be obvious that additional variety of effect may be obtained by arranging the threads in different colour patterns. For example, with odd ends and picks coloured dark and even ends and picks coloured light, section *a*, forms a solid dark area, section *c*, a solid light area, whilst a mixed colour effect results in sections *b* and *d*.

Cut effects produced by the use of special cutting threads

This is the more common method of producing the cut or sunk effect, and it is usually employed in cloths in which the threads are arranged in the proportion of 2 face to 1 back in warp and weft. In this system both the face and the back threads assist in forming the line, and it is necessary for the weaves to be arranged in precise order in relation to the threads to obtain the best results. In the first place at each cut two cutting threads between a pair of back threads are arranged to weave in 2-and-2 order with the face picks, and to oppose each other with their floats, as shown by the circles on the fourth and fifth ends and picks of the face plan given at G in *Figure 8.12*. Also each face float of two is arranged to include a face end or pick on each side of a back end or pick. The face weave is

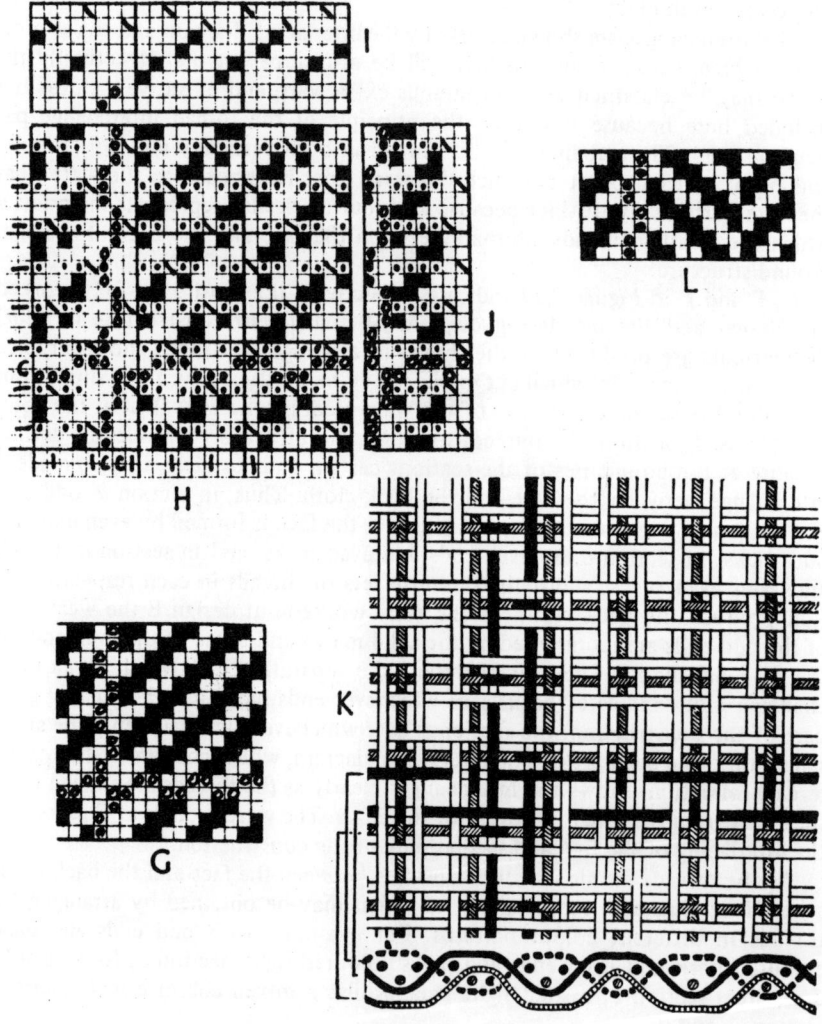

Figure 8.12

then placed, as far as possible, to support the cutting threads. Thus in G the float of one cutting thread is a continuation of the twill, while the other duplicates the float of the face weave. If the face weave is placed so as to oppose the cutting threads, as shown at L in *Figure 8.12*, which illustrates a defective plan, the line is made more open and not so distinct.

In constructing the complete design, which is given at H, the back thread between each face float of two is given a corresponding float, the cutting threads thus interweaving in 3-and-3 order, as shown by the circles on the sixth and seventh ends and picks of H. Each float passes under, or over, 1 face thread, 1 back thread, and 1 face thread. The plain weave also requires to be placed to support the cutting threads, as shown by the diagonal marks which are inserted to coincide with the centre of each float. In some cases, in order to increase the sunk effect, the cutting ends are woven as tightly as possible from a separate beam, in most circumstances, however, these ends weave tighter than the rest, and will create a distinct cut line even when placed on the same beam as the normal face ends. The draft for H on the lowest possible number of healds, is given at I, and the lifting plan at J.

The flat view given in the upper portion of K in *Figure 8.12* shows the interweaving of the ends and picks 1 to 18 of the weave H, the threads which assist in forming the cut being indicated in solid black. The solid line in the section given in the lower portion of K shows the interweaving of the first face cutting pick, the dotted line of the second, and the shaded line of the back pick which precedes them. The method of interweaving not only forms the sunk effect between the face cutting threads, but the face and back fabrics are tied very firmly together along the cuts.

9

Multi-layer Fabrics

TREBLE CLOTHS

In treble cloths there are three series of warp and weft threads which form three distinct fabrics, one above the other. Except for the ties, when a face pick is inserted all the centre and back ends are left down; when a centre pick is inserted all the face ends are raised, and all the back ends are left down; while when a back pick is inserted, all the face and centre ends are raised. The face ends and picks interweave with each other to form the face fabric, the centre ends and picks to form the centre fabric, and the back ends and picks to form the back fabric. By interweaving the centre ends or picks with the face and back picks or ends, the three fabrics are joined together, and the resulting cloth is equal in thickness and weight to the three single fabrics. Greater weight combined with equal fineness of appearance can thus be obtained in this than in the double system of construction. The weight of double woollen structures is frequently increased by excessively shrinking the cloth in the milling process, the chief disadvantages of which are that its elasticity, air permeability and clarity of effect are liable to suffer. This does not occur when increased weight is obtained by making the cloth three-fold, hence the treble principle can be advantageously employed in preference to the double system in adding weight to cloths which require little shrinking in the finishing processes.

Systematic construction of treble cloths

The method of constructing a treble cloth is illustrated stage by stage in *Figure 9.1* where, it will be noted, the weave in each of the three fabrics is not only the same but it is also started on the same footing. This creates the most favourable conditions for tying and should be used whenever identical weaves are required in all the layers.

A 2-and-2 twill is used in each fabric and this is represented at A, B and C which correspond to the face, centre and back cloths respectively and differ from one another only in the type of mark which is employed in each case.

Once the weave of each fabric layer is determined the construction can be commenced in the following order:

(1) Mark out around the margin of the design repeat the order of arrangement of the three series of threads. In the case of the example given at D the order for both the warp and the weft is 1 face (f), 1 centre (c) and 1 back (b). The minimum size of repeat for a 4-shaft weave construction in treble cloth is 12 X 12 but to show a sateen order of stitching the above example has been worked out over an area of 24 X 24.

(2) Insert the face weave on face ends and picks (solid marks), the centre weave on centre ends and picks (shaded marks) and the back weave on back ends and picks (diagonal marks). The different marks are used to simplify checking the design for correct placement of the weaves. This stage is shown at E in *Figure 9.1* and it may be noted that, as before, all marks indicate warp up unless otherwise stated.

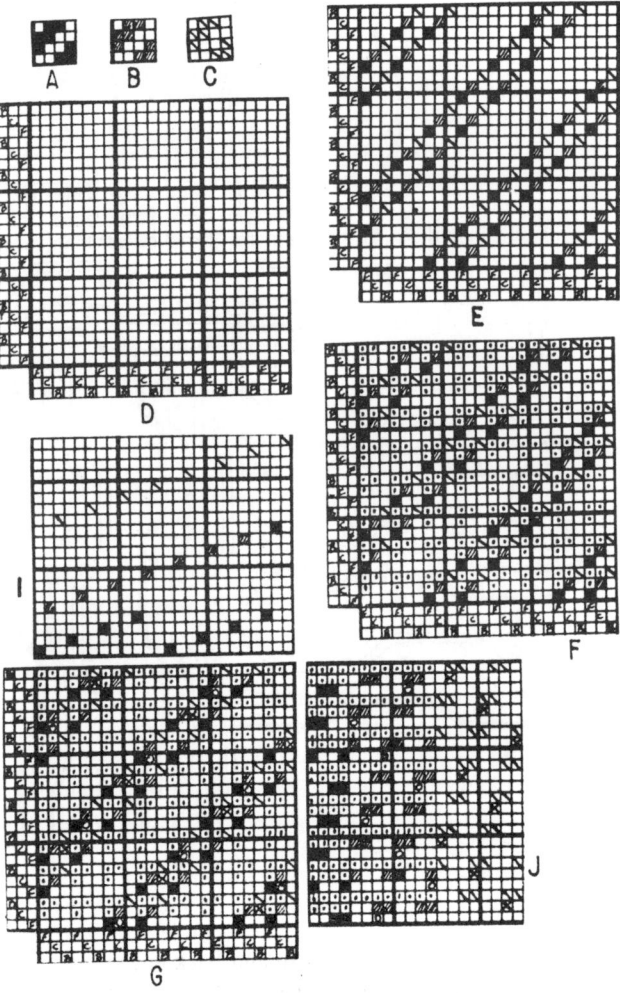

Figure 9.1

(3) Insert the separating lifts. In the treble cloths this is achieved in practice by lifting all the face ends on all the centre and back picks, and by lifting all the centre ends on all the back picks which is denoted by the dots in F. It must be borne in mind, of course, that the necessary corollary to this is

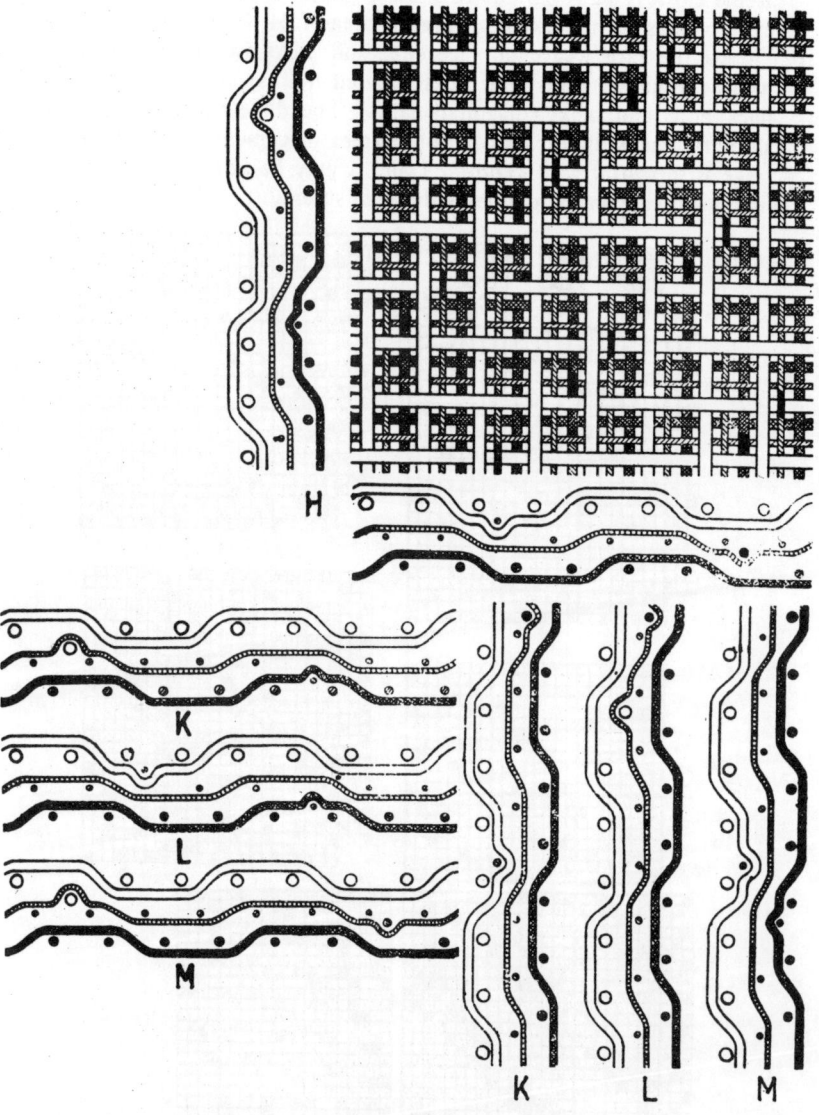

Figure 9.2

that on the face picks all the centre and back ends must be down, and on the centre picks all the back ends must also be down, but as the ends will automatically remain down unless marked, this latter situation is obtained by default.

(4) Introduce stitching marks. The treble cloths are usually self-stitched and the rules stated in connection with the tying of double cloths are equally applicable here. To reiterate the most important points: The stitching should be as regular as possible to prevent uneven tension; the ties should be adequately concealed on the face and on the back by corresponding face and back thread floats; the face and the back ends can only be used for stitching purposes when absent from the visible surfaces of their respective cloths. In the example given at G in *Figure 9.1* and in the interlacing diagram H in *Figure 9.2* a method of stitching is shown in which the three layers are united by raising the centre ends on the face picks at certain selected positions in an 8-shaft sateen order, and by similarly raising the back ends on the centre picks. The former lifts are indicated by the circles and the latter by the crosses in the design G whilst in the diagram H the stitch points are emphasised by the solid lines.

The fully worked-out design G represents a complete treble cloth structure with a suitable draft provided at I and a lifting plan, in which all marks indicate lifts, at J. The·sections given below and at the side of the interlacing diagram H in *Figure 9.2* indicate respectively the interweaving of the first face, centre and back picks and ends of the structure given at G and H. The sections show clearly that the two visible sides of the compound cloth are undisturbed by the ties due to their correct placement.

Methods of stitching

In the example given in *Figure 9.1* only one method of stitching has been shown in which the centre ends were lifted on the face picks and the back ends on the centre picks. It will be appreciated, however, that in a treble construction other possibilities exist, which may, in certain conditions, be more suitable than the common method indicated above. Thus, in the designs K, L and M in *Figure 9.3* the remaining three methods of tying, utilising the centre cloth yarns, are shown. All the three designs are identical with the construction G in *Figure 9.1*, apart from the method of stitching. In the design K the stitching is effected by dropping the face ends on the centre picks, indicated by the circles, and the centre ends on the back picks, represented by the crosses. In each instance, therefore, the tying consists of the cancellation of certain selected separating lifts. Construction L is stitched by raising the centre ends on the face picks (circles), and lowering them on the back picks (crosses). Finally, at M the tying is achieved by dropping the face ends (circles), and lifting the back ends (crosses) on the centre picks. The ties can be seen clearly in the sectional views in *Figure 9.2*, lettered to correspond, which show the first three picks and ends of each of the above three structures.

Although the first method, indicated at G in *Figure 9.1*, is the one most commonly employed, any of the other three may be preferable when the positions of convenient binding points in the face and back weaves are unsuitably placed in respect of the manipulations of threads required in the original system; or, if the relative thickness or quality of the yarns is such that the other methods result in a lesser degree of disturbance to the visible surfaces of the cloth. As the centre cloth never appears on the surface the position of the ties in that layer is of no particular importance in itself; however, efforts should be made to adhere, if possible, to the normal rules of stitching even in respect of the centre layer because any undue disturbance of its regularity is liable to result in a degree of distortion in the face or the back layer.

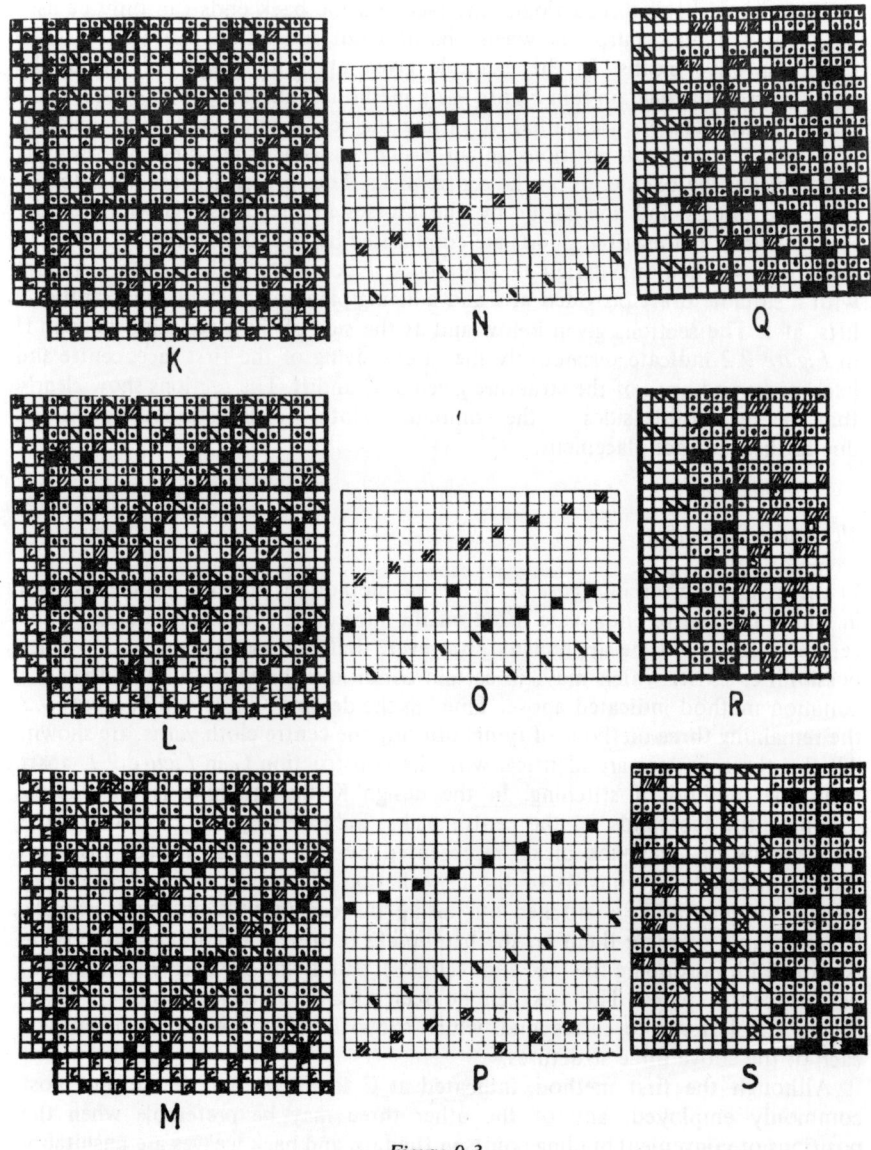

Figure 9.3

Apart from the four methods of stitching illustrated in the foregoing in which the threads of the centre cloth are involved it is also possible to tie a treble cloth together by either dropping the face ends on back picks, or, lifting the back ends on face picks, as in an ordinary double cloth. This form of stitching, in which the centre cloth merely acts as a wadding layer, is considered further on.

Beaming and drafting of treble cloths

From the point of view of the ease of warp yarn preparation and the ease of access to the weaving machine a single beam is more convenient than a double-beam or treble-beam mounting. When the materials and the weaves in each one of the three layers of a treble cloth are similar single-beam operation is possible. Some materials, notably the woollen yarns, are sufficiently accommodating to permit single-beam weaving even when there is a degree of dissimilarity between the yarn thickness and the weave in the various cloth layers. With other materials, in which such dissimilarities are encountered, it is necessary, however, to use two beams where two layers are similarly constructed and one is different, or three beams where all three structures are quite unlike one another. This is the case particularly when the warp take-up in each layer varies so that separate and independent regulation of each warp pay-off becomes necessary.

I in *Figure 9.1* and N, O and P in *Figure 9.3* illustrate four drafting arrangements commonly used in treble-cloth weaving. The main criteria are the regularity and the simplicity to facilitate the drawing-in of ends both during the entering and also following the breaks in weaving. As shown in the examples given it is usual to separate the healds into sets to correspond with each fabric layer but there is no specific reason why the treble cloths should not be woven in straight drafts if desired. It is sometimes claimed that the checking of the pattern chains is easier if the healds are separated into sets but as this is not a common occurrence it should not override the primary requirement of the regularity of end distribution. Whether the face, centre and back cloth sets are placed in the front, middle or back of the weaving machine is frequently a matter of choice. However, if the geometry of shed formation in a weaving machine is such that the front healds cause the least strain in the yarn then, undoubtedly, the weakest yarns, or, the most crowded and the most frequently interlacing sets of healds should be placed at the front.

In the lifting plans Q, R and S in *Figure 9.3* all the marks indicate healds up, the symbols used to denote a drop of an end for stitching in the corresponding designs having been deliberately omitted.

Construction of treble cloths with dissimilar weaves in the different fabric layers

The examples given in the preceding sections show that when the same weave is used in every fabric layer, and the threads are arranged in equal proportions, favourable conditions for tying are obtained by commencing the weave always in the same relative position. The weft and warp floats on the upper surface of the centre and back fabrics respectively are then directly below the warp and weft floats on the under-surfaces of the face and centre fabrics, hence there is no obstacle to the interweaving of the threads of one fabric with those of another. When different weaves are employed, however, such favourable disposition may not exist directly and a degree of experimentation may be necessary before the best relative starting position for each weave is determined. The construction of a fully worked-out design should not be commenced, therefore, before each of the constituent weaves is arranged separately in such a relationship with one another that the desirable coincidence of the warp and weft floats is obtained. In practice, when new constructions are attempted, the solution often lies in marking the weaves lightly on transparent design paper, superimposing one upon another and shifting them in turn until the greatest possible degree of coincidence between the warp-on-warp, and the weft-on-weft float is achieved. This, in fact, is most easily done in pairs because the two relevant relationships from the point of view of stitching are: (a) The position of the face weave in respect of the centre weave and (b) the position of the centre weave in respect of the back weave. As the face layer is normally joined to the back layer only through the intermediate agency of the centre layer the coincidental relationship between the floats of the face and the back cloth is of little significance. The two basic relationships are more difficult to establish without the aid of the transparent paper but they are clear enough in the designs A, B and C in *Figure 9.4* when studied in the relevant pairs. Thus, the face weave A which is a warp-faced satinette is placed in correct relationship with the 2-and-2 twill centre weave B; then, having determined the starting position of the centre weave, the back weave C which is also a warp-faced satinette is juggled with until the best relationship between it and the centre layer is established. The construction of a fully worked-out design can then be commenced. It must be realised, of course, that should it be entirely impossible to produce the required relationships between the pairs of weaves the weave of the centre layer, which is never visible, could be modified.

The fully worked-out design to correspond with the individual weaves A, B and C is shown at D and in the interlacing diagram E. The stitching is achieved by lifting the back ends on centre picks (circles) and the centre ends on face picks (crosses), and both sets of ties are distributed in a satinette order. The warp and weft sections below and at the side of the diagram E show that the tie placement in no way disturbs the appearance of the visible surfaces of the face and the back layers. The concealment of the back warp stitching lifts in the centre layer is, however, not perfect, as in some positions there is only one adjacent float instead of two at the stitch points. Although such an arrangement is slightly less than perfect it is quite acceptable in view of the fact that the aesthetic appearance of the centre cloth is of no importance—it is only its functionality that must remain unimpaired.

The diagrams F to K in *Figure 9.4* show the construction of a treble cloth in which the threads are arranged in the order of 1 face, 1 centre, 1 face, 1 back. A 4-and-4 twill is used for the face layer, and a 2-and-2 twill for the centre and back layers, the method of tying being the same as in the preceding example with the ties arranged in a twilled order. F is the face weave and is shown paired together with the centre weave G in an extended form to account for the disparity in their respective thread ratios. At H the same centre weave as at G is shown in the normal form paired together with the back weave I. It will be appreciated that as the thread ratios of the centre and back weaves are identical no need for an extension of either of them exists. The fully worked-out construction is given at J in which the circles indicate the stitching lifts of the back ends on centre picks, and the crosses, the stitching lifts of centre ends on face

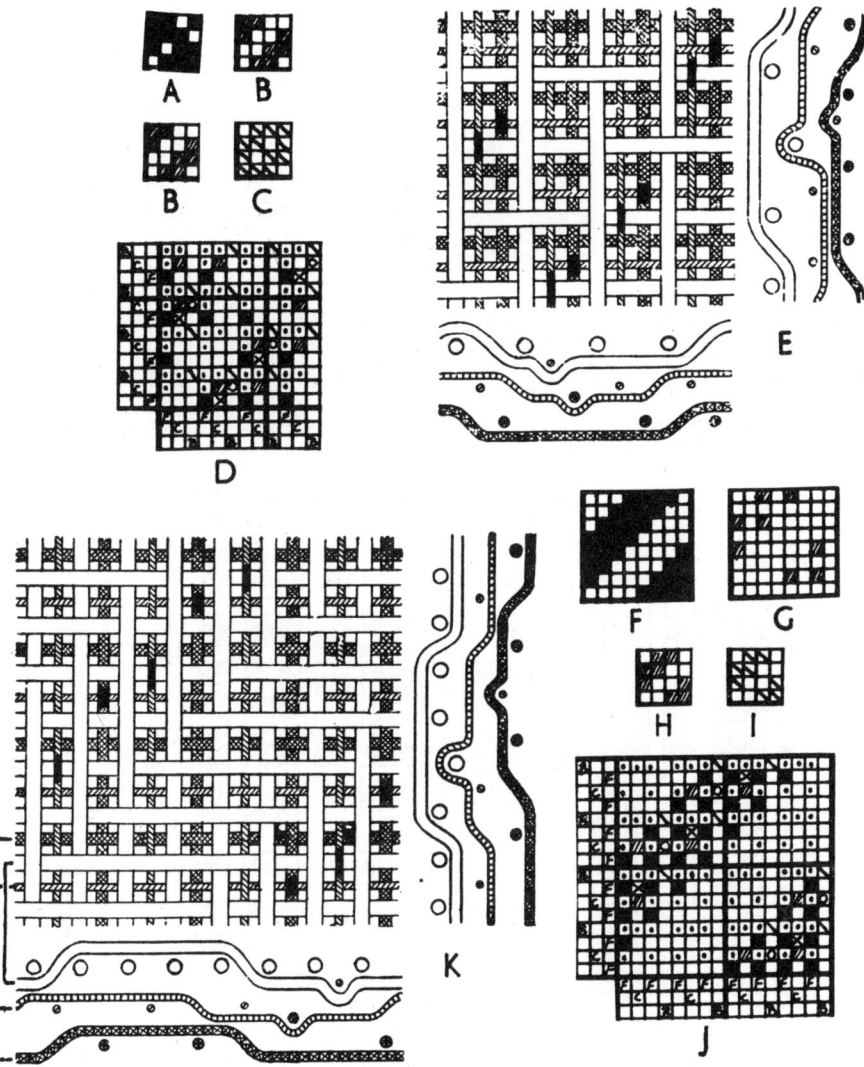

Figure 9.4

picks. In the interlacing diagram K the positions of the ties are clearly indicated by the solid black spaces. The warp section below the interlacing diagram shows the interweaving of the second, third and fourth picks of the structure, whilst in the weft section to the side of K the interweaving of the second, third and fourth ends is depicted.

In *Figure 9.5* the diagrams illustrate the construction of a treble cloth in which there are 2 face and 2 back threads to 1 centre thread, the arrangement being 1 face, 1 back, 1 centre, 1 face, 1 back, as indicated at O. Five-thread satin and sateen weaves are used for the three fabrics, the face layer being a warp satin and the centre and back layers the weft sateen. This in effect means that

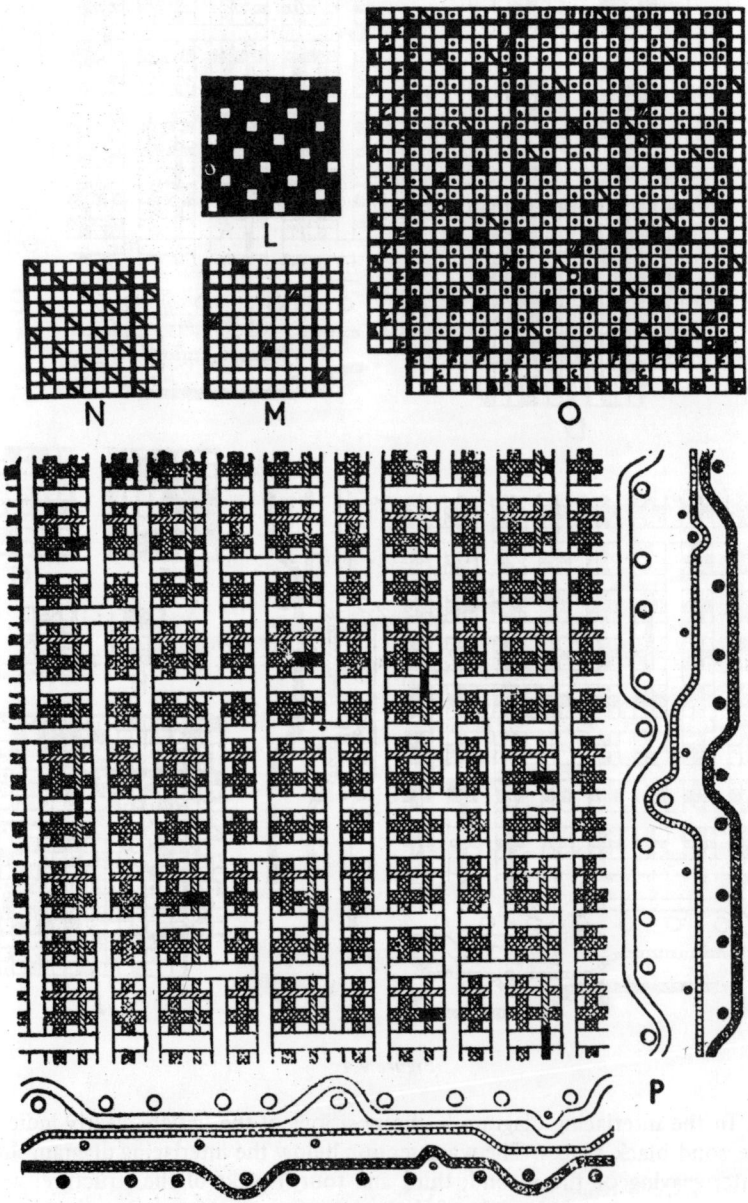

Figure 9.5

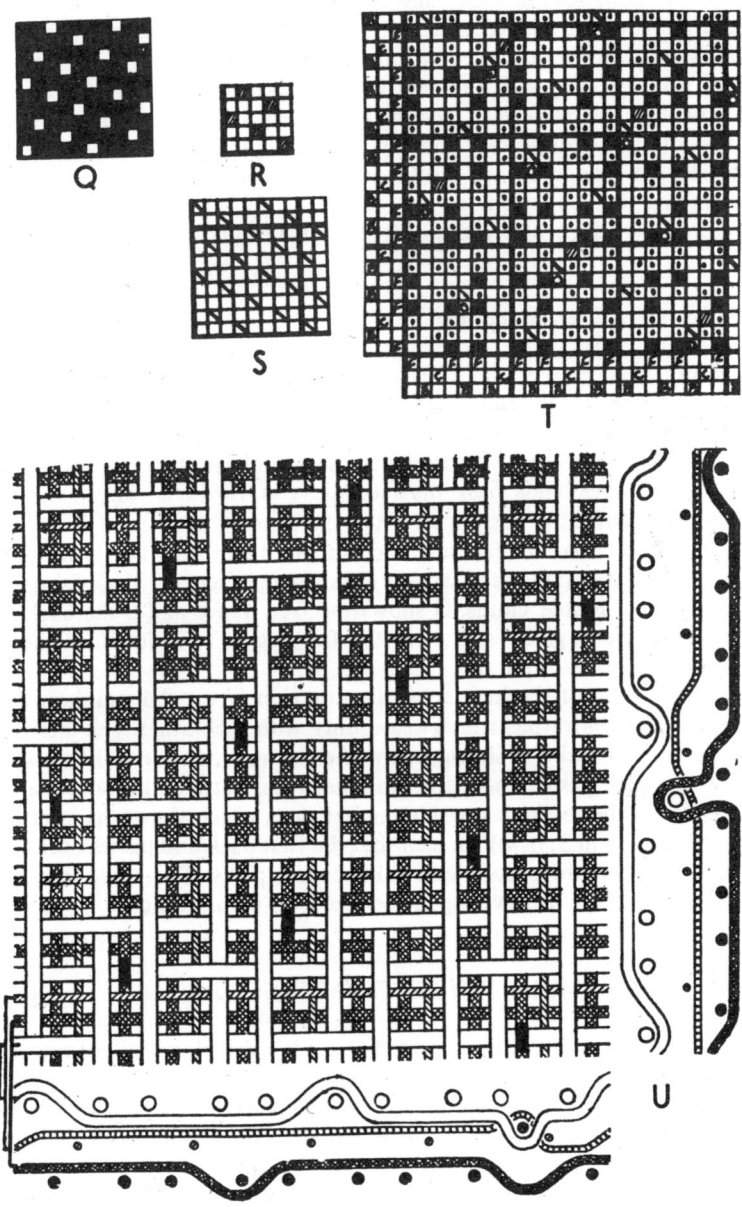

Figure 9.6

the fabric is reversible as the visible surfaces of both the face and the back cloth are composed of warp floats. The tying is effected by lifting the centre warp threads on face picks, as shown by the circles at O, and by dropping the centre ends on the back picks as indicated by the crosses. This method is the most suitable one for stitching reversible warp face constructions and is particularly applicable to structures in which the centre warp is not much thicker than the face and back yarns, whereas the centre weft is much coarser. Following the previously established practice the three constituent weaves L, M, and N are arranged side by side, with the centre weave extended to compensate for its lower ratio, for the purpose of locating the correct stitch points. It will be noted from the interlacing diagram P and the accompanying cross-sectional views that the ties in the face and back cloths are adequately covered by two adjacent corresponding floats of warp and are, therefore, perfectly placed. Furthermore, the continuity of the centre yarn floats also remains undisturbed.

Use of the centre layer as wadding

The design T in *Figure 9.6* and the corresponding diagrams show a method of uniting the three fabrics which is different from any of the foregoing. In this case the centre threads do not interweave with either the face or the back threads, but are used purely in forming a wadding cloth, the tying being effected by raising the back ends over the face picks in 10-thread sateen order. The system can be advantageously used when the centre yarns are of lower quality and much thicker than the face and back yarns. In arranging the positions of the weaves and ties it is only necessary to consider the face and back fabrics, as in the self-stitched double cloths. Q shows the face weave and S the back weave. The centre weave is given at R, and the full design at T, in which the circles indicate the positions of the ties.

An examination of the interlacing diagram U in *Figure 9.6* will show that the centre ends and centre picks interweave only with each other. The interweaving of the picks 1, 2, and 3 is shown below the flat view, and it will be noted that the first face pick passes under the ninth back and between the fourth and fifth centre ends. In the section alongside the flat view, which shows the interweaving of the ends 1, 2, and 3, it will be seen that the first back end passes over the fifth face pick between the second and third centre picks. The tying may also be similarly effected by the lowering of the face ends on back picks.

MULTI-PLY BELTING STRUCTURES

Solid woven multi-ply beltings are today produced mainly for conveyor work. At one time they were also made extensively for power transmission but with the introduction of the V-belt drives the flat belting is now used for driving only in exceptional circumstances. Where it is retained it is usually in the form of a narrow belt and is, therefore, woven by the narrow fabric techniques similar to those used for the construction of safety belts for the aircraft and motor-car industries.

Conveyor belts are produced in widths of 0·4 m to 1·5 m and in thicknesses varying from 3 mm to 10 mm. The thinner belts are of a two-ply construction the number of plies increasing progressively with the increase in the belt thickness so that to produce the thickest belts six-ply or seven-ply structures are required. Almost any type of material can be conveyed and the goods handled range from rock and coal to grain, dusts, foodstuffs and bagged or packaged articles. To suit the wide variation in the materials different finishes are applied although the p.v.c. coated belts are by far the most common on account of their toughness, non-flammability and easy cleaning qualities. If required the belts can be run at very high speeds reaching for some materials 210 m/min, or they can be operated at 6 to 10 m/min for such operations as sorting or picking. They represent a very economic method of carrying a large volume of material over short distances and flat belts about 1 m wide can convey as much as 200 to 250 t/h. When suitably troughed a similar belt would convey double the above tonnage at a modest power requirement of 25 h.p. when running on plain bearings or 18 h.p. when mounted on roller or ball bearings. Apart from being able to carry materials along a horizontal plane the belts can operate over considerable gradients which for some goods can be as steep as 25°. For very acute inclines the conveyor can be modified into an elevator belt by having buckets bolted on to its surface.

It will be clear from the foregoing that to produce a worthwhile article for the purpose the architecture of the construction must be very carefully thought out. The difficulty exists in the fact that there is a clash of requirements for a conveyor belt—on the one hand, it should be rigid enough to carry considerable weight of materials without undue sagging between supports, on the other, it should be sufficiently flexible to permit easy bending when running over pulleys or when troughed. The stresses suffered by a belt upon bending are due to the forces of expansion and contraction between the outer and inner faces of a belt and are defined by the diagram and the general formula given in *Figure 9.7.*

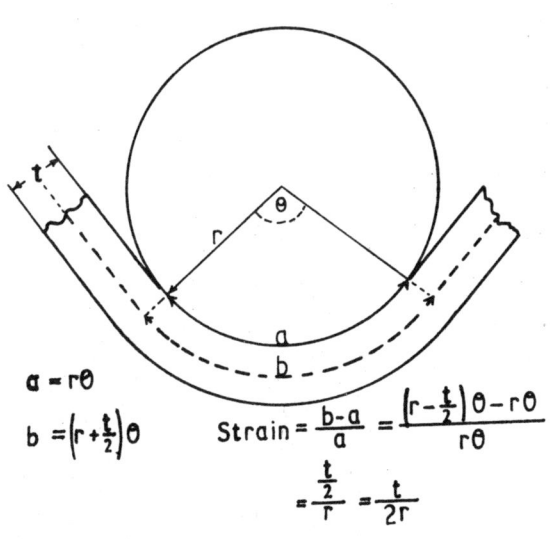

$$a = r\theta$$

$$b = \left(r + \tfrac{t}{2}\right)\theta$$

$$\text{Strain} = \frac{b-a}{a} = \frac{\left(r - \tfrac{t}{2}\right)\theta - r\theta}{r\theta}$$

$$= \frac{\tfrac{t}{2}}{r} = \frac{t}{2r}$$

Figure 9.7

From the illustration it becomes apparent that the best solution of the divergent and opposite requirements may be achieved by a compromise, constructing the central layers, which suffer no length deformation, more rigidly than the outer layers. The construction of a belt must also be such that it will not permit undue stretching under load.

Early constructions did not conform to the ideas of a planned engineering approach which were not developed until later and the belts frequently consisted of a multi-ply (3 to 6) plain weave stitched right through the plies in a manner shown by the weft section A in *Figure 9.8.* As all the layers were constructed in the same weave and with the same density of yarn spacing the outer plies suffered very considerable deformation stresses upon bending. Failure of a belt of this type occurred prematurely unless very large diameter pulleys were used. With the improved knowledge of the behaviour of belts under loads and with better understanding of the role of the construction in achieving superior belt performance many other designs were tried from which certain basic types were developed.

In one type of belting which is made in three-ply or four-ply structures the outer faces consist of twill weaves, such as 2-and-1 or 2-and-2. This permits denser setting of the warp thus allowing greater number of yarns to sustain the load and inhibit stretching because the load is distributed over a greater number of units and also on account of a lower inherent crimp in the warp yarns. The looser structure of the twill also allows greater freedom of flexing than is possible with a tight plain weave interlacing. The neutral central axis of the belt which, as stated before, does not undergo the same strain as the outer layers is

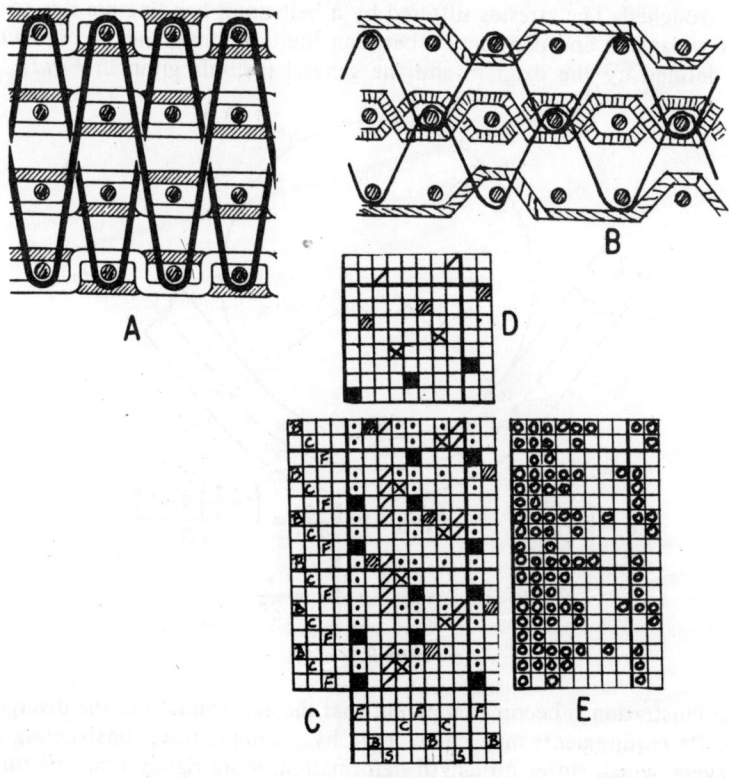

Figure 9.8

made more rigid by using in it a plain weave structure. This, apart from improving the load-bearing characteristics of the belt, also improves the fastener holding properties—another important aspect in an endless belt operation.

A three-ply structure of the type described is illustrated in the form of a weft section at B in *Figure 9.8*, and as a fully worked-out design at C. In the section only one warp end of the 3-thread twill is shown in the face and back layers to preserve the clarity of the structure. The stitching yarns, indicated in solid lines, produce a hinge action by which a freedom to expand or contract in the outer layers is not restrained but their binding is sufficiently effective to prevent any layer to creep in respect of another which is undesirable as it promotes the tendency to ply separation. In the design C the weave of each layer is indicated by the distinctive marks, the separating lifts by the dots, and the operation of the stitchers by the diagonal marks. The ends and picks are designated by the letters F, C, B, and S at the margins of the design which stand respectively for face, centre, back, and stitcher. The draft for the structure is shown at D and the lifting plan at E in *Figure 9.8*.

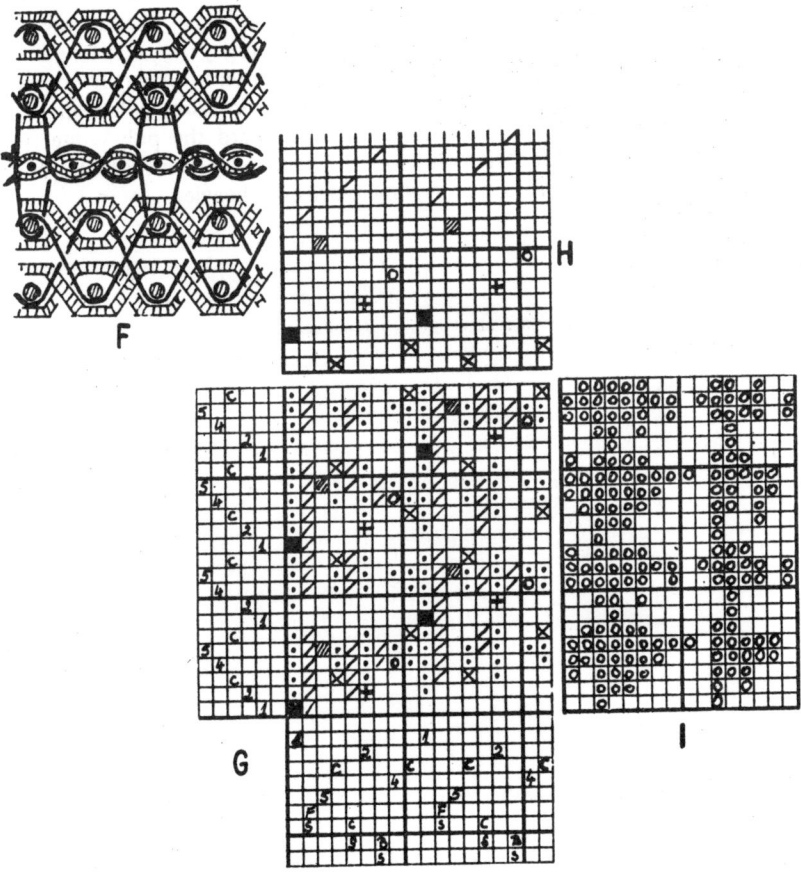

Figure 9.9

A different construction consisting of five layers is shown in the weft section at F in *Figure 9.9*. In this fabric the rigidity and inhibition of stretch is provided by the fine, densely set plain weave layer along the neutral axis of the belt. The other four layers each consist of a coarse plain weave construction which can flex reasonably easily due to an ingenious system of stitching. The binding yarns (marked in solid black) operate on a two-tier base radiating from the centre outwards. The binding is so arranged that the belt retains complete cohesion even when the extreme outer layers (1 and 5) are completely worn. The fully worked-out design for this structure is given at G in which the distinct marks indicate warp lifts in each ply and the dots and diagonal marks refer again to the separating lifts and the stitching weave respectively. The ends and picks which form the different plies are designated in the margins by 1 (the top layer), 2, C (the centre layer), 4 and 5 (the bottom layer) and the stitching ends are marked FS, CS and BS for face, centre, and back stitcher respectively. A suitable draft for this structure is provided at H and the lifting plan at I.

Cloths of the belting type are woven in heavy weaving machines using positive tappet shedding with either conventional shuttle or single rigid rapier forms of weft insertion. The yarns may be all cotton but at present the most common materials are heavy multifilament polyamide yarns combined with cotton during doubling or used as composite core spun yarns. Apart from adding to the bulk the presence of cotton is desirable to improve adhesion to the various coating agents and to act as insulation to guard the polyamide fibres against damage due to heat. The counts of the yarns vary considerably ranging from 95 tex to about 1000/3 tex or in some cases even heavier.

10

Figured Double and Treble Cloths

Figured double and treble cloths represent a principle of construction which is widely employed to produce considerable ranges of ornamental fabrics in different materials and weights for such divergent uses as dress fabrics, fancy waistcoats, overcoatings, curtainings, upholstery materials and rugs. An inexhaustible structural variety is possible and in this chapter it is intended to show only the main classes and to establish the principles of construction involved.

Most of the structures of the above type are a development of the cloth interchange principle dealt with in Chapter 8 in which a number of simple figured effects falling within the scope of the dobby shedding mechanisms were given. The basic principle of layer interchange remains the same and whether a dobby or a jacquard is employed the separating lift is the governing factor determining which layers at any given part of the construction shall occupy the top and the bottom positions. A treble interchanging cloth is only slightly more complex as it involves interchange at three different levels but these are as much under the control of the separating lifts as are the two layers in a double cloth.

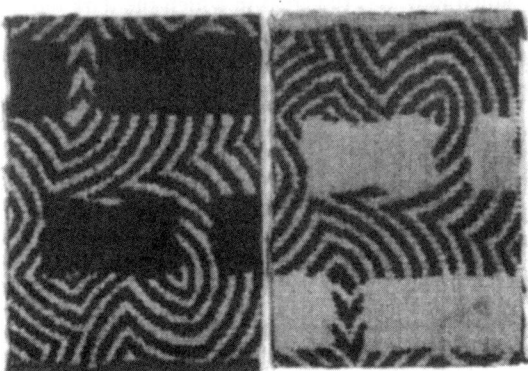

Figure 10.1

To produce a distinctive design the threads of one fabric require to be different from those of the other in respect of colour, material or thickness whilst the number of threads in each layer may be in either equal or unequal

173

proportions. Many interchanging fabrics are produced in reversible styles, as shown in the fabric in *Figure 10.1*, in which a light figure on a dark ground on one side corresponds with a dark figure on a light ground on the other side.

At one time the figured interchanging cloths were extensively produced in special jacquard harnesses and mountings designed to increase the figuring scope of the coarse pitch machines or to simplify the painting of the designs and the card cutting. The traditional mountings applicable to the manufacture of the interchanging cloths such as the sectional harness tie are described together with the other special jacquards in Appendix I. In this chapter it is assumed that the fabrics are produced on fine pitch ordinary jacquards using condensed and simplified painting and modern card-cutting machinery as outlined in Chapter 1.

FIGURED INTERCHANGING DOUBLE CLOTHS

Simple interchanging structures

The least complex structures in this group consist of two layers which differ in colour and which, by alternating between the face and the back according to a predetermined plan, produce a reversible design in two colours. An example of this type of construction is given in *Figure 10.1* which shows both sides of the cloth side by side. The weaves employed in each layer in that particular cloth are plain and this, in fact, is the most frequently used weave as it permits the production of a compact and serviceable fabric at lower settings than any other weave. However, apart from the question of the economics of production, there is no limitation as to the type of weave which could be employed in the interchanging double cloth and many other interlacings are used. It is also possible to have one layer constructed in one weave and the other in a different weave.

The construction of the simple interchanging cloths on design paper is illustrated in stages in *Figure 10.2*. At A a simple motif is shown in a condensed and simplified form which indicates the areas in which the dark cloth (solid squares), or the light cloth (blank squares), is displayed on the surface. The degree of condensation adopted here is by 2 in both directions so that each vertical row equals two ends, one dark and one light, and likewise, each horizontal row corresponds to two picks, one dark and one light. In finely set fabrics it is possible to condense the design by 4 in each direction without losing too much detail, especially when plain weave is used in both layers. It will be appreciated that in normal jacquard designs much more ornate figures will be produced than the dice effect given at A, but whatever the effect the motif is as representative of the constructional features of an interchanging double cloth as the most elaborate figured design.

If it is required to produce the design A in plain weave then each 2 × 2 area of the solid squares is representative of the detailed weave shown at B in *Figure 10.2*, and each 2 × 2 area of blank squares is representative of C. It will be noted that at both B and C the dark ends weave plain with the dark picks and the light ends weave plain with the light picks, the only difference between the two being the occurrence of the separating lifts. Thus, at B, dark ends are raised over light picks ensuring that the dark cloth occupies the top or face position, and at C, light ends are raised over dark picks which makes the light coloured cloth layer assume the surface position with the dark layer underneath it. The interchange is shown clearly by the warp section D which represents the first horizontal row of A, and by the fully worked-out design E. Referring to the fabric given in *Figure 10.1* it will be appreciated that despite the complexity of the figure each dark portion of the cloth is produced exactly as at B, and each light coloured portion exactly as at C in *Figure 10.2*.

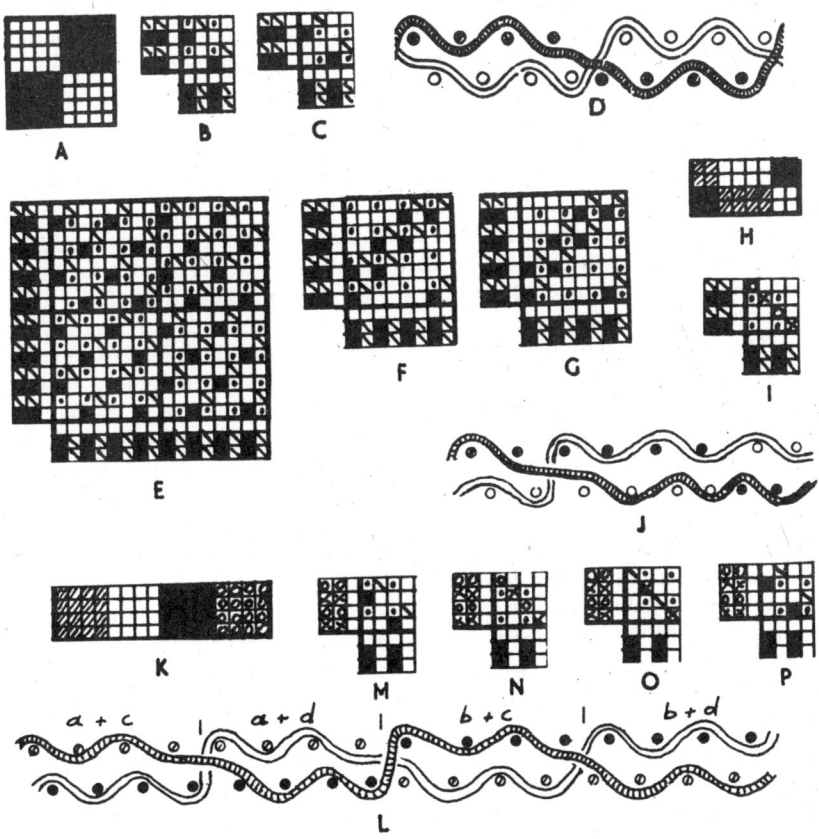

Figure 10.2

As the motif A is representative only of the disposition of the two differently coloured layers between the face and the back of the cloth and not of any specific interlacing it can be used also to illustrate the construction of interchanging twill weave fabrics. Thus, if it is required to produce the design A in a 2-and-2 twill weave, then each 4 × 4 area of solid squares in the motif is representative of the detailed weave given at F, and each 4 × 4 area of blank squares equals to that at G in *Figure 10.2*. It will be noted that, again, the only difference between F and G is in the positions of the separating lifts which are as at B and C respectively.

Apart from varying the weaves in the fabric as a whole, it is also possible to produce interchanging constructions in which different weaves are used in each cloth layer. If, for example, it is intended to have the dark layer in plain weave and the light coloured layer in 2-and-2 twill then the solid portions of A will be representative of B, and the blank portions of A will correspond to G in *Figure 10.2*. Although, theoretically, any two weaves could be combined in the manner suggested above, in practice it is necessary to consider the possibility of cloth distortion if the two weaves have considerably different frequencies of interlacing. In such situations, and particularly when all the ends come from the same beam, there is a tendency of one layer becoming much slacker thus forming distinctly puckered areas. This, when not wanted, as in overcoatings, represents a fault; however, there are many other ornamental structures in which the puckering of one layer in respect of the other is deliberately magnified as a special feature.

In the foregoing examples solid two-coloured effects were produced by inter-weaving dark with dark and light with light ends and picks. Using the same colour arrangement it is, however, possible to obtain a three-coloured effect in the cloth by interweaving in selected areas dark ends with light picks and *vice versa*. Thus, in addition to the solid dark and solid light coloured areas, mixed colour areas can be formed which, due to the close juxtaposition of the differ-ently coloured threads, assume a hue intermediate between the one and the other. This is indicated by the motif H in *Figure 10.2*, in which the cross-hatched portion represents the mixed colour area of the design. Assuming that the weave is plain throughout, the solid portions of H will weave as B, the blank portions as C and the cross-hatched portions as I in which the plain weave is produced between the dark ends and the light picks. As the dark ends are lifted on dark picks the top cloth layer in this portion will, therefore, consist of a plain weave composed of dark ends and light picks whilst a similar bottom layer consists of light ends and dark picks as shown clearly by the warp section J which corresponds to the first horizontal row of the condensed design H.

As a further elaboration of the simple interchanging cloth, designs in four different hues can be obtained if the two colours of weft are different from the two colours of warp. Thus, assuming that the warp colours are a and b, and the weft colours c and d the four-colour effect is due to the combination of a + c, a + d, b + c, and b + d as indicated by the motif K and the warp section L in *Figure 10.2*. The fully worked-out constructions to correspond with the different portions of K are given at M, N, O and P respectively. Although the different areas are not as clearly defined as when they are composed of solid colour they often produce pleasingly subdued effects. The colour range in this form of construction and also in the preceding ones can be considerably ex-tended, if desired, by planting and chintzing.

Combination of fine and coarse fabrics

Figured effects can also be formed by interchanging a fine fabric layer with a coarse one as illustrated by the cloth in *Figure 10.3*. The design is given in a condensed form at A in *Figure 10.4* in which the blank areas represent the portions in which the fine cotton/polyester blend yarn fabric forms the surface, whilst the cross-hatched areas indicate where the coarser worsted cloth is displayed on the face. Plain weave is used for both layers and the threads are arranged in the proportion of 2 fine to 1 coarse in both the warp and the weft directions. For this reason the degree of condensation used in the design A is 1 : 3, i.e. each vertical and horizontal row in the design equals three threads—two fine and one coarse. The fully worked-out weave representative of a 2 × 2 area of the blanks is given at B, whilst that for a similar area of cross-hatched squares is shown at C. The warp cross-section at D corresponds to the beginning of the sixth horizontal row of A.

Figure 10.3

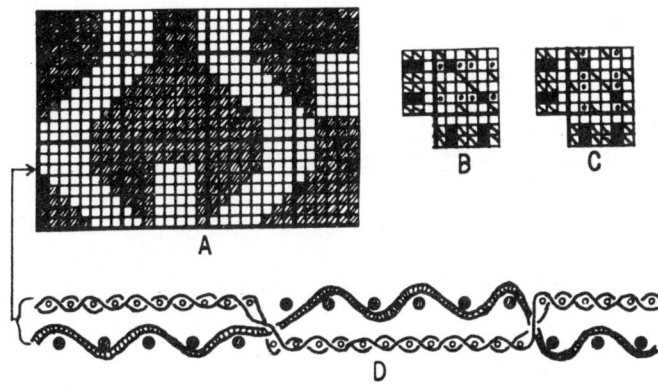

Figure 10.4

In constructions in which there is a very considerable disparity in the thickness of the fine and the coarse threads the ratios between the two sets of yarns can be increased to 3 : 1, 4 : 1, or even 5 : 1. However, as the very coarse layer tends to distort the threads of the very fine layer at the point of interchange the uses of such extreme combinations are confined to a very small range of effects verging on the bizarre.

Combination of double weaves with warp and weft float

In addition to forming figure by interchanging the fabric layers, as described in the foregoing, the cloths can be further embellished by floating the weft or

Figure 10.5

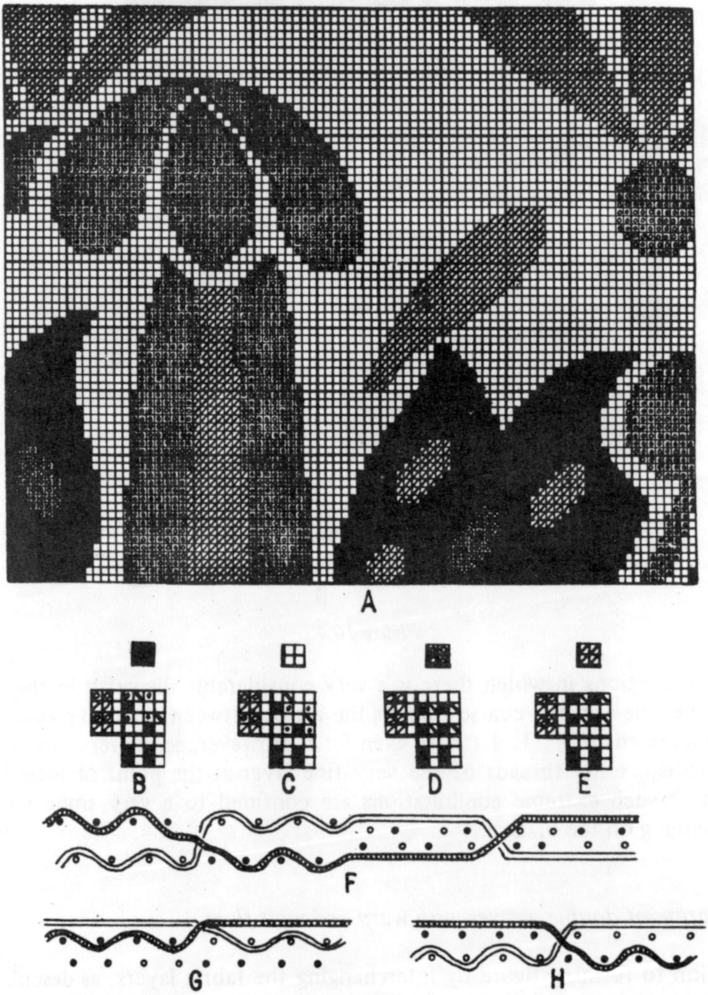

A

B   C   D   E

F

G   H

Figure 10.6

warp threads. In *Figure 10.5* a cloth is represented in which two colours of warp combine with two different colours of weft to form two interchanging plain cloths each in a mixed colour effect and, additionally, each weft is floated in places to form a solid coloured figure on its own. When one weft floats on the surface the other makes a corresponding float on the back, the two warps being contained in between the weft floats.

A portion of a simplified design (condensed by 2) is given at A in *Figure 10.6* with the detailed weaves for the four different structural areas of the cloth worked out at B, C, D and E. The warp cross-section at F shows the order of thread interlacing in each of the four areas.

At G and H in *Figure 10.6* two different sections are provided to show alternative methods of construction which may be adopted for the float portions of the figure. At G one weft floats on the surface supported by the plain weave cloth produced by the interweaving of the second warp and weft whilst the first warp forms a corresponding warp float area on the back. This method produces a firmer construction than that used at A and may be more suitable for upholstery fabrics whilst the former may be preferred for curtainings. At H yet another possibility is presented in which the weft float is supported by a corresponding float of its own warp whilst the other warp and weft weave plain underneath. This method offers the advantage of utilising the warp placed directly below the floating weft for stitching the pick in on the surface should the weft float be judged excessively long. When the floating warp is placed in the position indicated in the section H the occasional stitching-in is carried out without disturbing the plain foundation cloth.

Although in the foregoing examples only the weft float was considered it will be appreciated upon the examination of the sections F, G and H in *Figure 10.6* that in each case additional ornamental features could be readily produced by reversing the positions of the warp and weft floats to obtain warp float figures on the face of the cloth.

The cloth represented in *Figure 10.5* has been made for a heavy curtaining fabric to the following particulars: 1 thread 30/2 tex cotton, 1 thread 28 tex filament rayon in warp and weft; 28 ends and picks per cm.

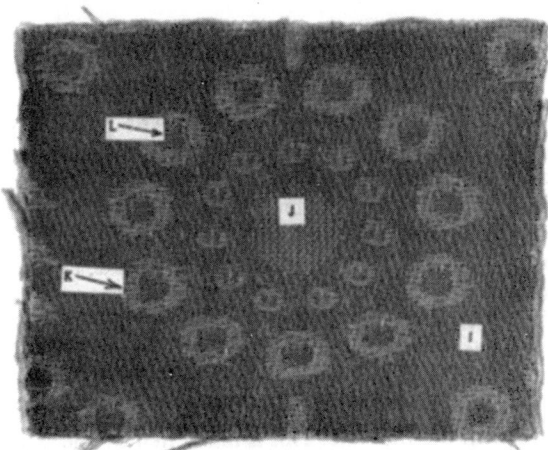

Figure 10.7

In *Figure 10.7* an entirely different curtaining fabric is shown which is produced in the, so-called, folk weave style which is characterised by a comparatively open construction. It also incorporates weft float figure areas which are produced exactly as in the cloth given in *Figure 10.5* but its main interest lies in the rather unusual combination of weaves. Thus, the ground, marked I in *Figure 10.7*, is produced by the darker ends and picks in a 5-shaft satin weave with the reverse of that in the lighter ends and picks in the bottom layer; portions of the figure designated J are produced in plain weave with the lighter threads forming the face cloth and the darker ones the back cloth; other portions, K, weave in a 5-shaft sateen in light threads on the face with the reverse weave in the dark threads on the back, and at L the dark weft floats on the face whilst the light weft floats on the back with both warps in between.

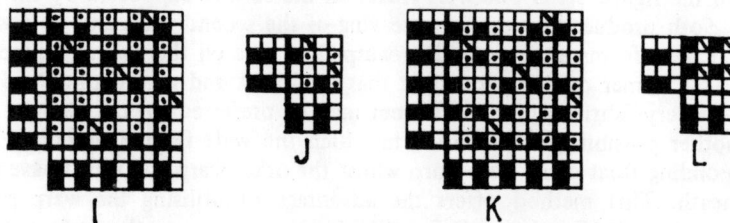

Figure 10.8

A perfectly reversible fabric is produced for which the detailed weaves are given in *Figure 10.8* designated I to L to correspond with the lettering in *Figure 10.7*. The two sets of threads are exactly the same in thickness and differ only as to colour, the arrangement being 1 dark, 1 light in both directions. The warp and the weft consist of identical 74/2 tex cotton yarns and the settings are 26 ends and 22 picks per cm.

Cloque or crepon effects

In *Figure 10.9* the face and the back of two different fabrics are shown, illustrating the appearance of a cloque texture in which a waved or cockled surface is produced. The effect may be due to the weave structure, or the use of yarns with different shrinkage properties, or to both. The possibility of using the difference in contraction between the two layers, due to the employment of tightly and slackly interlaced weaves, to create distortion has been mentioned earlier. The effect which causes cockling of the slack fabric layer can be magnified if the warp in the tightly interlaced cloth layer is drawn from a heavily tensioned beam and the warp in the slack cloth is placed on a lightly tensioned beam. Where only a slight degree of cockling is required in the layer containing the excess length of yarn this method may be sufficient in itself. Where, however, a pronounced cockle is desired the utilisation of differential shrinkage properties in yarns is necessary, the shrunk cloth layer remaining straight whilst the one not subject to shrinking becomes corrugated or cockled due to the excess length in its yarns compared with the first one. This is illustrated by the schematic diagram at A in *Figure 10.10*.

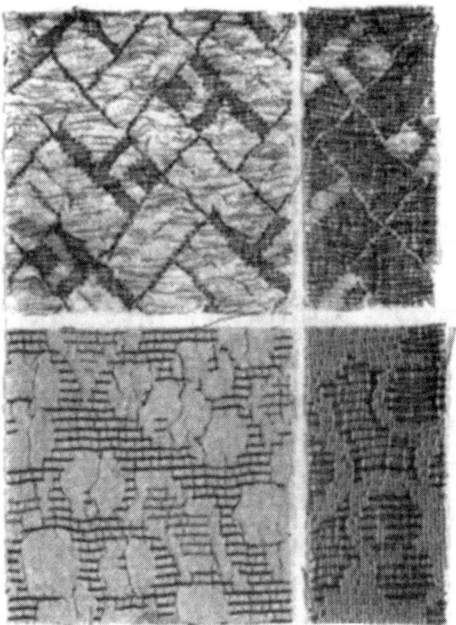

Figure 10.9

The differential shrinkage properties are best considered in connection with pairs of materials one member of which is affected and the other unaffected by a particular agency. Thus, if one layer in an interchanging double cloth consists of woollen yarns and the other of cotton then in a milling process the woollen layer shrinks and the cotton one cockles as it is less affected by the milling action than the former layer. Similarly, a cotton and a polyamide layer when immersed in a caustic soda solution will result in the shrinkage of the cotton component and cockling or wrinkling of the polyamide one. In another method of achieving differential shrinkage one layer composed of crepe twisted yarns will shrink more than the second layer consisting of low twist yarns. In addition, it is also possible to utilise in a similar manner the heat sensitive properties of some synthetic materials which have the capacity to shrink appreciably when exposed to heat whilst making the other cloth layer from materials which remain unaffected by exposure to higher temperatures.

All of the above methods can be employed for the production of cloque effects and the choice frequently depends on what particular pairings of materials are the most desirable for any specific purpose. The two fabrics represented in *Figure 10.9* are both of the light dress type in which the cockled appearance was achieved by the use of high twist yarns in one layer and low twist yarns in the other. The high twist fabrics are in both cases very open and form a net-like construction through which the finely set structure composed of the low twist yarns is clearly visible. The open structure is achieved by the special arrangement of the threads which in the upper fabric in *Figure 10.9* is: 3 low twist, 1 high twist in the warp and 4 low twist, 1 high twist in the weft; and in the lower: 4 low twist, 1 high twist in the warp and 6 low twist, 2 high twist in the weft.

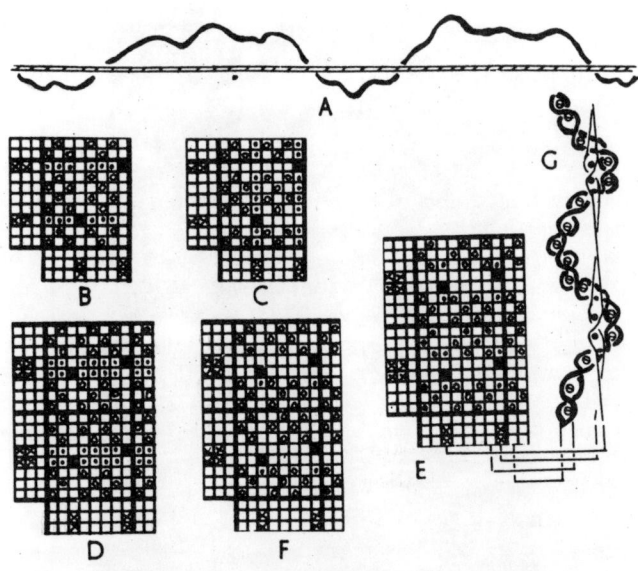

Figure 10.10

Plain weave is used throughout in each layer. Fully worked-out weaves are shown in *Figure 10.10*, those at B and C referring to the upper and those at D, E and F to the lower fabric in *Figure 10.9*. In each of the detailed weaves the high twist threads are indicated by the crosses at the margins of the designs. At B the construction is shown in which the finely set low twist yarns weave plain on the face with the high twist yarn producing the open plain weave structure on the back. C shows the reverse situation. At D and E the constructions in the second cloth are shown which correspond respectively with B and C in the first cloth except that at E the cloth made from the high twist yarns is brought to the face in only narrow horizontal portions. This is done in order to create a ground consisting of a series of wefts as an added feature illustrated clearly in the weft section at G. At F in *Figure 10.10* another effect is given which forms a secondary feature in the lower cloth in *Figure 10.9* and which consists of an even thinner band of high twist yarns on the face, this band being only one pick wide on the surface. Thus, two grades of welt are formed in the ground areas, a pronounced one at E and a discreet one at F.

Combination of double cloth with warp or weft float on single cloth ground

The fabric illustrated in *Figure 10.11* shows a compact upholstery fabric of very rich appearance which embodies five different constructional areas. The ends are all the same consisting of 17/2 tex filament viscose rayon, 84 ends per cm; in the weft arrangement is 1 pick 34 tex filament viscose rayon, 1 pick 140 tex condenser spun cotton, 29 picks per cm.

The ground is composed of a single cloth produced in a warp-faced rib structure indicated at B, and also in the weft section at A in *Figure 10.12*. In the design B the coarse cotton picks are indicated by the solid marks at the margin of the repeat. It will be noted that the rib weave is such that it splits the weft into two layers the top layer consisting of the fine rayon and the back layer, of the coarse cotton picks. The back is comparatively loosely bound as the warp

Figure 10.11

makes only half the number of interlacings on the back compared with the face which, due to the density of end spacing, is completely covered by the warp. The section A shows the interlacing of the first and the second ends which indicate clearly the manner in which the warp is made to separate the weft into layers.

In constructions which involve the combination of single cloth with double cloth areas the warp-faced rib is the weave commonly employed for the single

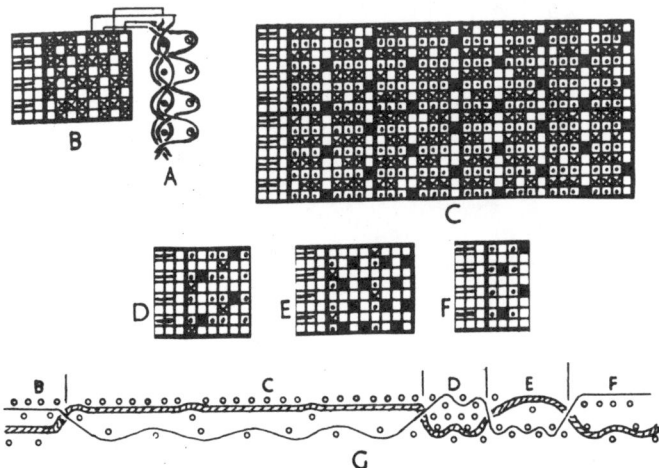

Figure 10.12

cloth areas because it can easily accommodate the large number of ends per unit space without distortion. The weave also provides a very hard wearing surface and can be adapted to produce attractive ground structures when the warp is arranged in two-colour combinations in almost any proportion such as 1 and 1, 2 and 1, 3 and 1, etc. A number of rib weaves which could be usefully employed for this purpose are given in Chapter 6.

The figured portions in the cloth given in *Figure 10.11* are all constructed on the double cloth principle. The main effect is produced in a warp satin construction in which three out of four ends act as face ends and produce the face satin in conjunction with the coarse weft whilst the fine weft weaves plain underneath with the remaining ends as indicated at C in *Figure 10.12*. Auxiliary features in the figure are produced in (1) a plain weave on the face in which one-third of the ends weave plain with the fine picks whilst another third weave plain with the coarse picks on the back and the remainder act as wadding ends in the middle, as shown at D; (2) another plain weave on the face in which only a quarter of the ends weave plain with the coarse weft on the face thus permitting this weft due to low warp cover to show clearly through, whilst the remaining ends weave plain with the fine picks on the back, as given at E; and (3) a weft float where the fine filament weft floats on the surface supported by half the ends directly underneath it, whilst the other half makes a plain structure with the coarse picks on the back, as indicated at F. All the five structural areas of this fabric are shown also by means of a warp section at G in which two picks are given and each different area is lettered to correspond with the fully worked-out designs B to F.

Combination of double cloth with extra threads for wadding or figuring purposes

In interchanging double cloths extra threads are sometimes used for the sole purpose of adding bulk. As the incorporation of such threads in a fabric,

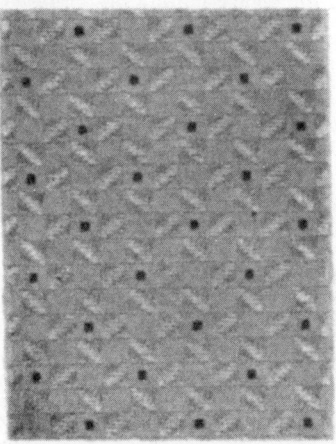

Figure 10.13

however, invariably results in additional costs it is normal to utilise them in a dual capacity, both as wadding and also in a figuring capacity. In jacquard weaving the extra material is usually introduced in the form of weft to avoid the inconvenience of re-tying the harness or the need for elaborate casting-out schemes. In dobby styles it is frequently preferable to use extra warp yarns which permits higher rate of production.

The latter style is represented by a fancy waistcoat fabric in *Figure 10.13* in which the extra warp forms a simple spot effect by floating on the face of a fine plain weave layer when it occupies the top position. The double cloth consists of a fine plain weave layer which alternates with a coarser twill layer. Where the extra warp is not required to form a spot on the face it is made to float on the back. Apart from showing one method of combining a double cloth with extra warp figuring threads the fabric in *Figure 10.13* is also representative of a combination of a fine with a coarse cloth layer. A complete repeat of the construction is shown in the simplified and condensed motif at A in *Figure 10.14* in which the solid squares represent the extra warp spots. Where the fine yarns weave plain on the face the coarse yarns also weave plain on the back as indicated

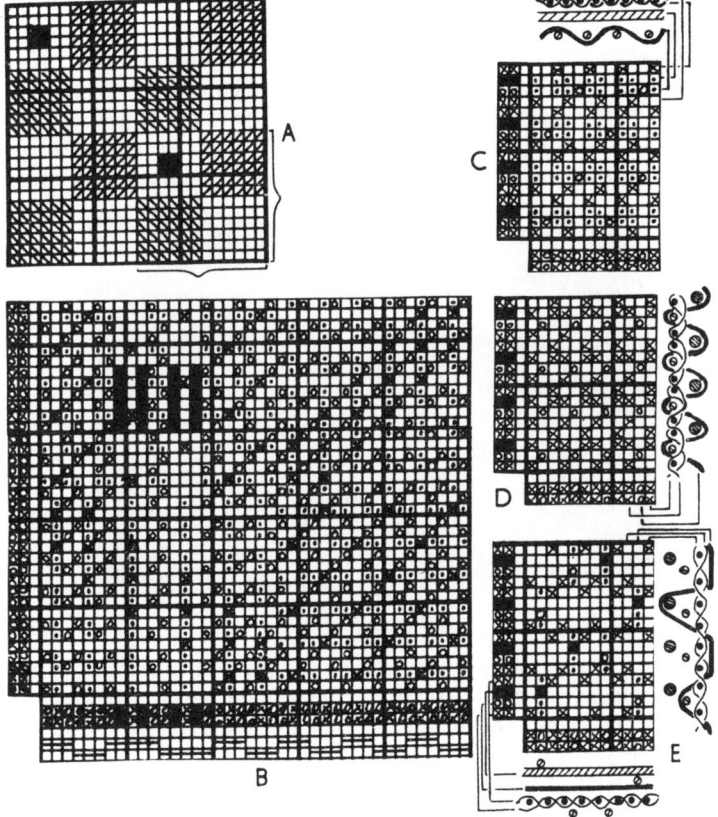

Figure 10.14

by the blanks at A. The diagonal marks show the areas in which the coarse yarns produce a 6-end twill on the face. The direction of inclination of the diagonal marks is representative of the direction of the twill which in either variant is backed by the fine plain cloth. One quarter of the repeat, as marked by the brackets at A, is given in full at B in which the fine ends and picks correspond to the circles, the coarse ends and picks to the crosses, and the extra ends to the solid marks at the margins of the design. It will be noted that the warp is sleyed-in three ends/dent except where the extra ends are introduced, these being additional to the two fine and one coarse ends which are combined within each split. The extra ends are raised over both the face and the back picks when required to form the figure on the face but remain down on all the picks in between.

In *Figure 10.15* a heavy non-reversible curtaining fabric is shown in which extra weft yarns are used for wadding and, at intervals, also for figuring purposes. The warp in this fabric is arranged—2 ends of filament rayon, 28 tex, to one end of two-fold cotton, 33/2 tex, 40 ends per cm; and the weft—2 picks of filament rayon (same as warp) to 1 pick of two-fold cotton 50/2 tex, and 1 pick of condenser spun cotton 150 tex, 32 picks per cm. The ground consists of a double plain cloth in which the rayon yarns produce the face, the two-fold cotton yarns the back, the heavy condenser cotton yarns acting as the wadding materials between the two layers. The figure is composed of two structurally different areas. In the first, a sunk effect is formed by interweaving the yarns in a single cloth warp rib weave, and in the second, the wadding weft is brought to the surface where it is loosely stitched where required by the cotton warp whilst the fine rayon yarns weave plain underneath. The detailed weaves for each of the

Figure 10.15

three different areas are shown respectively at C, D and E in *Figure 10.14* with the salient features of each structure clearly indicated in similarly lettered cross-sectional diagrams. The filament yarns are indicated by the crosses, the two-fold cotton yarns by the circles and the heavy condenser weft by the solid marks at the margins of the designs. All marks in the designs represent warp up.

FIGURED INTERCHANGING TREBLE CLOTHS

More elaborate cloths of a similar character to the double plain styles are woven with three series in one direction and two series in the other direction, or with

Figure 10.16

Figure 10.17

three series in both directions. *Figure 10.16* illustrates a curtaining fabric of the last type in which the weft is arranged 1 pick yellow, 1 pick blue, and 1 pick white; and the warp, 1 end red, 1 end green, and 1 end white, all staple viscose rayon. Seven effects are formed in the design, as indicated by the different marks in the corresponding condensed plan given in *Figure 10.17*, which illustrates the method of designing for the texture. The weaves are so arranged that the yellow and blue wefts are frequently brought to the surface, full use thus being made of the contrasting colours.

The various weaves that are combined are illustrated in full at A to G in *Figure 10.18* in which the weave marks indicate warp. Linked with the face threads of the weaves a plan on 2 X 2 squares is shown in which the marks coincide with the corresponding effect in *Figure 10.17*. The positions which the threads occupy in the different sections of the cloth, represented by the respective weaves A to G in *Figure 10.18* are indicated in the following list:

Table 1

Weave	Face	Back	Centre
A.	First weft and first warp	Second weft and third warp	Third weft and second warp
B.	Second weft and second warp	Third weft and third warp	First weft and first warp
C.	Third weft and third warp	Second weft and first warp	First weft and second warp
D.	First weft and third warp	Second weft and first warp	Third weft and second warp
E.	Second weft and third warp	First weft and first warp	Third weft and second warp
F.	First and second wefts and first warp	Third weft and third warp	Second warp
G.	First and second wefts and second warp	Third weft and third warp	First warp

Plain weave is produced on both sides of the cloth, but in F and G two wefts are intermingled on the face, so that there are no weft threads in the centre. In

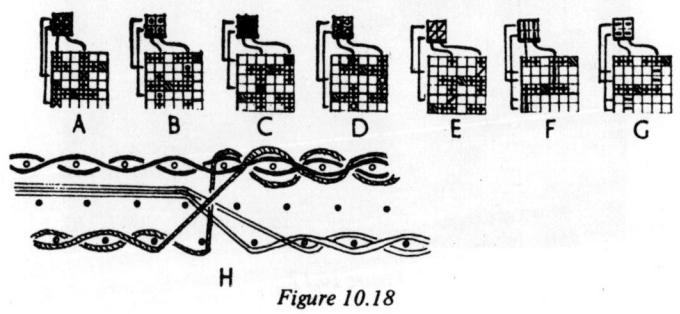

Figure 10.18

A, B, C, D and E there are both weft and warp threads in the centre which, however, do not interlace with each other, but simply serve as wadding threads, the picks passing over the ends. The warp section H shows, on the left, the effect A, to which areas B to E are structurally similar, and, on the right, the effect F which is similar to G.

11

Tapestry Structures

The classical tapestry was usually a structurally simple fabric in which the highly figured and elaborate design was produced by the placement of coloured threads within well-defined areas of the cloth. The warp was normally comparatively fine and of a neutral hue to prevent interference with the multi-coloured wefts in which the design was developed. The ends were operated in the plain weave order and into each shed small portions of weft would be placed, each short length conforming in colour with the painted design positioned directly behind the warp threads. Each equivalent of a pick in a normal cloth would, therefore,

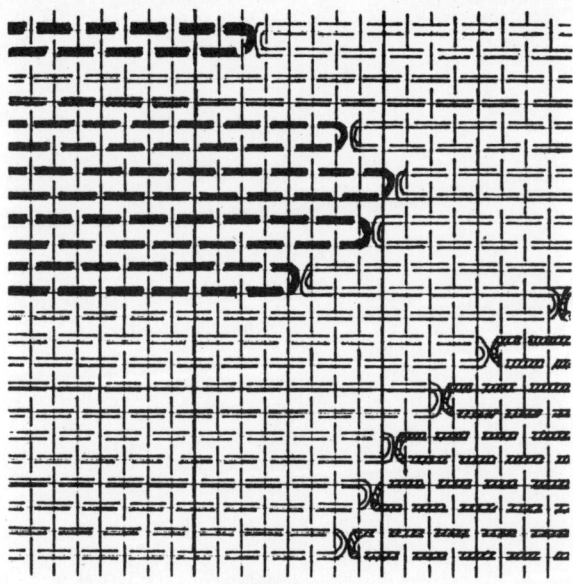

Figure 11.1

consist of discontinuous lengths of weft and the design would be built up, row by horizontal row, with the differently coloured lengths of weft occupying increasing or decreasing spaces in conformity with the changes in the dimensions of the figures as indicated by the interlacing diagram in *Figure 11.1*.

The warp sheet in the hand tapestry loom was frequently placed vertically and in broad tapestries several weavers would be stationed side by side, each responsible for a vertical strip of design of a given width, and each armed with a set of small hand shuttles containing the requisite colours of weft. The cloths produced in this manner usually consisted of single-repeat, large picture panels used mainly as wall hangings, and similar techniques were employed in tapestry weaving in most of the ancient civilisations. The highest peak of development of this art form in Western Europe was probably reached between the fifteenth and the seventeenth centuries in France and in the Low Countries where some of the most famous contemporary artists frequently engaged in the production of designs for tapestry panels. A facsimile reproduction of a sixteenth century tapestry picture in a miniature printed form is provided in *Figure 11.2.*

Figure 11.2

Although the making of tapestries in the manner described above is still practiced in various countries as a studio activity, the modern, machine-produced tapestry fabric has little structural affinity with the classical picture panel. The similarity between the two exists in that the figure in both is due to the display of colour but structurally many of the modern tapestry upholstery fabrics represent the most complex of the compound woven constructions. They may consist of several figuring warps and wefts as well as of stitching and ground yarn elements in both directions. They are produced on fine pitch jacquard machines for which the designs are prepared in the usual condensed form (see Chapter 1) the degree of condensation depending on the size of the structural unit. Thus, a tapestry cloth arranged with 3 figuring warps to 1 face and 1 back stitching warp, and 2 figuring to 1 stitching weft would be condensed by 5 warp-wise and by 3 weft-wise. The more complex of the structures frequently utilise the full capacity of modern card-cutting machinery for accommodating the different weave areas.

The tapestry construction has been used for hangings, sofa rugs, upholstery work, table covers and carpets. At present, fabrics of this type are mainly employed for the upholstery purposes for which the hard-wearing quality of the structure is particularly well suited. The figuring elements may consist of cotton, wool or man-made staple yarns; the ground warp, if present, is almost invariably a two-fold cotton yarn, whilst the stitching elements are frequently fine, two-fold cotton or filament nylon. The stitching yarns are an important element in the construction as they ensure cohesion and wear resistance of the fabric by preventing the formation of long floats whilst they may also add to the structural variety which can be enhanced by deliberately altering the appearance of similar colour areas by the changes in the order of stitching. Despite performing a vital role the stitching yarns must be unobtrusive and they must not interfere with the colour values of the main figure areas. For these reasons they are usually very fine (12 to 17 tex) and may be either dyed black or be entirely transparent as is the case with the fine filament nylon yarns.

SIMPLE WEFT FACE TAPESTRIES

These represent the least complex of the tapestry structures and consist of a ground warp the lifts of which determine the disposition of the figuring wefts, the stitching warp, and a number of figuring wefts. The wefts, of which there are usually between two and four, interchange between the face and the back of the cloth thus producing a design in the number of colours equal to the number of different wefts. It will be appreciated from the examples which follow that

Figure 11.3

in a 2-weft tapestry it takes two picks to complete a horizontal row of the design; in a 3-weft structure—three picks, and in a 4-weft—four picks. Thus, if other conditions remain equal, a 2-weft tapestry is produced at twice the rate of a 4-weft fabric with an obvious advantage in respect of the cost of production.

The stitching warp operates in a regular order stitching the weft floats both on the face and on the back so that no long floats are formed in any part of the fabric. Occasionally, the ground warp is also used for figuring on the surface thus adding another ornamental feature without increasing the cost of production.

Two-weft tapestry structures

Figure 11.3 represents a fabric in which the design is due to the interchange of two differently coloured wefts while a portion of the corresponding design given at A in *Figure 11.4* illustrates the simplified and condensed method of painting adopted for these structures. The warp is arranged 1 stitching end to 1 ground end, although in the heavier structures there may be 2 ground ends to 1 stitching.

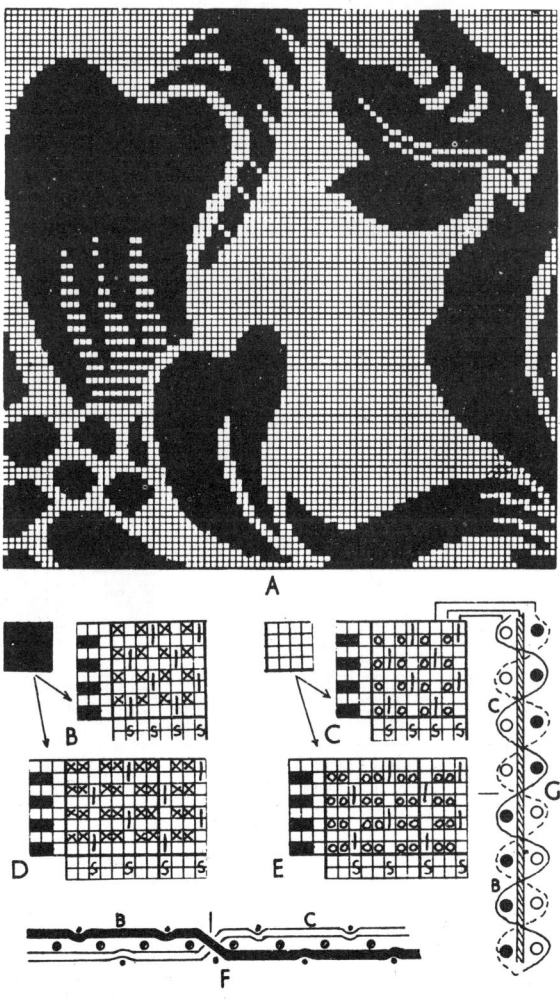

Figure 11.4

In the latter case the 2 ground ends are usually operated as one, being placed in a decked mail eye (q.v.) so that the jacquard operation is identical with the former case and only an expansion of the repeat size is obtained.

The detailed weave for the parts of the design painted solid in A is given at B, whilst C illustrates the weave of the blank portions of A. The crosses and the circles in the weaves B and C represent the lifts of the ground warp and the vertical lines indicate the lifts of the stitching ends. D and E in *Figure 11.4* correspond respectively to B and C and show the construction of the double ended version of the structure. The warp section through the fabric given at F shows clearly the interchange of the figuring wefts whilst the operation of the stitching ends becomes obvious upon studying the weft section at G.

Figure 11.5

Although structurally the 2-weft tapestries are quite simple, considerable differences in their appearance can be obtained by changing the orders of arrangement of ground and stitching warp ends or by changing the order of operation of the stitching ends. The fabric in *Figure 11.5* represents a simple 2-weft tapestry with a distinctly ribbed appearance achieved by arranging the warp in the ratio of

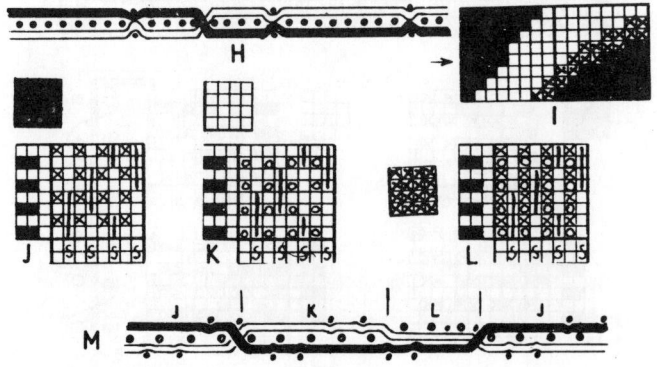

Figure 11.6

8 ground ends to 2 stitching ends which is shown by the warp section at H in *Figure 11.6*. In *Figure 11.7* another 2-weft tapestry is given where the difference in the appearance is due to the order of operation of the stitching ends which in

this cloth work in a broken twill weave order. In addition this fabric also illustrates the use of the ground warp as a figuring element. A portion of a condensed design for the latter construction is shown at I in *Figure 11.6* where it will be noted that a third colour, represented by the crosses, is used to denote the areas in which the ground warp is raised to the surface. The detailed weaves for each different area of figure in this cloth are given at J, K and L with the warp section through the fourth row of the design I being shown at M in *Figure 11.6.*

Figure 11.7

As the ground warp cannot be stitched, there being no weft stitching elements, its use is confined mostly to outlining the weft figure for the purpose of greater emphasis where it can enhance the general appearance without creating an excessively long float. If the ground warp is not used for figuring then the 2-weft tapestry cloth is perfectly reversible, a dark figure on a light ground being formed on one side, and a light figure on a dark ground on the other side. Sometimes, however, one of the wefts is chintzed (q.v.) in order to develop some portions of the figure in different colours, in which case the cloth is not reversible, because the chintzed weft produces horizontal bands of colour in the ground on the underside. The cloth in *Figure 11.7* represents an example of a chintzed development of the figure in a 2-weft tapestry structure. All the fabrics given in the foregoing show typical upholstery cloths but exactly the same interlacings can be used to produce rugs and carpets, the only difference being in the coarseness of the yarn elements used. The structures particularly well suited for the production of such rugs are those given at D and E in *Figure 11.4* which are sometimes enhanced by the extensive use of the chintzing techniques.

Three-weft, and four-weft tapestry structures

These fabrics represent a further development of the 2-weft tapestry from which they differ mainly in the number of the differently coloured figure areas which can be achieved. When one of the wefts is displayed on the surface the remaining ones weave together on the underside and for this reason the standard 3-weft, and 4-weft constructions are not reversible which will be seen clearly by reference to the sectional diagrams given at A, B, E and F in *Figure 11.y.* A and B show respectively a warp and weft section of a 3-weft tapestry whilst E and F illustrate the corresponding sections of a 4-weft tapestry.

Figure 11.8

The fabric illustrated in *Figure 11.8* is an example of a 4-weft tapestry composed as follows: 2 stitching ends of 14/2 tex cotton, 2 ground ends of 60/2 tex cotton, 24 ends per cm. The weft is 170 tex staple viscose (low twist, long staple) yarn, 6 picks of each colour per cm. Although the actual cloth sample contains four wefts the above particulars are equally applicable to fabrics containing fewer or greater numbers of wefts as long as they are constructed with 6 horizontal design rows per cm. A portion of the condensed design suitable for this cloth is given at C in *Figure 11.9* in which four colours are used to denote an area in which each of the four different wefts is displayed on the surface. The degree of condensation is by 4 weft-wise and by 2 warp-wise so that each horizontal row of the design represents a complete structural row of the fabric and each vertical row represents one ground end and one adjacent stitching end. The stitching ends operate continuously 4 up, 4 down as indicated clearly in the section F and when the cloth is woven face side up the ground ends are raised on all the wefts which are not required at that point on the face and remain down on the weft which does form the face float. As this method of weaving results in very heavy lifts it is more usual to weave such fabrics face side down in which case, on average, only one-quarter of the total number of ground ends is up on any given pick. A fully worked-out structure for the seventeenth and eighteenth horizontal rows of the design across the vertical rows 17 to 40 is given at D and shows the detailed weaves for each differently coloured area of the condensed design when the cloth is woven face side down. At D the lifts of the stitching ends are denoted by the vertical lines whilst the lifts of the ground ends are represented by the marks used in the design to show where each weft is displayed on the surface except that crosses are employed to show the lifts of the ground ends on the weft represented by the blanks in the condensed design. The warp section at E fully conforms to the first four rows of the design D as indicated by the brackets.

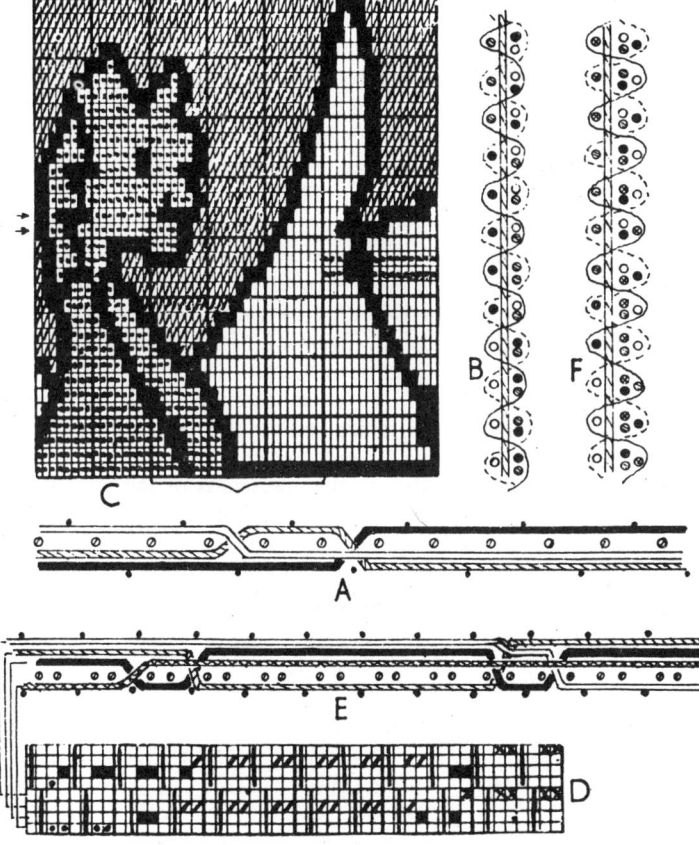

Figure 11.9

In 3-weft, and 4-weft tapestries effects can be produced simply in three or four colours by displaying the different colours of weft on the surface as shown in the fabric given in *Figure 11.8,* in addition, however, all the structural modifications considered in respect of the 2-weft tapestries can also be employed. Thus, the ground warp can be brought to the surface as an additional figuring element and various orders of stitching can be applied to the different colour areas. It is also possible to stitch the same colour areas in different orders thus changing the texture of one in respect of another. Sometimes two figuring wefts are brought to the surface together resulting in a mixed colour area. This technique, however, is used infrequently as it produces a somewhat indeterminate effect.

Three-weft, and four-weft reversible tapestries

In upholstery fabrics there is no reason to produce reversible structures because only the face side of the fabric is displayed to view. In rugs, however, both sides of the fabric may be utilised and for this purpose the 3-weft, and 4-weft tapestry structures may be arranged so that either surface is equally pleasing aesthetically. In such fabrics the figuring wefts are arranged in three layers so that when one weft forms the figure on the surface another colour forms an identical figure on the back and the remaining weft colour or colours act as wadding yarns in the centre. The stitching ends are not affected by this system and it is only the ground ends that determine which colour of weft is to be placed in which layer. Thus, when it is intended to display a figuring weft on the face all the ground ends in that area are dropped, if the second weft is the one chosen to appear on the back then all the ground ends are raised in that area, and to contain the wefts unwanted either on the face or on the back half the ground ends are raised, say, all the odd ground ends, so that such wefts remain quite invisible in the centre layer between the raised odd, and the lowered even ground ends. The result of this type of manipulation is shown clearly in the warp sections given at A and B in *Figure 11.11* which respectively refer to a 3-weft, and a 4-weft reversible tapestry structure.

The fabric shown in *Figure 11.10* is an example of a heavy 4-weft rug construction made as follows: 1 stitching end of 74/3 tex cotton, 2 ground ends of 320/2 tex staple viscose rayon, 10 ends per cm; weft-400 tex woollen yarn, 6 picks of each colour per cm. Due to the coarseness of the weft the stitching warp is very highly crimped, sometimes exceeding 100 per cent, whilst the ground ends lie almost straight.

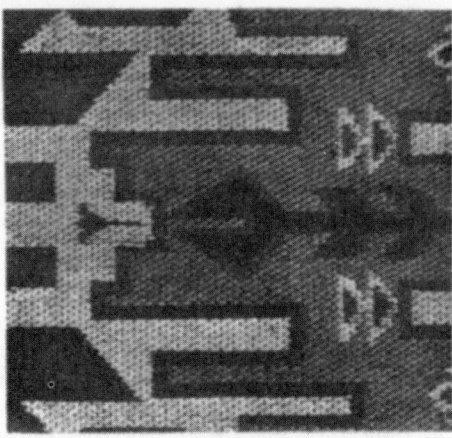

Figure 11.10

C in *Figure 11.11* shows a portion of the condensed design painted in four different sets of marks to denote the four colours of weft with the detailed weaves for each area being shown at D, E, F and G. It will be noted from the detailed weaves that, numbering the wefts in the order in which they are inserted, the following scheme is produced:

Weft No. 1 on the face, weft No. 3 on the back, wefts No. 2 and 4 in the centre.

Weft No. 2 on the face, weft No. 4 on the back, wefts No. 1 and 3 in the centre.

Weft No. 3 on the face, weft No. 1 on the back, wefts No. 2 and 4 in the centre.

Weft No. 4 on the face, weft No. 2 on the back, wefts No. 1 and 3 in the centre.

Thus, the wefts 1 and 3 supplant one another between the face and the back and so do the wefts 2 and 4. Although in the scheme shown the odd and the even

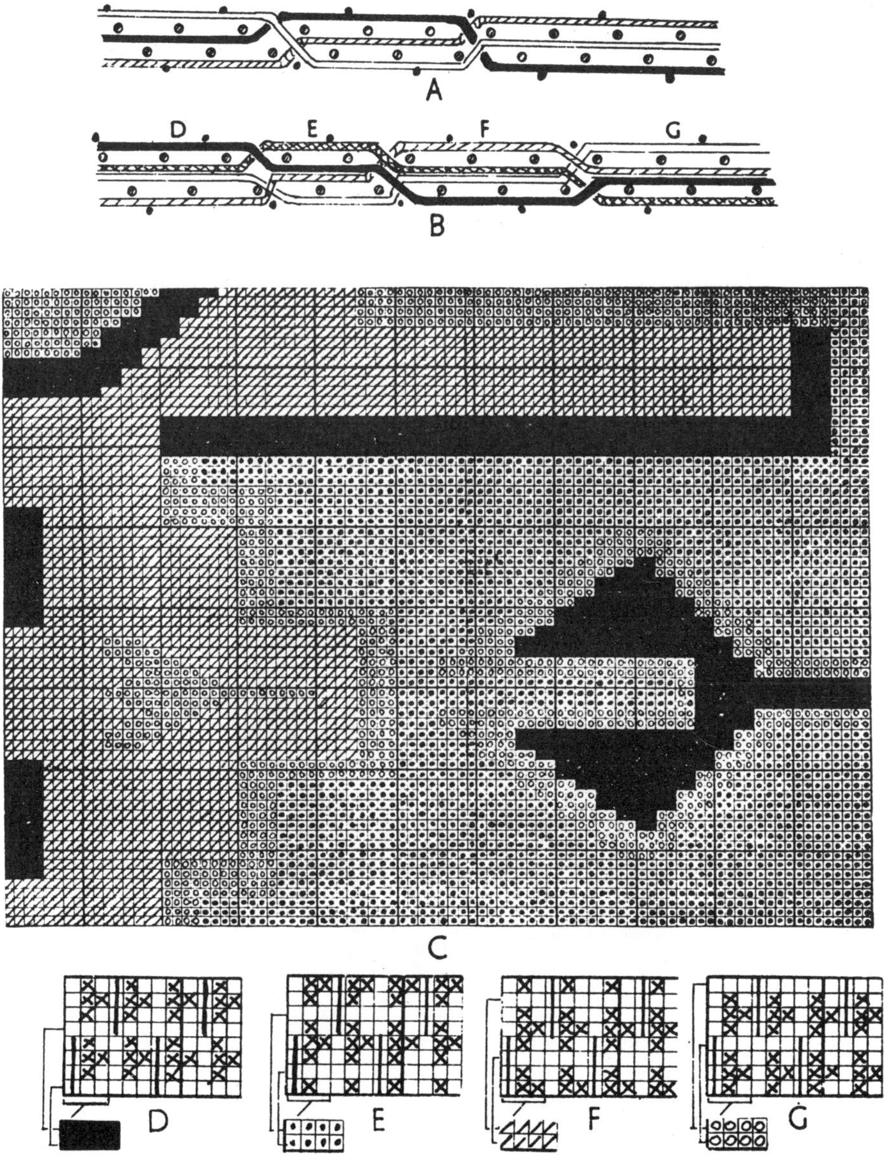

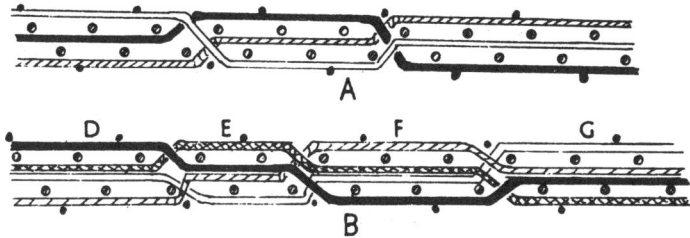

Figure 11.11

picks are paired together this is in no way compulsory and any one colour of weft can be made to supplant any other with equal facility. The interchange of the wefts can be followed by studying the warp section at B which corresponds to the scheme given above.

In the example given the stitching ends operate 4 up, 4 down, on an alternate basis as shown by the vertical lines in plans D to G and due to their distribution the face and the back of the cloth appear to be stitched in a plain weave order. If, however, the ends were distributed in the order of 4 ground, 2 stitching then a ribbed appearance could be achieved similar to the one illustrated in *Figure 11.5*. As this results in longer weft floats it is reserved mostly for the lighter weights of rugs which are not expected to undergo such severe wear as the heavier construction illustrated in *Figure 11.10*.

REPP-STITCHED WEFT FACE TAPESTRY STRUCTURES

Due to their complexity these structures were at one time woven on special harness mountings which in addition to healds also included lifting rods. An example of this form of harness, devised to simplify the painting and cardcutting of the designs, is given in Appendix I. At present this type of tapestry in common with other similar structures is produced on fine pitch jacquards and the employment of modern card-cutting machinery permits the use of simplified methods of design painting. The effect is most commonly produced in three figuring wefts with the ground warp elements contributing a fourth colour area on the face and a fabric of this type is represented in *Figure 11.12*. The tightly

Figure 11.12

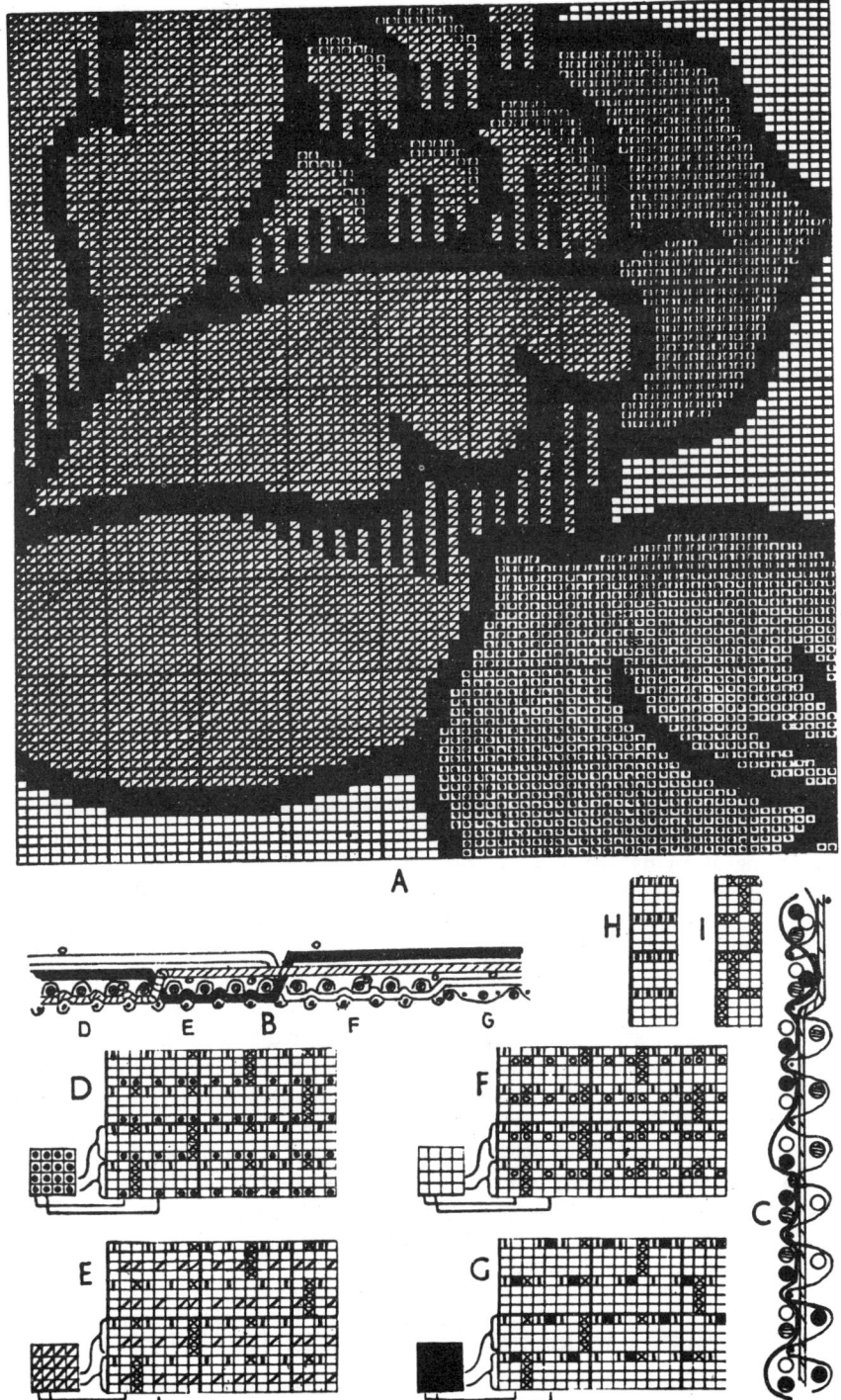

Figure 11.13

stitched and finely ribbed surface is due to the use of a face stitching warp in conjunction with a stitching weft yarn. The close binding of the face results in a very superior upholstery cloth in which the figuring wefts will not slip or roll due to rubbing. The back of the cloth, which is separated from the face by the ground warp, is loosely stitched by special back stitching ends to accommodate the excess material without straining the face and to permit easy flexing of the fabric to conform to the shape of the furniture. In the cloth illustrated in *Figure 11.12* the following yarns and settings have been used: Warp arrangement—1 ground end, 1 face stitching end, 1 back stitching end, 1 ground end, 1 face stitching end, 24 ends per cm; all the warp yarns are cotton, the count of the ground ends being 48/2 tex, and of the stitching ends 14/2 tex; a three-beam mounting is necessary due to considerable differences in the crimp of the different warp yarns, which is respectively 3 per cent, 27 per cent and 8 per cent for the ground, the face, and the back stitchers. The weft is arranged 3 figuring picks to 1 stitching pick, 12 picks of each weft per cm; the figuring yarns are cross-bred or lustre worsted, 98 tex, and the stitching weft is the same as the warp stitchers, i.e. 14/2 tex cotton.

As the ratio of the ground ends to the face picks is in the proportion of 48 : 60 per unit space the count of design used for the simplified design should be 8 × 10. The ground ends are used in emphasising the outline of the ornament and in shading the figure, as shown by the solid marks in the simplified design A in *Figure 11.13*, whilst the main ornamental features are produced by the three figuring wefts, as represented by the dots, diagonal marks and blanks respectively.

The cloth is woven face side down to avoid heavy lifts and is thus represented by the warp section at B in *Figure 11.13* in which the four different figure areas are lettered D, E, F and G to correspond with the detailed weaves of each area given alongside. As each horizontal row of the simplified design is representative of four picks, 3 figuring and 1 stitching, it is necessary to cut four cards from each row, the order of cutting for every weft in each differently operated design area being given by the fully worked-out weaves at D to G. The order of interlacing of the face and back stitching warps is also given separately at H and I in *Figure 11.13* from which it will be noted that the face stitcher is dropped on all the figuring wefts and raised on the stitching pick, whilst the back stitching warp is raised to stitch the back of the fabric in a satinette order dropping on the stitching picks in a reverse satinette order to bind the figuring picks which float on the back to the ground structure. From the fully worked-out plans it can also be noted that the back stitchers are not permitted to show on the face (underside, as woven) of the cloth and are, therefore, raised above every figuring pick of weft when it forms the face effect even though it appears according to the basic weave given at I that they should be down. The operation of the warp stitching elements is clearly shown in the weft section at C. The marks in the detailed weaves all indicate warp ends up in weaving the cloth face side down. It will be appreciated that if it is desired to weave the cloth the right way up blanks instead of marks should require to be cut.

COMBINED WARP AND WEFT TAPESTRY STRUCTURES

The cloth in *Figure 11.14* illustrates a complex tapestry structure in which the effect is achieved primarily by the use of several differently coloured figuring warps. Although the warp in this type of structure represents the main figuring element, figuring wefts are also used to enhance the diversity of patterns which can be obtained. The upholstery fabrics in this class of construction produced at present contain between three and five series of figuring warp ends and two figuring wefts. The warps are comparatively fine and when operating on the face do not cover the picks completely thus making it possible for the weft to be visible. The two figuring wefts are quite coarse and are usually in two strongly contrasting colours—one very dark and the other very light. Being visible, each weft is capable of influencing to a considerable extent the colour of a given warp which rests upon it such that the same warp colour area looks quite different depending upon which weft forms the background to it. To give an illustration, it will be appreciated that when a blue warp operates on the background of a fawn weft the hue of this portion of the design differs vastly from another area in which the same blue warp is backed by a brown weft. Thus, using the above method in which both wefts form an effect in conjunction with any one of a number of warps it would be possible to produce 10 distinct colour areas with five figuring warps and two constrasting wefts. In addition, it is possible to produce other areas in which each weft figures on the face independently. Being much coarser than the warp it will cover the surface completely without permitting any adulteration of its own colour. Further increase in the diversity of colour ornamentation is possible by colour planting in the warp and this is frequently resorted to. The figuring yarns are closely bound by the warp and weft stitching elements which help to achieve a distinctly ribbed, hard wearing surface.

Figure 11.14

A good quality cloth as shown in *Figure 11.14* may be produced with the following particulars: Warp—1 stitching end, 12/2 tex cotton; 3 to 5 figuring ends, 40/2 tex spun viscose staple; 108 ends of each kind per 10 cm; weft—2 figuring picks, 115 tex condenser spun cotton; 1 stitching pick, 17 tex filament rayon; 92 picks of each kind per 10 cm.

In a construction embodying 1 stitching and 3 figuring ends, and 1 stitching and 2 figuring picks the natural degree of condensation in the preparation of a simplified design would be by 4 warp-wise and by 3 weft-wise, one small square of design paper thus representing a complete structural unit. The number of different colours used in a simplified design depends on the number of colour and structural areas produced. In the example worked out in *Figure 11.15* eight different colours have been used and these are shown in the plans A to H. The simplified plans are connected by lines with the corresponding detailed weaves in which the marks indicate warp up. The first, second and third figuring warps are respectively interwoven with the first figuring weft at A, B and C, and with the second figuring weft at D, E and F, whilst in G the first, and in H the second figuring weft is brought to the surface. The vertical marks in the detailed weaves indicate the lifts of the stitching ends, and distinctive marks, which correspond to those shown in the simplified plans A to F, represent the lifts of the figuring ends in forming the surface effect, whilst the diagonal marks show the other lifts of the figuring ends.

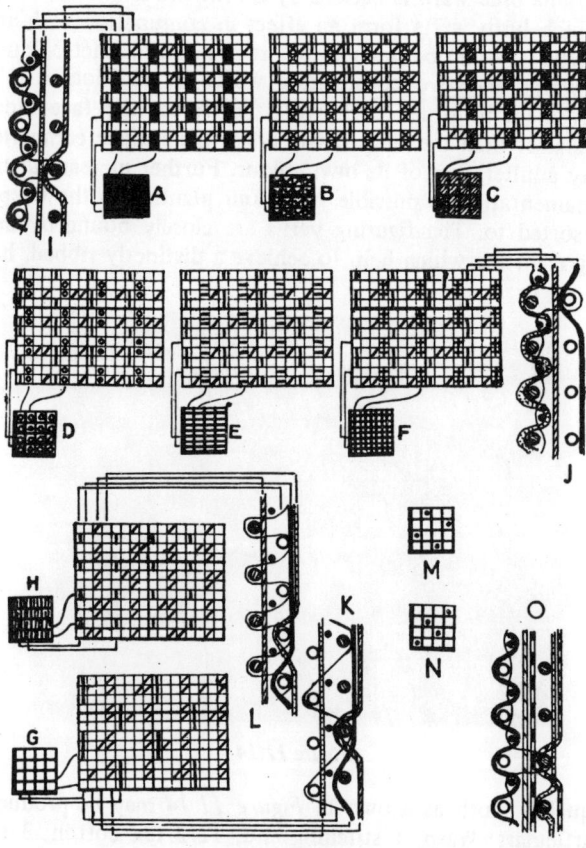

Figure 11.15

In each weave A to F the figuring warp required on the surface floats over both figuring picks and under the stitching pick thus making a continuous 2 up, 1 down interlacing. One of the remaining figuring warps acts as wadding and separates the wefts into a face and a back layer being operated in a continuous

1 up, 2 down order. The third figuring warp floats on the back and is loosely stitched into the body of the fabric in a satinette order as given at M or N. The stitching warp is down on the face figuring weft and up on the back one except when it is dropped to stitch through to the back which is done in a satinette order corresponding to the one given at M or N but operating in reverse. The stitcher is also invariably raised on each one of the stitching picks thus being responsible for the characteristically ribbed appearance of this class of structure. The manipulations of the various elements are clearly shown in the weft sections at I and J which correspond with the detailed weaves at A and F respectively. It will be appreciated that all the areas A to F are structurally similar and the differences in them are due entirely to colour. This structural similarity is also independent of the number of figuring warps used so that in a fabric containing, say, five figuring warps the face structure would be exactly as that given in the sections I and J, and only the number of wadding and backing yarns per vertical row would increase as depicted in the section shown at O.

The detailed weaves at G and H in *Figure 11.15* and the sections K and L show the figuring weft float areas in the fabric. When one of the figuring wefts is floated on the surface one figuring warp acts as wadding being placed directly underneath the face weft and above the other figuring weft and the stitching weft. The remaining two figuring warps float on the back and are stitched in to the figuring weft on the back in a satinette order, one as at M and the other as at N. Again the two areas are structurally similar the difference depending on the colour of figuring weft which floats on the face. In G and H, however, the structures are different because each illustrates a different method of stitching the float of the figuring weft. The differences are particularly clearly shown in the corresponding sections at K and L where it can be observed that whilst the first weft at K is bound rather loosely in an alternate order by the stitching warp operating 3 up, 3 down, at L the second weft is repp-stitched by means of 1 up, 2 down operation of the stitching warp. The differences in the method of binding result in a sufficiently different surface appearance to permit their utilisation as additional effect areas; thus, if required, the cloth described above could be painted in 10 colours, six of which would correspond to the warp face effects as at A to F, whilst the other four would refer to the areas of weft float in which each of the two figuring wefts would be displayed first in a loosely bound, and then in a repp-stitched form.

Although in the foregoing only one principle of figuring has been explained it will be readily appreciated that in view of the presence of a large number of yarn elements in the warp and in the weft a considerably greater structural diversity could be achieved. Thus, the tapestry structure could be combined with areas of wadded double cloth or treble cloth, other effects could be formed by floating the figuring elements without stitching in the manner of extra warp or extra weft and so on. In respect of the diversity of effect it is worthwhile to note that the filament rayon weft stitcher, which is one of the elements in the cloth described in detail above, has been employed in a deliberate attempt to enhance the general appearance of the fabric. This weft, which in the figuring weft float areas is entirely invisible, in the warp face areas is visible to a minute extent in the 'furrows' between the ridges formed by the figuring elements (note sections I and J). Due to its high sheen it provides, at certain angles of reflection, a shimmering depth which sets off the duller figure areas perfectly. Thus, with a degree of ingenuity and some knowledge of materials available it is possible to make even the comparatively insignificant and utilitarian elements of the construction contribute to the aesthetic appearance of a cloth.

12

Gauze and Leno Structures

In gauze and leno weaving certain ends—termed crossing ends—are passed from side to side of what are termed standard ends, and are bound in by the weft in these positions. The crossing and standard ends may be arranged with each other in various proportions, as 1-and-1, 1-and-2, 1-and-3, 2-and-2, 2-and-3, etc., but an essential condition is that each group of crossing and standard ends must be placed in one split of the reed. A crossed system of interlacing can be obtained when all the warp is brought from one beam, and in some cases this is essential in order to produce the desired effect. Very frequently, however, effects with such a difference in the take up are produced that it is necessary for the two series of ends to be brought from separate beams. The warp may consist entirely of crossing and standard ends, or stripes of these may be combined with stripes in which the ends interlace in the ordinary manner so as to form plain, twill, figure, etc. It is also possible for the crossing and standard ends to form the crossed interlacing alternately with straight interlacing in any required order. There is, therefore, almost unlimited scope for the production of variety of effect in striped, checked, and figured fabrics by combining gauze or leno with practically any other system of interweaving. Where the crossed interlacing occurs, an open, perforated structure may be formed, or the crossing ends may be interwoven in zig-zag form on the surface of a more or less compact ground texture. Nearly all kinds of yarns and yarn combinations can be used, but in open perforated structures particularly the threads should be as smooth, as uniform in thickness, and with as little loose fibre on the surface as possible. As the warp yarns in leno weaving are subjected to a higher degree of friction and greater stresses, their average strength should be higher than the minimum required for normal weaving.

The fabrics produced by this method are employed for curtainings, shirtings and for blouse and dress materials as well as for various industrial uses such as filter cloths, screens and sieves. Their great merit lies in a very considerable stability of the interlacing combined with its open nature. The size of the interstices can be determined precisely and will remain stable and uniform even under a degree of pressure.

The yarns used most frequently in the manufacture of these fabrics are cotton, spun rayon staple, cotton/polyester blends, filament polyamide and polyester, glass, and occasionally silk. Due to the friction associated with this system of weaving, yarns susceptible to static electrification should be well protected either by lubrication or by other techniques of static elimination or prevention.

THE PRINCIPLE OF LENO STRUCTURE

The terms leno and gauze are used somewhat indiscriminately but generally it is accepted that whilst leno may be applied to all structures in which some ends are transferred from one side to the other of the standing or standard ends, the term gauze is reserved for open structures produced in a plain or similar simple leno interlacing.

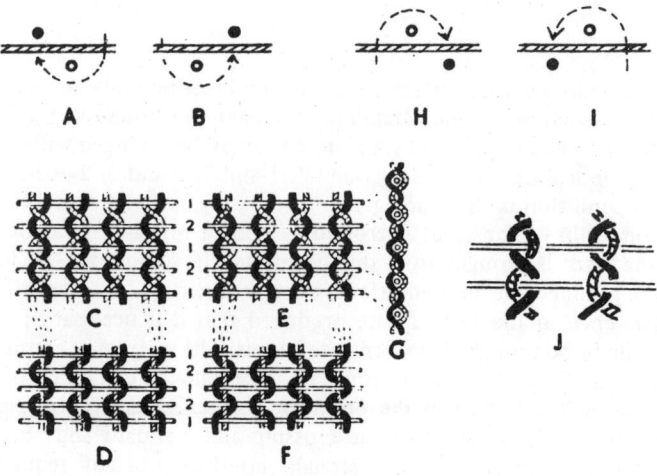

Figure 12.1

Although there exist a number of mechanically different systems to achieve the necessary lateral movement of one thread in respect of another the resultant structures produced by each of them may be identical. The thread manipulations required to produce the simplest structure of this type known as the plain leno or gauze are depicted in *Figure 12.1*. At A and B respectively cross-sectional views of two successive sheds are given in which at A the crossing end (black) forms the top shed on the left of the standard (white), and at B on the right of the standard end. Thus, in plain leno, using the bottom douping system illustrated at A and B, the crossing end is up and the standard end down on every pick but in between each successive shed the crossing end crosses under the standard end prior to each lift and the weft is held between the half-twists of the crossing end. The interlacing diagrams C and D show the appearance of the plain leno structure, the former obtained when one beam is used and the distortion of both the sets of ends is equal, and the latter achieved when the crossing ends are placed on a lightly tensioned beam and, therefore, bend prominently, and the standard ends lie straight being placed on a heavily tensioned beam. At E and F in *Figure 12.1* two other structures are given which correspond respectively with C and D but differ from them in that the alternate vertical rows of the leno in the former two are point drafted. This in effect means that whilst in one row the crossing end on a given pick crosses from left to right, in the next row the crossing end on the same pick crosses from right to left. Diagram G shows a section of a plain leno structure cut through the weft whilst the appearance of a plain leno cloth is illustrated in *Figure 12.2.*

Figure 12.2

The cross-over of the crossing end may occur under the standard end as shown at A and B, which, as stated, is termed bottom douping, or it may occur over the the standard end which is termed top douping. The latter case is illustrated by the two successive shed diagrams given at H and I in *Figure 12.1* which show a situation exactly the reverse of the one depicted at A and B. At H the crossing end forms the bottom shed to the right, and at I to the left, of the standard end which on each pick remains in the top shed. Thus, in top douping the crossing ends are down and the standard ends up on every pick but as the crossing end is transferred alternately from one side of the standard to the other between each pair of sheds the weft is held securely in the half-twists of the crossing end. The interlacing diagram at J shows that the structure produced in this way is fully comparable with C except for the reversed position of the crossing and the standard ends in relation to the weft—the two could not be distinguished if one of them were to be turned over. As bottom douping is mechanically more convenient it is much more commonly encountered and most of the examples given subsequently have been worked out on the bottom douping principle.

The basic sheds of leno weaving

From the diagrams and the descriptions given in the foregoing it will be apparent that the crossing ends may be required to form sheds either on one or on the other side of the standard ends. In order to achieve this each crossing end must be controlled by two healds, the mails of which are placed one on either side of the standard end weaving in conjunction with the given crossing end. In order to prevent unnecessary see-sawing of the crossing yarns against the standard yarns upon the alternate shed formation the dual heald control over the crossing end is not direct but through an intermediary of a third element known as the doup. Thus, for the proper control of its operation each differently working crossing end requires three shedding elements. The crossing ends may be drawn through the warp beam to the front of the loom either on the left or on the right of the standard ends with which they combine to make the leno rows. If it is assumed that they are drawn on the left-hand side then each lift of the crossing end on the left will be quite a normal lift, as no transfer to the 'wrong' side of the standard is involved, provided that the heald which holds the crossing end on the other side of the standard releases its hold on the crossing end temporarily. This is ensured by the doup, and the shed on the left thus formed is illustrated schematically at A in *Figure 12.3* and is known as the *open shed.* Making the crossing end lift on the right involves first pulling it across from the normal side, i.e. the side on which it was drawn through, to the 'wrong' side which again is accomplished with the aid of the doup. This shed, known as the *crossed shed,* is illustrated at B in *Figure 12.3* and requires additionally the operation of an easer, an auxiliary element, which gives-in temporarily an extra length of yarn required during the formation of the crossed shed. Summarising: On open sheds it is necessary to raise the heald controlling the crossing ends on the normal side of the standard and the doup; on crossed sheds it is necessary to raise the heald controlling the crossing ends on the abnormal side of the standard, the doup and the easer. Although in the example given above the left has been designated as the normal, and the right as the abnormal side the situation would be reversed if the crossing ends were originally drawn through from the beam on the right-hand side of the standard ends instead of on the left.

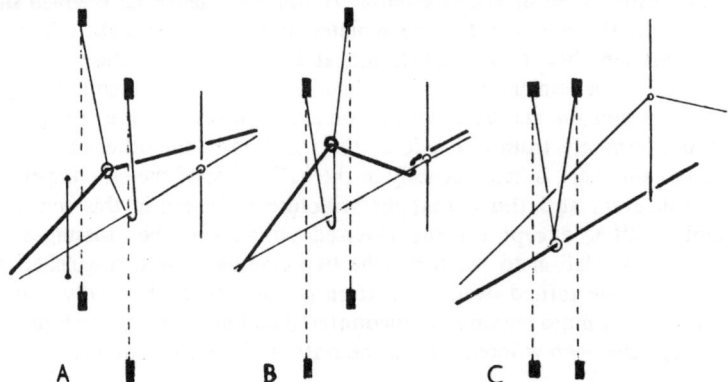

Figure 12.3

On both the leno sheds the standard end in bottom douping remains down if, however, on some picks of the construction it is required to lift the standard end to form the upper shed line whilst leaving the crossing end down then the standard heald is raised and all the elements which control the crossing end are left down. The lift of the standard end which is termed the *plain shed* is illustrated at C in *Figure 12.3*.

Methods of producing the leno structure

The method of control of the crossing end illustrated in schematic diagrams in *Figure 12.3* has at one time been used in hand-loom weaving and in slow-speed power-loom weaving to produce leno centre selvedges but is not applicable in modern high-speed weaving machinery. Equally so, the method which depended on the use of string or twine doups is at present employed only in exceptional circumstances and cannot be any longer considered as a major system of leno fabric production. For this reason it is dealt with in Appendix I together with the other traditional mountings. The main methods of production of leno and gauze structures in current use may be listed as follows:

(1) Flat steel doups with an eye.
(2) Flat steel doups with a slot.
(3) Gauze and tug reed mechanisms.
(4) Eyed needle and slider frame devices.
(5) Rotating bobbin and geared disc mounting.

The first four systems given are considered subsequently in the order in which they are listed. The last system is only mentioned because it represents an entirely different principle of twisted thread formation obtainable in weaving but as it is only applicable to the construction of leno selvedges it cannot be regarded as a cloth or a design-forming element.

In addition to the main leno-forming elements most methods of leno production require an easer the function of which has been already mentioned. In some circumstances where open or semi-open shedding motions are employed another auxiliary mechanism may be necessary and this is known as the shaker or the jumper. The device is described later—at this juncture it will be sufficient to realise that it is necessary to bring the standard heald level with the leno healds between the sheds when the transfer of the crossing end from one to the other side of the standard end is taking place.

LENO WEAVING WITH FLAT STEEL DOUPS WITH AN EYE

The simple type of flat steel doup in which the doup needle has an eye at the top is somewhat limited in terms of figuring capacity and is, therefore, mainly employed in the production of industrial fabrics in the plain leno weave as shown in *Figure 12.1*. However, it can also be used successfully for simple ornamental fabrics and is particularly useful for effects in which ordinary cloth is combined with leno in stripe formation.

The flat steel doup assembly or unit consists of a doup, shown at D in *Figure 12.4*, and two lifting healds designated H1 and H2. Each lifting heald is composed of two identical flat steel strips joined together at a point marked W by a spot weld. The doup or doup needle has an eye at the top through which the crossing end is drawn and it sits in a sliding fit between the flat strips of the lifting healds H1 and H2 thus bridging the gap between them. Each lifting heald is controlled by its own heald frame in the usual manner through the steel rods fitting at A whilst the doup needles are threaded upon their own rods at B. The two rods B are spring loaded through a double yoke arrangement to ensure the return of the doup after each lift. Each lifting heald is capable of raising the

212

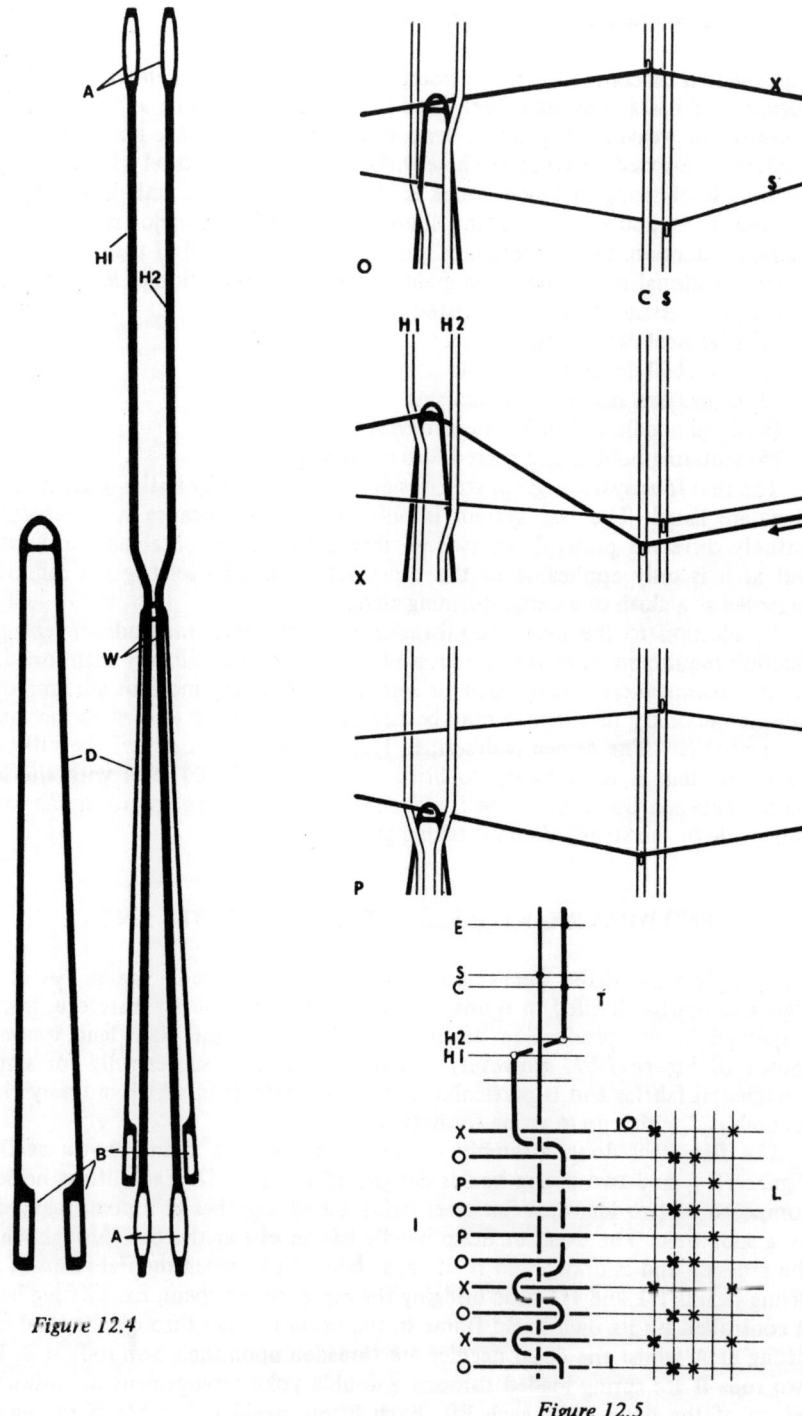

Figure 12.4

Figure 12.5

doup due to the presence of the spot weld in the flat strips and in the production of plain lenos the elements H1 and H2 lift alternately.

The formation of the leno sheds in this assembly is illustrated at O, X and P in *Figure 12.5* which respectively show the open, crossed, and plain sheds. It will be noted from these diagrams that the leno assembly is mounted at the front followed some distance behind by the heald controlling the crossing end and finally by the standard heald. The gap between the leno assembly and the first of the ordinary healds is intended to reduce the angle at which the crossing ends lift during the formation of the crossed sheds and the best results are obtained when the distance of the gap is approximately 10 cm.

In the draft T in *Figure 12.5* it will be noted that the crossing end is drawn on the right side of the standard which means that any sheds formed by this end on the right will be open sheds and any sheds formed on the left of the standard will be crossed sheds. The diagram O depicts an open shed which is achieved by the lift of the right-hand side lifting heald, H2, and the crossing end heald, C, the H2 element being responsible for raising the doup. The formation of the crossed shed is shown in the diagram X in which the left-hand side element of the doup assembly, H1, is raised. This action transfers the crossing end to the left-hand side of the standard and is assisted by the simultaneous operation of the easer, indicated by the arrow on the right of the diagram, to provide the extra length of yarn necessary to compensate for the cross-over. The heald C on crossed shed remains down. In fact, the crossing-end control heald C could be entirely dispensed with and both sheds could be achieved by the doup lifting action alone but this is rarely practised now since it was established that without this heald the sawing action of the crossing end against the standard end and standard heald upon the formation of crossed sheds is much more severe. Formation of the plain shed is depicted by the diagram P and involves the lift of the standard heald S only, all the crossing-end control elements remaining down.

In the draft, T, in *Figure 12.5* the crossing end is shown to be controlled by the easer, E, the heald C and the two lifting heald elements of the leno assembly, H1 and H2. The standard end is controlled by the standard heald alone. At I, one full repeat of the construction consisting of 10 picks, is represented by means of an interlacing diagram. In it each shed is designated by a symbol O, X or P to indicate what type of shed is formed. The correct designation of the sheds in leno weaving is important as it determines which shedding elements are pegged or cut on each pick. Thus, it will be noted from the lifting plan L that on pick 1, which is an open shed, H2 and C are pegged; on pick 2—a crossed shed— H1 and the easer are pegged and on pick 6—a plain shed—the standard heald is pegged. The doup lifts on every pick on which either one of the two lifting healds, H1 and H2, is raised.

Upon examination of the interlacing diagram I it will be seen that in one repeat of the depicted construction there are five open sheds and three crossed sheds and this illustrates a point of some importance. In leno weaving the formation of a crossed shed always creates an additional strain and, therefore, other conditions being equal the side on which there are fewer lifts of the crossing end should always be designated the crossed shed side. The diagram I illustrates the appearance of the construction when it is produced with two beams, the beam carrying the standard ends being the more heavily tensioned. However, the same construction could be equally well produced, with a single beam, only in such a case both sets of ends would be displaced similarly producing a cloth with a different appearance.

Weaving with more than one leno assembly

Using a doup with an eyelet, each different order of manipulation of the crossing end requires an additional leno assembly. The method of mounting of the additional assemblies is similar in as much as all the leno assemblies are mounted at the front, then follows a gap after which are mounted the crossing end healds in a consecutive order, then the standard healds, also in consecutive order so that the standard healds working in conjunction with the first leno assembly come first, those working with the second assembly come second, and so on. As the crossing sheds between the differently patterning leno rows occur usually on different picks each assembly normally requires its own separate easer bar each of which must be individually controlled. It is obvious that each additional leno assembly results in a greater complexity of the draft and of the weaving process and for this reason more than two leno assemblies are rarely employed on any one loom. A construction produced with the aid of two leno assemblies or units is illustrated in *Figure 12.6* and is described in the following section.

Point draft or counter leno

By referring to *Figure 12.5* it will be seen that by drawing in the crossing end in a neighbouring leno row on the left of the standard instead of on the right and crossing it to the right it would be possible, using the existing elements, to produce a novel structure in which the alternate leno rows would work in a mirror image order. The point draft or counter leno is often employed because it permits the creation of new and different effects with a considerable economy in the number of the shedding elements used.

Figure 12.6 illustrates quite an ornate tie cloth construction produced on the counter leno principle with the use of only two leno units and six ordinary healds. The actual appearance of this cloth made from filament polyester yarns is shown at A in *Figure 12.7*, whilst at B the same cloth is given enlarged four times.

The diagram I in *Figure 12.6* provides a pictorial representation of two weft repeats and one warp repeat of this cloth from which it will be seen that the construction consists of four leno groups, each with one crossing and two standard ends, but only two leno assemblies. This is achieved by making each pair of neighbouring leno groups operate on the counter leno principle thus utilising the same lifting unit twice over in a point draft fashion. The crossing ends 1 and 4, numbered at the bottom of I, are both raised on exactly the same picks but whilst 1 makes open sheds on the right, and crossed sheds on the left of the standard ends, 4 makes open sheds on the left and crossed sheds on the right. In this way two crossing ends are made to converge towards, and diverge away from, each other forming a cell-like structure. The crossing ends 7 to 10 work similarly but on different picks and between them utilise the second leno assembly. The full draft is shown at T from which it may be noted that two easers, E1 and E2, are used and the two sets of ends come from separate beams to permit full sideways deflection of the crossing ends whilst keeping the standard ends straight. The order of pegging or cutting of the various shedding elements is given in the lifting plan L.

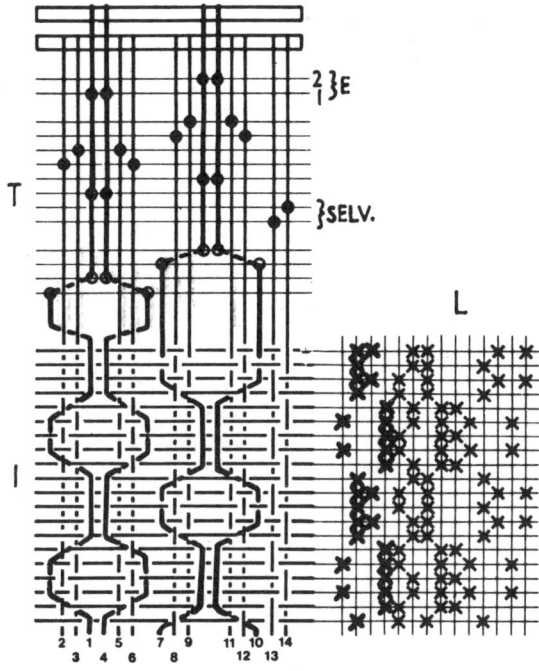

Figure 12.6

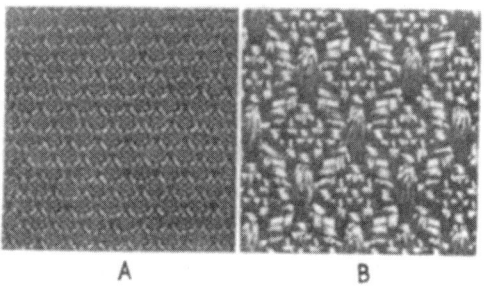

Figure 12.7

The ends 13 and 14 which are shown in the construction in *Figure 12.6* are not part of the leno structure but represent the selvedge ends. They are included to illustrate the most advantageous point at which the staves which control the selvedge ends can be placed. It may be noted at this juncture that whilst the standard healds may occasionally be used for the selvedge end control in leno weaving, particularly when they are operated in a continuous plain or other simple weave order it is generally preferable to use separate healds for the selvedges. When separate healds are used, as in the example given, then they should be placed in front of all the other ordinary healds and behind the leno units exactly as shown in *Figure 12.6*. In this way the 10 cm gap between the leno assemblies and the other healds is partially utilised without the slightest interference with the structure as these selvedge staves, obviously, do not carry any mails in the centre of their span.

In the structure in *Figure 12.6* as in the previous structure the open shed side has again been designated as the side with the greater number of lifts of the crossing end and it may be noted that the crossing end lifts five times on the open, and only twice, on the crossed shed side in each weft repeat. Although in the example given two leno units have been employed it will be readily appreciated that a considerable variety of effect in counter lenos can be achieved with only one unit and this type of work is exemplified by the left-hand side portion of I in *Figure 12.6* in which the cell structure achieved by ends 1 to 6 utilises only one leno unit.

Special lifts of the standard ends

As the doup with an eyelet is somewhat limited in the figuring capacity and the multiplicity of the leno units increases the difficulty of weaving, structural variation is sometimes achieved with considerable economy by an ingenious order of lifts of the standard ends.

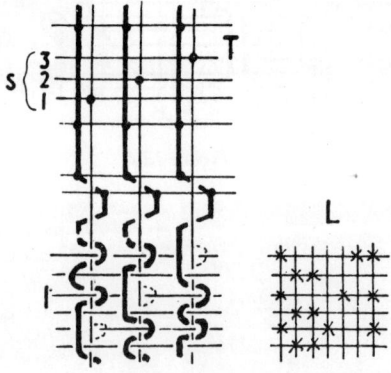

Figure 12.8

At I in *Figure 12.8* a structure is shown in which a twilled leno effect is produced which at first sight appears to require three leno assemblies. On closer scrutiny, however, it becomes clear that it can be obtained with a single leno unit and the use of three standard healds. The two lifting healds of the leno assembly operate alternately forming open and crossed sheds as in the plain leno. On certain selected crossed sheds, however, the standard end is also lifted which ensures that the crossing end cannot remain in the crossed position because there is nothing to retain it there and will return to the open shed side due to tension thus creating a long float. The temporary crossed shed positions of the crossing end are indicated in the diagram I by the dotted lines. The draft for this structure is given at T and the lifting plan at L in *Figure 12.8*. The frequency at which the trick lifts of the standard ends occur must be judged carefully otherwise a loose structure is liable to result.

Russian cords

Figure 12.9 shows a leno and plain weave stripe fabric termed a Russian cord. The structure is quite solid and consists usually of a number of ends weaving plain in a stripe arrangement interspersed by ends in a strongly contrasting colour which produce the leno stripe. The leno stripe contains one or two thick standard ends or a larger number of fine standard ends over which the crossing end passes from side to side on succeeding picks. The weft in the leno stripe is entirely concealed, the surface being formed by a distinctly bulging cord effect. The face side of the cloth is given in the interlacing diagram at R in *Figure 12.9* but the fabric is normally woven face side down with a bottom douping arrangement. It will be noted from the diagram that the crossing ends form crossed and open sheds on alternate picks and, therefore, the two lifting healds of the leno assembly work exactly as in the plain leno. Cords as wide as 6 mm have been successfully woven in flat steel doup assemblies using the simple doup with an eyelet.

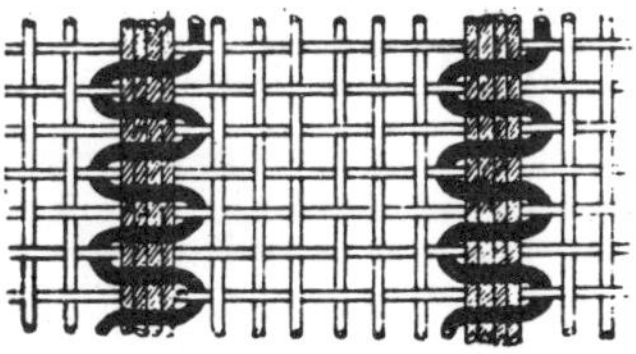

R

Figure 12.9

Fabrics of this type are comparatively densely wefted which causes the oscillating crossing ends to create a solid wrapper effect. Heavy cord lines are thus produced which, on account of the contrast in colour with the ground warp and weft, appear to be formed in extra weft. The crossing warp requires to be very much longer than the standard warp, and in the example is about four times as long, but the proportionate lengths vary according to the reed, the picks of the cloth, and the bulkiness of the standard ends. Variety of effect is sometimes given to these styles by having the standard ends different in colour from the crossing ends, and ceasing to form either the crossed or the open sheds for a number of times in succession. The thick standard ends are thus left uncovered by the crossing ends for a space, the latter lying straight in the cloth and being practically concealed by the former, hence the continuity of the coloured line is broken by spots of another colour.

Simple net lenos

The term net or spider leno is commonly applied to doup styles in which the crossing ends are mostly floated on the surface of the cloth, and are interlaced so as to form waved lines. The effect formed by the crossing ends is usually a chief feature of the pattern, and these ends, therefore, require to be of special material, colour, or thickness so that they will show in clear contrast with the ground. Each group of standard ends generally forms a compact ground structure across which the doup ends are traversed, the latter ends being really introduced on the extra warp principle. An open appearance is, however, sometimes given to a fabric by suitably missing splits between the groups of ends.

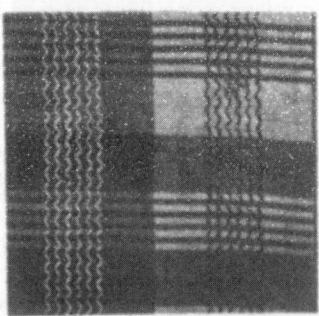

Figure 12.10

Figure 12.10 illustrates a style in which the crossing ends are all drafted in the same direction across four standard ends. A portion of the structure, showing how the threads interlace, is represented at I in *Figure 12.11*, while the draft is given at T, and the lifting plan at L. The standard ends, as is very frequently the

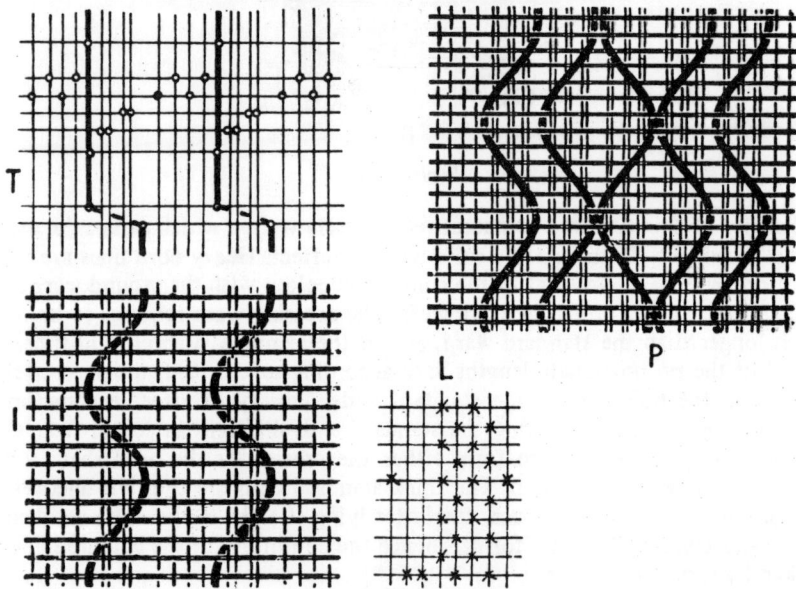

Figure 12.11

case in these styles, work two ends together in plain order throughout, so that they will spread out as much as possible; and they are also kept as straight as possible in the cloth, so that the maximum amount of traverse will be given to the crossing ends. *Figure 12.11* illustrates the production of the effect wrong side up with a bottom doup, the crossing ends being lifted on one pick in every five. The order of lifting fits with the plain interweaving of the standard ends; thus, where a crossing end is raised, the double standard end next to it is left down, and the former is held by the weft against the latter. *Figure 12.10* illustrates a good method of colouring the ground of a net leno style. Thus, light crossing ends are introduced on a dark ground and dark crossing ends on a light ground, while to correspond with the vertical lines formed by the crossing ends narrow horizontal lines are formed by light picks on a dark ground, and by dark picks on a light ground. The fabric represented in *Figure 12.12* shows a modification of the last style produced by point drafting, a method which can be used very effectively for giving prominence and variety to the zig-zag interlacing. A flat view, showing the interweaving of the threads in the broad leno stripe, is given at P in *Figure 12.11*, in this case the effect being shown face side up although normally it would be produced face side down using the bottom douping system.

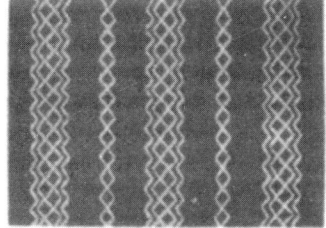

Figure 12.12

As each group of crossing and standard ends must be placed together in one split of the reed, the denting of these styles is a very important feature. For example, in the fabric represented in *Figure 12.12* there are 22 ends in each plain stripe which are dented two per split, while in the narrow doup stripe there are two groups of ends which must be placed in two splits, and in the broad doup stripe five groups, which must be placed in five splits. The narrow doup stripe, however, occupies the width of 12 ends, or six splits of the plain stripe, and the broad doup stripe the width of 26 ends or 13 splits. As a general rule, the effect can be produced by suitably missing splits between the groups of ends, the double standard ends, in working plain, readily springing out and filling up the spaces created.

Simultaneous bottom and top douping

The system in which the two ends which form a leno weave are both drawn through the eye of a doup needle, one operating as a bottom doup and the other as the top doup, is particularly useful in weaving yarns which are susceptible to sideways deflections. In ordinary forms of leno weaving formation of the crossed shed inevitably results in severe angular deflections of the crossing ends and a degree of rubbing between the yarns or between the yarn and the healds. Some materials, such as fibre glass or low twist woollen yarns, would be difficult to weave in those conditions which can be appreciably alleviated by the use of the double-doup mounting.

The principle of operation of this type of mounting is illustrated at X, O and P in *Figure 12.13* from which it can be observed that the crossing end is mounted in the leno assembly operating as the bottom doup whilst the standard end is

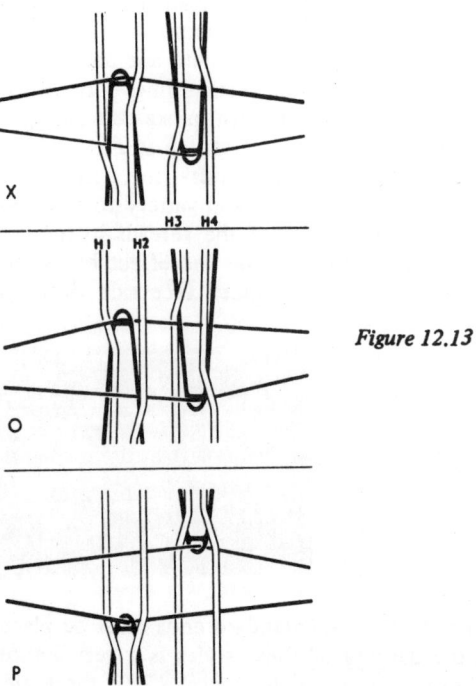

Figure 12.13

mounted in the top doup leno assembly. As opposed to the normal system where the crossing end is displaced in respect of the standard, in the double-doup mounting the two ends are mutually displaced one in respect of the other by the opposite movements of the two respective doups. At X the crossed shed is formed by the crossing end moving up (H2 lift) on the right of the standard as the standard end is simultaneously moving down on the left of the crossing end (H3 drop). Exactly the opposite situation occurs at O in which H1 is raised and H4 is dropped. The position of the lifting healds during the formation of plain sheds is shown at P. It will be realised that the bottom doup needle is spring-loaded in the downward direction so that it is pulled down after any lifts by either H1 or H2 whilst the top doup needle is springloaded in the upward direction being pulled up after each downward pull by either H3 or H4.

Although the above system offers the advantage of reduced bending or deflection of the yarns it is rather limited in its figuring capacity. For greatest ease of weaving the number of the crossed sheds in the weft repeat of the weave should be balanced by an equal number of the open sheds. This, however, does not preclude the plain lenos which are frequently produced on the double-doup systems or such simple fabrics as those illustrated in *Figures 12.14* and *12.15*. The construction illustrated in *Figure 12.15* is suitable for cellular woollen blankets and shows the possibility of using the double-doup mounting for the production of point or counter leno effects. In both the above figures the

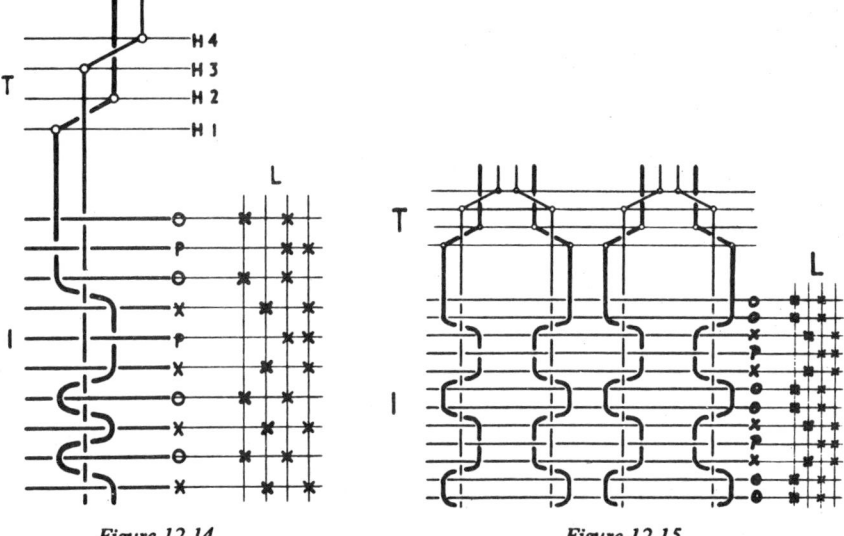

Figure 12.14   Figure 12.15

interlacing diagrams are designated by the letter I, the drafts by T and the lifting plans by L whilst the types of shed formed on each pick are marked by X, O and P referring respectively to the crossed, the open, and the plain sheds. The actual appearance of a cellular blanket is shown in *Figure 12.16.*

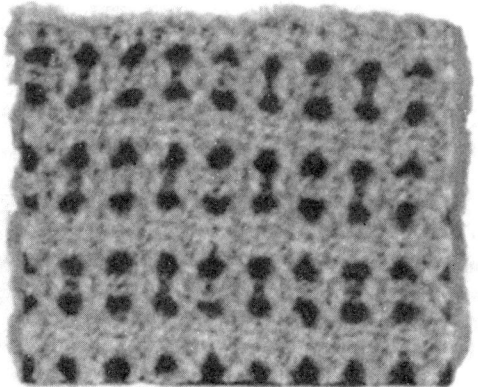

Figure 12.16

The mounting offers an additional mechanical advantage in that, due to an equal amount of displacement in both sets of ends, there is no need for an easer motion and neither is there any necessity for a shaker device.

LENO WEAVING WITH FLAT STEEL SLOTTED DOUPS

When the doups with an eyelet are used any differences in the order of interlacing between vertical rows of the structure require normally the use of additional leno assembly frames. This, as has been stated, adds to the difficulties of weaving and for this reason is avoided if possible. The slotted doup offers a possibility of obtaining different interlacings in the neighbouring rows of leno with the use of only one leno assembly frame and is, therefore, employed in all figured leno constructions.

The slotted doup needle is shown in *Figure 12.17* at D where it will be seen that whilst one leg of the needle is open the other one is solid. The needle could be reversed either way so that the slotted or open leg could be placed on the right or on the left. The leno unit or assembly with the slotted doup is illustrated side by side and the principle of operation of this unit is identical with the one given in *Figure 12.4* in connection with the simple eyelet doup. The difference between the two lies in the fact that when the lifting heald, H2, of the assembly is raised the crossing end does not need to rise with it. It may remain in the bottom shed by running down into the slot if an additional control heald, through which it is also threaded, remains down as well. If there are, say twelve separate control healds through which the adjacent crossing ends are drawn then twelve different interlacings on the slotted side are possible and the control healds must, therefore, be regarded as figuring healds. The slotted side is the open shed side of the structure, therefore, the choice of whether a lift does or does not take place exists only on the open shed side. No such choice exists on the crossed shed side because when the lifting heald, H1, lifts the doup it bars the slot and the crossing end must lift to the top with the doup. As already stated the position of the slot may be reversed and in addition to the freedom of choice of the lifts on the slotted side there is also the possibility of point draft or counter leno operation.

The principle of weaving lenos with slotted doups is illustrated in *Figure 12.17* where a construction is shown in which a crossing end operates in conjunction with two standard ends—an arrangement frequently employed in the production of marquisettes. The two standard ends usually weave plain but may be operated in other weaves if desired. The crossed shed is formed as before with the lifting heald, H1, of the leno assembly up and the easer operated to release a length of yarn as indicated at X, the leno figuring heald F remaining down. The normal open shed shown at ON is also formed in the same way as in the eyed doup, i.e. with the lifting heald of the leno unit, H2, up and the crossing end control heald (or figuring heald) F up. At OA the open shed is shown upon which the option to lift the crossing end to form a leno binding is not exercised because the heald F has been retained in the low position and the crossing end is thus permitted to slide down the slot. If necessary, plain sheds in which all the elements controlling the crossing end are down and one or both standards are up can also be made and such a shed is shown on pick 7 at I in *Figure 12.17* but as there is no advantage in this the interruption of leno binding is normally produced by lowering heald F on open sheds as illustrated at OA and at I on pick 5. The construction given in the interlacing diagram at I with the draft at T and the lifting plan at L has been selected to illustrate all the possible shed formations in slotted doup weaving and, apart from the two sheds already referred to, it will be noted that crossed sheds are formed on all the even picks, normal open sheds on picks 1 and 3 and an open shed combined with a trick lift of the standard end on pick 9. On the latter pick due to the lift of the second standard end crossing end binds across only one standard instead of two as on picks

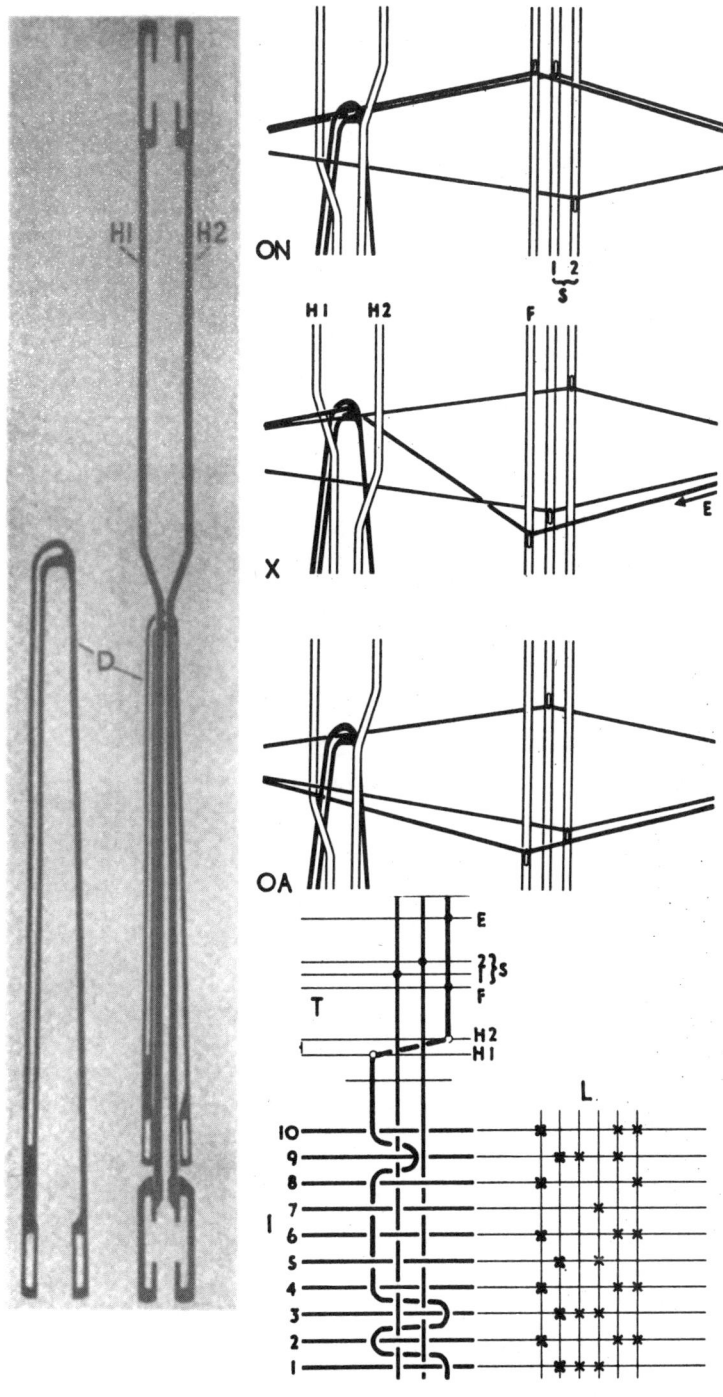

Figure 12.17

1 and 3. Trick lifts of the above type could, of course, be operated in reverse on the crossed sheds.

Frequently, in weaving lenos on slotted doups the two lifting healds of the leno assembly, H1 and H2, are operated alternately throughout and structural variations are obtained by suitable manipulations of the leno figuring healds. This has the advantage of never lowering the doup to the bottom as between sheds it only descends half way down before it is taken up again so that reduced spring loading can be applied to the doup return motion with the resultant reduction of strain and wear on the leno assembly. On the other hand if it is required to make several sheds in succession the lifting heald H2 and the doup can stay up for as long as necessary with only the leno figuring healds working and this also results in less wear.

Simple figured effects

The slotted doups when combined with a large number of figuring healds lend themselves particularly well to the production of twilled or diamond effects in

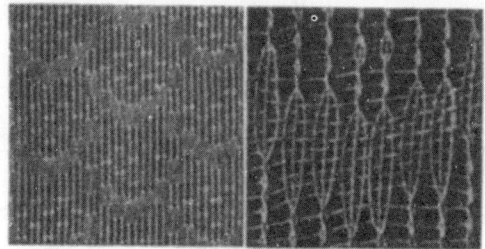

Figure 12.18

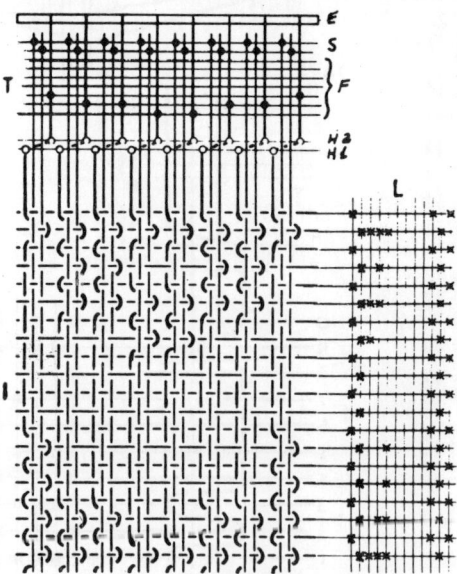

Figure 12.19

open curtaining fabrics. A cloth of the above type is shown in *Figure 12.18*—actual size on the left, enlargéd on the right. It repeats over 24 leno groups, each containing three ends, one crossing and two standard, and requires only one doup assembly, seven leno figuring healds point drafted in pairs, and two standard healds. The size of the full repeat weft-wise is 32 picks. A portion of the repeat which corresponds with the enlarged illustration in *Figure 12.18* is shown in the interlacing diagram I in *Figure 12.19*. The two lifting healds of the leno assembly operate up and down alternately as indicated in the lifting plan, L, forming crossed sheds on every even pick and permitting open sheds to be formed on every odd pick. However, as will be noted, the figuring healds F are dropped on some open sheds thus ensuring that these do not take place. As a result the crossing end on those occasions stays down and remains on the crossed shed side of the standards producing the figured effect which is clearly visible. The standard healds, S, operate a continuous plain weave and the easer bar, E, is raised on every crossed shed as usual. In the draft, T, only the first three figuring healds are shown drawn through but the full draft can easily be constructed by reference to the cloth in *Figure 12.18*.

Structures produced with two crossing ends per slot

In addition to the effects in which one crossing end works in conjunction with two or more standard ends attractive effects are frequently produced in 2-crossing-2 styles in which two crossing ends are threaded through each doup

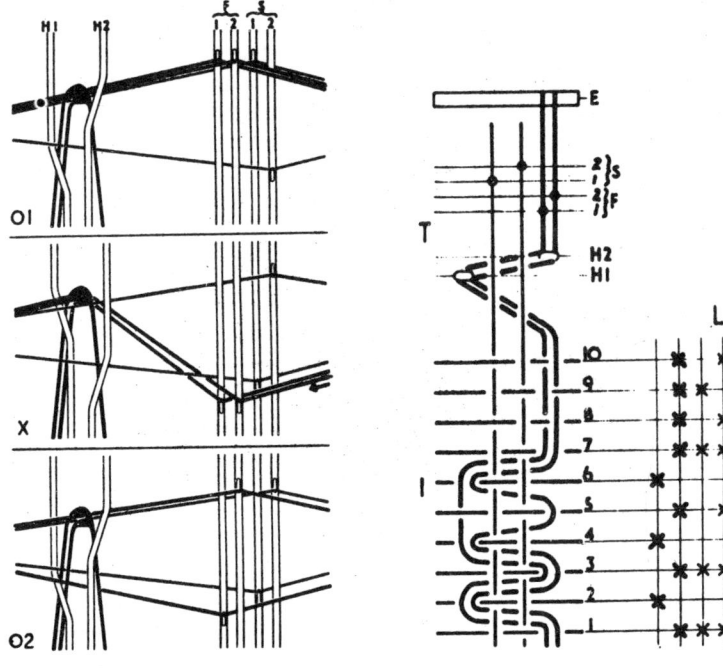

Figure 12.20   Figure 12.21

slot. The operation of the doup assembly in such a structure is shown at X, O1 and O2 in *Figure 12.20*. At X a crossed shed is formed and both the crossing ends in the slot must lift to form this shed. On the open shed side, due to the control exercised by the figuring healds, the crossing ends may both be lifted as at O1, or only one may be lifted, the other remaining down as at O2, or indeed, if so required, both may remain down.

In the interlacing diagram shown at I in *Figure 12.21* various interlacings are shown. Crossed sheds are produced on picks 2, 4 and 6 and straightforward open sheds with both crossing ends lifting on picks 1, 3 and 7. On pick 5 a special effect is produced because one crossing end makes an open shed in an orthodox manner whilst the other stays down and because it is down between two crossed sheds it remains on the crossed shed side so that one crossing end is on the side of the standards and one on the other side. On picks 8, 9 and 10 the crossing ends make open sheds alternately which in effect results in a bar of plain weave across the cloth. The standard ends operate in a continuous plain weave but in common with other leno structures this operation could be interrupted to produce trick lift effects or weaves other than plain in conjunction with the crossing ends. The draft at T and the lifting plan at L indicate the manner of operation of the shedding elements, the healds being designated as in the previous diagrams.

Use of slotted doups for point draft or counter lenos

The actual and enlarged views of the cloth in *Figure 12.22* represent a cellular shirting or blouse materials frequently made on the point draft leno principle. The technique of counter leno weaving in the slotted doups does not differ from that explained in connection with the eyed doups apart from the freedom which exists on the open shed side due to the presence of the slot. In the first fabric illustrated by the interlacing diagram I in *Figure 12.23* this freedom of patterning is not utilised and, in fact, the construction at I, which represents the cloth illustrated in *Figure 12.22*, is so simple that it could have been

Figure 12.22

produced on the ordinary eyed doup. It is given mainly to show the various drafting possibilities which exist when two doup assemblies are used. The need for the two doup assemblies in this case exists not as a result of the complexity of the structure but solely due to the density of the cloth. The interlacing diagram I and the lifting plans L1 and L2 represent the fabric as it is woven,

K   *Figure 12.23*

which is face side down. Due to considerable displacement of the crossing ends two beams are required and these are designated B1 and B2.

At T1 and T2 two drafts are illustrated. Draft T1 is the normal draft in which the crossing ends in the leno assemblies cross the standard ends to the left and to the right in alternate leno groups. All the crossing ends are drawn through the same figuring heald. The three standard ends which together with one crossing end form a leno row or group are drawn point draft fashion over two standard healds and, as the crossing ends which belong to the first and the second leno assembly make the crossed sheds on the same picks, one easer is sufficient. In the alternative draft, T2, the two leno assemblies are drawn so as to permit one set of crossing ends to form an open shed as the other set makes the crossed shed. In this way the tension across the warp is equalised better and more even running of the loom can be expected. Due to this arrangement two figuring healds and two easers become a necessity. The standard ends are straight drafted over three standard healds. The space between the leno assemblies and the figuring healds is, as usual, occupied by the selvedge frames, V, in both drafts.

To indicate that much more elaborate effects can be produced on the point draft system with the slotted doups than the one given at I in *Figure 12.23*, another construction is illustrated at K. This structure is portrayed with the aid of design paper which represents a quicker technique of designing than the interlacing diagram. The lifts of the standard ends are shown by the double vertical marks, the crossed sheds by the solid marks and the open sheds by the crosses whilst the floats of the crossing end between the lifts are represented by the thick continuous lines. Using bottom douping the cloth would be woven face side down as indicated in the lifting plan P. It will be noted that in the design a separate vertical row has been allocated to every crossing end on each side of a group of four standards under which the crossing end floats. Two vertical rows of design paper are required for each crossing end as space must be available to show the lifts of the crossing ends on the open *and* on the crossed shed side of the standards.

In the draft M the shedding elements are marked according to the previously established designations and from the design it is clear that quite elaborate and large repeats, in this instance 30 ends × 36 picks, can be achieved with a single point drafted doup assembly if the density of the cloth is not very high.

The double-slotted flat steel doups

This form of doup shown in *Figure 12.24* permits further extension of the patterning capacity of the flat steel doup and ensures particularly efficient locking of the threads within a leno row thus making the unit especially valuable for very open figured leno fabrics.

The leno assembly with the double-slotted doup has two crossing ends each threaded through its own slot. A number of standard ends, usually not less than two, are drawn on the standard healds and pass between each pair of lifting healds H1 and H2. When one lifting heald of the leno unit is raised one of the two crossing ends compulsorily forms a crossed shed whilst the other is at the same time free to make an open shed. As both ends are also drawn each through its own figuring heald the freedom to produce an open shed exists for either of them. The drafting in the assembly is such that when both the crossing ends are raised on one side of the group of standards with which they form a leno row one of the crossing ends is in the crossed, and the other in the open shed position. This means that for the double-slotted assembly two easer bars are a necessity. Both the crossing ends may be raised together on one or the other side of the standards or each can be made to operate on the opposite sides of the group of standard ends.

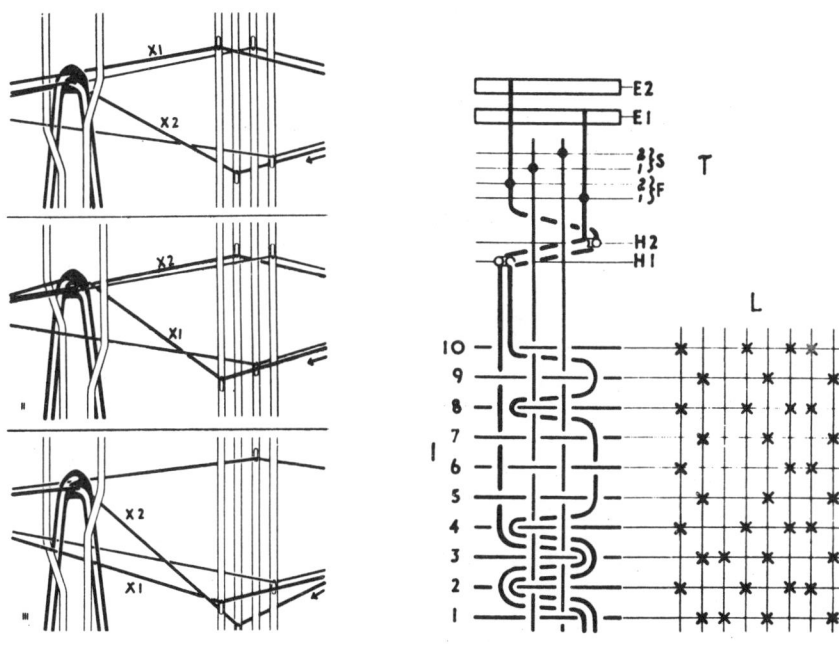

| Figure 12.24 | Figure 12.25 |

The essential manipulations of this assembly are shown at I, II, and III in *Figure 12.24* where at I both crossing ends are raised on the right of the standard ends, crossing end 1 in the open and crossing end 2 in the crossed shed; at II both are raised on the left but now with X1 in the crossed end X2 in the open shed; at III X2 is up in the crossed shed but X1 remains down. The effect of these manipulations is shown in the interlacing diagram I in *Figure 12.25* where on picks 1 and 3 operation of healds as given at I is necessary; on picks 2, 4, 8 and 10 the healds operate as at II; on picks 5, 7 and 9 as at III; and on pick 6 the situation opposite to that shown at III takes place. The method of drafting for the double-slotted doups is indicated at T and the lifting plan for the construction is given at L.

A more elaborate structure produced on a double-slotted doup assembly is shown in the actual and enlarged views in *Figure 12.26* for which a partial interlacing diagram, I, with the draft T, and lifting plan L to correspond are given in *Figure 12.27*. In this structure two leno ends are drawn in each slot and these operate in conjunction with four standard ends so that each leno row consists of eight ends. In the leno working portions of the design the four crossing ends cross the four standards on every pick thus drawing the eight ends together very closely. The figure is produced by permitting the eight ends to spread apart in selected areas by operating them in a double-ended plain weave.

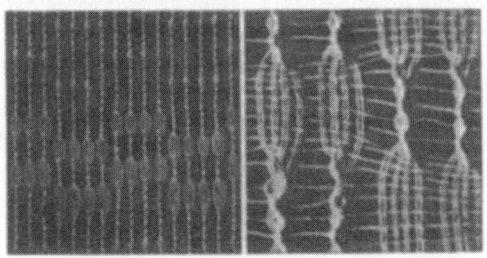

Figure 12.26

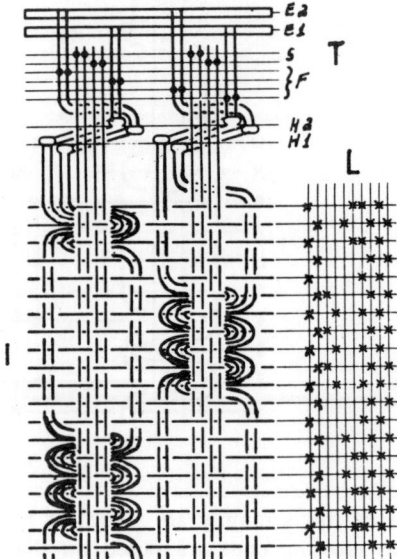

Figure 12.27

As shown in *Figure 12.26* quite a large figure repeat is achieved with the use of only one double-slotted doup assembly, four leno figuring healds, two standard healds and two easers.

It will be noted from the two examples given that a feature of the double-slotted doup operation is the alternate lifting of the two lifting healds of the leno assembly. In this way on every pick one of the two crossing ends forms a crossed shed whilst the other is free to be raised or dropped by its figuring heald as desired.

EQUALISATION OF YARN TENSION IN OPEN AND CROSSED SHEDS

The crossing ends in leno weaving may at any given pick form crossed, open or plain sheds (q.v.). The latter two types of shed do not place any extraordinary strain upon the ends but the formation of the crossed shed involves the transfer of the crossing end and its lift on the wrong side of the standard. To form a clear shed at the front the crossing end must be so placed that it does not lift the standard ends under which it has crossed. This placing is achieved normally by the control heald which on crossed sheds is down holding the crossing ends behind the doup well clear of the standards which in effect results in an acutely angled back shed. The combination of the acute back shed with the lateral movement of the crossing end would extend the crossing end severely and to prevent this, on the formation of a crossed shed, an extra length of yarn should be provided. This is achieved by the easing action which may be negative or positive in nature.

Negative easing action

One of the simplest devices for providing an extra length of yarn during leno weaving is a spring loaded equaliser bar. This is, as a rule, placed behind the standard healds in the manner shown in *Figure 12.28* and is particularly useful in tappet looms as it saves the use of a crossing end control heald frame. The bar is operated by the varying tension in the crossing ends and is shown at A in *Figure 12.28* in a high position to which it is pulled by the crossing ends when they form an open or a crossed shed although it will be realised that the position

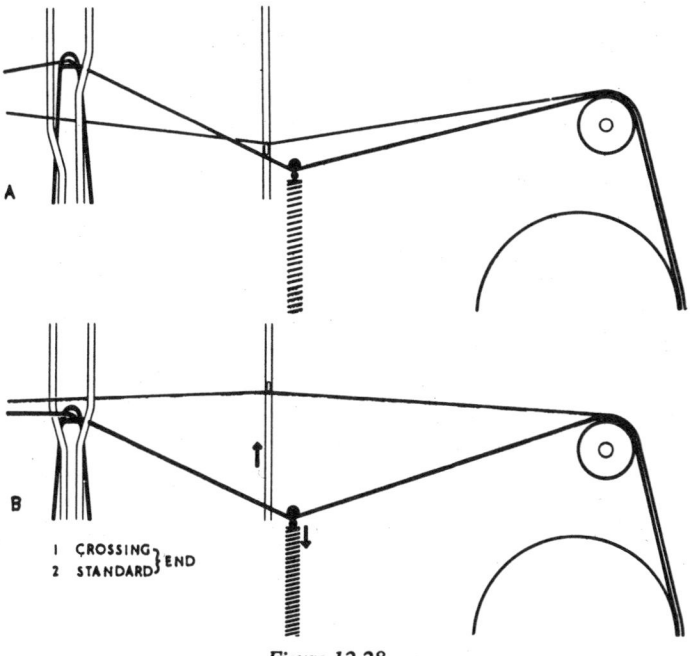

Figure 12.28

of the bar will be marginally higher at crossed than at open sheds. At B, the position of the bar is shown between the sheds when the doup descends relaxing the tension on the crossing ends with the spring taking up the slack by pulling the bar down. If there are a large number of standard heald frames the positioning of the equaliser bar behind them may cause the crossing ends to descend down so low that they may, at the back, brush against the top driving shaft of the loom. To avoid this the equaliser bar may be placed in front of the standard frames maintaining, however, the recognised distance of 10 cm between itself and the doup assembly. This situation arises invariably in jacquard lenos where the depth of the figuring harness makes it necessary to place the bar in front.

In high density or coarse lenos some difficulties are occasionally experienced with the twisting together of the neighbouring crossing ends on the equaliser bar. When this occurs the solution is provided by the substitution of an equaliser heald frame for the bar. This is placed in front of the standard frame, is also negatively controlled by springs but prevents the twisting of ends by having each crossing end drawn through a separate mail eye. As in respect of the bar the oscillations of the equaliser frame are controlled by the increase and relaxation of tension in the crossing ends. Both the devices operate within slides which maintain their horizontal displacement and are also provided with stops which limit the maximum permissible fall and rise of the device.

Faller rollers represent another type of a negative easing device which is sometimes used. In this case the weight of the roller takes up the slack in the crossing ends which upon shed formation pull the rollers up to provide themselves with the extra length of yarn and between sheds permit the rollers to fall down within their slides. Obviously, the weight of the roller must be adjusted correctly according to the amount of tension which it is intended to impose.

The negative devices are suited mainly for the production of open, lightweight leno fabrics and can operate only in conjunction with the eyed doups. For slotted doups positive easing action must be used.

Positive easing action

Positively acting easer bars are normally mounted in the proximity of the back rest, usually slightly above or below it. All the crossing ends are passed over the bar which at the required point is moved forwards either by a dobby or by cam action to release a predetermined length of yarn. As the extra length is required only temporarily, after each release which coincides with the crossed shed, the extra length is taken back either by a spring return action or by a cam.

In *Figure 12.29* a commonly used dobby-controlled device is illustrated. The easer bar, E, fulcrumed at F, is attached to a dobby jack connection, J, via the long arm, L. In the diagram O, during the formation of an open shed—and also on plain sheds—the easer remains in the normal position with the arm L held against the stop, P, by the stout return spring, R. In the diagram X, i.e. on crossed sheds, the arm L is raised by the dobby jack connection which causes the easer to swing forward thus delivering a required length of yarn. Ample adjustment of the movement is possible ensuring that for each structure just sufficient length of yarn is provided, as an excess would result in a slack shed. A slightly different arrangement operating, however, on the same principle is used when the crossing ends are placed on a separate beam mounted above the back rest.

In constructions such as plain gauze in which all the crossing ends form open and crossed sheds on alternate picks the easer bar may be operated by a cam fixed to the bottom shaft. The diagram given in *Figure 12.30* illustrates such a motion which is of particular value in tappet looms. The diagram shows the

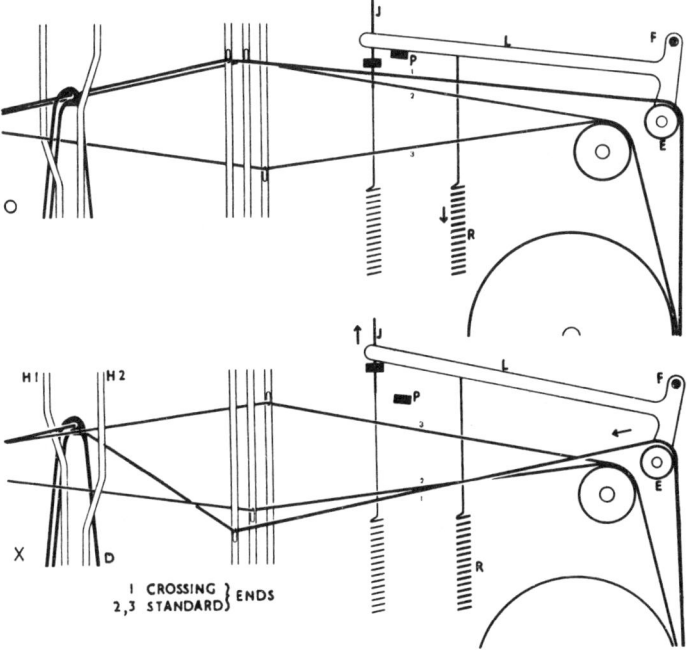

Figure 12.29

crossed shed in which the easer bar, E, is forward, delivering the extra length of yarn. On the following open shed this extra length will be taken back by virtue of the action of the increasing diameter of the cam, C, upon the anti-friction roll, A, attached to the easer bar lever L. The lifting of the lever L will force the easer back. The spring, R, ensures that contact between the cam surface and the anti-friction roll is maintained throughout the rotational cycle.

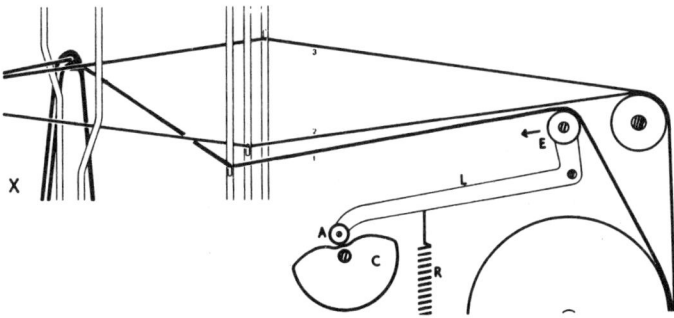

Figure 12.30

In structures in which one set of crossing ends forms crossed sheds on different picks from another set two independently controlled easer bars may be necessary. This arises when more than one leno assembly is used or when double-slotted doups are employed. In *Figure 12.31* a two easer arrangement is shown operating

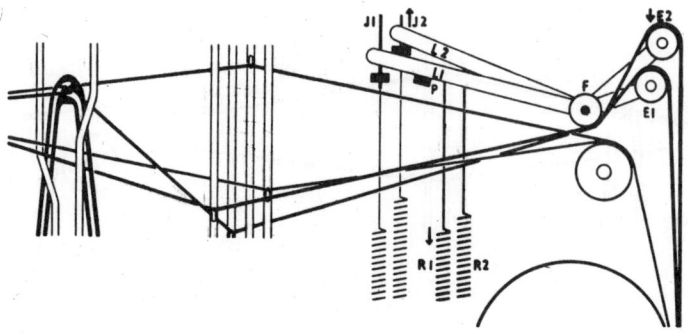

Figure 12.31

in conjunction with a double-slotted doup assembly. In this system the two easers are operated alternately because the two sets of crossing ends form the crossed sheds on alternate picks but as each easer bar is controlled by a different dobby jack any order of operation can be achieved. The action of this type of easer is exactly as described with reference to a single dobby-controlled easer given in *Figure 12.29*.

The shaker device

As stated at the beginning of this chapter the shaker or jumper device is required when lenos, in which the crossed sheds follow the open sheds (and *vice versa*) on succeeding picks, are woven in open or semi-open shedding mechanisms. In bottom douping the function of the shaker is to bring the standard ends half way up the shed between the sheds to permit the cross-over by the crossing ends without undue friction or tension on the standards (in top douping the standards would be brought half-way down). The reason for this action becomes apparent by reference to the plain gauze or leno weave where the crossed and open sheds occur alternately. In this structure the crossing ends are up and the standard ends down on *every* pick the cross-over taking place between the picks. Between the picks the crossing end is half-way down the shed having been taken down by the doup which follows the falling heald of the leno assembly before at that point the doup needle starts rising again to follow the rising heald, the two operating up and down alternately in plain leno weaving. However, the standard end is never pegged to rise from the bottom shed and would, therefore, never reach the level at which the cross-over can be accomplished if it were not for the shaker.

It will be clear that when one or a number of plain sheds separate the crossed from the open sheds no shaker is necessary because the doup is at the bottom and a cross-over can be effected without difficulty. There are, however, other cases in which the shaker can be dispensed with even though successive crossed and open sheds do occur and these can be listed as follows:

(1) When simultaneous top and bottom douping assemblies are used to produce the leno weave.

(2) When there are two or more standard ends in each leno group which weave plain, as in this instance all the standard ends are in mid-shed positions between picks as well as the crossing ends and the cross-over can be effected without any difficulty.

It will be obvious, of course, that if closed shed mechanisms are used the shaker is entirely unnecessary because all the ends after each shed are at the same level.

The half-lift of the standard heald is obtained in various ways, either by cam action or by a special attachment to a dobby which most makers now provide for the purpose. The attachments operate in a fixed manner and produce the shaking movement in between every pick whether it is required or not. The additional vibration of the standard ends is obviously detrimental therefore the shaker device is used only if absolutely necessary. Whatever means are employed to achieve the shaking movement it should be remembered that the connections must be made in such a manner as not to interfere with the normal full lifts of the standard heald for the purpose of ordinary interlacing.

JACQUARD LENOS

Apart from Madras muslin, which is dealt with separately in a section which follows, jacquard lenos fall into three distinct groups of structures:

(1) Open structures produced on the 1-crossing-1 principle in which the leno groups weave plain leno in the ground and also produce a plain weave and warp or weft float figure in selected areas. These are usually produced with the aid of one slotted doup assembly placed in front of the jacquard harness.

(2) Figured marquisette type fabrics produced on the basis of 1-crossing-2 in which in addition to a slotted doup assembly two standard healds are mounted at the back to control the standard ends which weave plain throughout.

(3) More solid 2-crossing-2 structures in which open leno areas alternate with brocade figuring thus achieving the effect of an opaque figure on a semi-transparent ground. This type of structure is woven without any heald frames using individually controlled slotted doup assemblies in which each lifting arm of every assembly is connected to a separate harness cord at the top and its own lingo at the bottom. The return movement of the doup needle itself is also achieved by individual lingo weighting.

As all the jacquard lenos are usually woven in quite open settings comparatively small jacquard sizes are adequate in most instances. Frequently, the centre closed shed jacquard motions are employed being particularly well suited for the leno structure.

One-crossing-one styles

A typical jacquard mounting for this style of leno is illustrated by means of the comber-board diagram at D in *Figure 12.32*. The crossing and the standard ends are all drawn through the figuring harness in a transposed draft order. The draft transposition is a matter of convenience which simplifies the construction of the detailed weaves for the purpose of card cutting. It will be noted from the inter-lacing diagram, C, that in the figured portions of the design the standard ends are the odd ends and the crossing ends are the even ends. Therefore, unless the standards are drawn on the odd long rows and the crossers on the even ones, as in the transposed draft, the transposition of the design itself in vertical pairs needs to be carried out in the figured areas for the purpose of card cutting. The use of the transposed draft avoids the need for that operation. The doup assembly frames are placed the usual distance of 10 cm in front of the harness and the lifting healds H1 and H2 are raised on alternate picks being controlled from the hooks in a spare row. The operation of the doup assembly is shown to the left of the detailed weave plan B in *Figure 12.32* from which it will be seen that in the arrangement illustrated crossed sheds are produced on even picks and open sheds on odd picks. As slotted doups are used, all crossing ends must lift on the crossed shed side on every even pick, but they are free to be manipulated at will on the open shed side by the figuring harness.

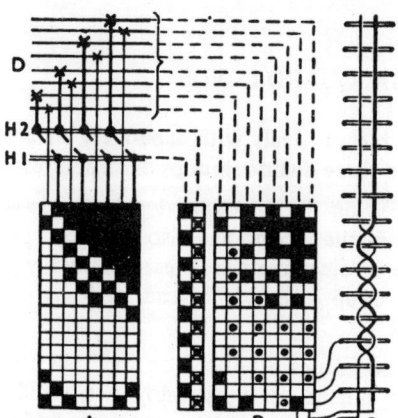

Figure 12.32

The simplified design at A indicates the method of painting—the warp float is indicated by the solid marks and each area of warp float figure is, as is usual in leno brocades, surrounded by an area of plain weave, the purpose of the surround being to spread out the standard and the crossing ends to make a fuller float figure. Without it the ends would tend to cling together in pairs in which they were twisted together in the plain leno ground. The leno ground in the design is left blank. The detailed cutting plan is shown at B from which it will be observed that the crossing ends on odd picks are raised by the harness to produce the open leno shed, the lifts being indicated by the dots. The harness on the even picks of the ground area must remain down because that is when the crossed shed is produced by the lift of the H1 element of the doup assembly. In the figure and plain weave areas every mark of the simplified design is cut except for the marks which indicate even end lifts on even picks; at these points the crossing (even) ends are raised to the top shed by the frame H1 automatically. The detailed weave for the figure and plain weave areas is indicated by the solid marks in plan B. The diagram C represents the interlacing of the last leno group in the portion of the design given. As every crossing end in this arrangement makes the crossed shed on alternate picks the easing operation can be controlled by means of a cam.

One-crossing-two styles

Figured marquisette styles are frequently produced in jacquards assisted by a heald mounting. The arrangement is shown by the comber-board draft given at D in *Figure 12.33*. In front of the harness there is a slotted doup leno assembly operated exactly as described for the 1-crossing-1 styles above. The crossing ends are additionally drawn through the jacquard figuring harness but the standard ends are not jacquard controlled and are drawn through a pair of ordinary healds placed at the back of the harness and operated in plain weave order from two of the spare hooks in the jacquard. Thus the two standard ends which in

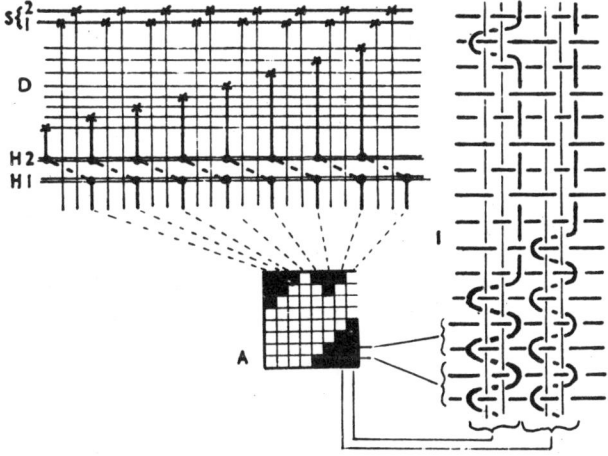

Figure 12.33

conjunction with the one crossing end form one leno group weave plain uninterruptedly. As the crossed sheds are formed also continuously on alternate picks the patterning consists of areas in which the freedom to form the open shed is exercised and which, therefore, produces the orthodox 1-crossing-2 leno, and areas in which the option to make the open shed is not taken up resulting in a three-ended plain weave. The two different areas are shown by the interlacing diagram I which represents a small portion of a repeat. In the leno portion the three ends of the leno group are drawn very close together resulting in an open fabric appearance; in the plain weave areas the three ends spread out assisted by the fact that the outside ends of each group of three weave the same tabby and thus tend to migrate together overcoming the natural tendency to remain apart induced during weaving by the dents of the reed which separate them. As only the crossing ends are jacquard controlled the painting of the design is very much simplified being condensed by 2 weft-wise and consisting of marks to indicate the 1-crossing-2 leno portions as shown at A. The construction is implemented as follows—on the picks on which the crossed shed is produced a blank card is inserted as all the crossing ends remain down in the harness but are raised by the H1 frame of the doup assembly; on the picks on which the open shed is formed the ends indicated by the solid marks are cut because they are the ends which are raised to form the open shed and, therefore, the leno part of the design, the ends indicated by the blanks being left down. The simplified design A corresponds to the interlacing diagram I in which the manipulations described above can be easily traced. It will be appreciated that the blank or uncut card inserted on crossed shed picks is 'blank' only in respect of the figuring harness—it will contain appropriately cut holes in the spare rows for the operation of the lifting heald frames of the doup assembly and the standard heald frames. As in the previously described system, in which the crossing sheds also occurred regularly on alternate picks, the easing operation can be cam controlled.

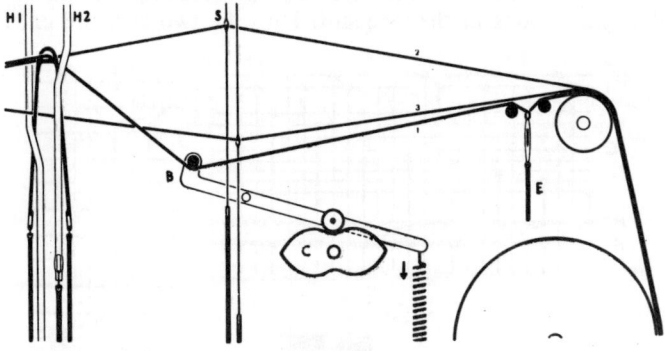

Figure 12.34

A different mounting used in 1-crossing-2 styles is represented in *Figure 12.34* which also shows the form of easing action suitable for such mountings. In this method no heald frames are involved and each doup assembly consisting of eyed doups is an individual unit in which the lifting healds H1 and H2 are controlled by separate jacquard hooks. In this way each crossing end can be made to operate in a different sequence from its neighbour. Some may be raised on the crossed shed side and some on the open shed side for several picks in succession, others may be left down to produce a warp float on the underside. The standard ends are also individually controlled by the jacquard harness which means that in this mounting full structural variety can be introduced with warp and weft float brocaded figure, areas of three-ended plain weave as in the previous mounting, and areas of leno weave which in themselves can vary considerably.

One system of control is represented by the comber-board diagram at A in *Figure 12.35* which represents one short row of harness. It will be noted that one short row consists of six ends but requires eight hooks and needles as the crossing end lifts must be controlled separately on the crossed and on the open shed side. At B a small portion of design is shown which contains an area of warp and weft float figure bordered by the plain weave and an area of leno weave. The design corresponds exactly to the interlacing diagram at I. In the design each leno group of three ends occupies four vertical rows of design paper which correspond with the needle numbers as given at A. All marks in the design indicate warp up—the solid marks show the lifts of the standard ends, the crosses refer to the lifts of the crossing ends on the crossed shed side and the circles to lifts of the crossing ends on the open shed side—and all the marks are cut in a straightforward manner. Simplification of the design may be carried out by colour coding the different structural areas in the usual way. In view, however, of the comparatively small size of the repeats in leno weaving—small in terms of the number of ends and picks involved—simplification in this instance frequently causes more trouble than it is worth resulting in a large number of detailed weaves of varying repeat size which for the purpose of automatised cutting must be brought into the common size denominator. Very often, to curtail the preparatory processes, simplification results in the reduction of the variety of effect produced.

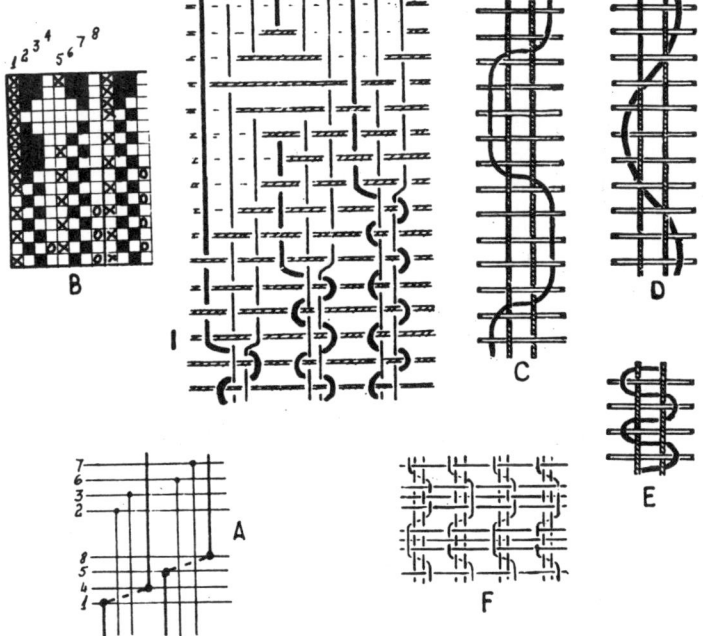

Figure 12.35

The system of control described permits full flexibility of operation in the ordinary and in the leno portions of the design. A rich variety of leno effects can be used and, if desired, all of them could be contained within one repeat. Diagrams C, D, E and F in *Figure 12.35* show four frequently employed effects in 1-crossing-2 leno styles some of which will clearly differ from others in the amount of crimping introduced and the designer's skill must be employed to the full to ensure that within a repeat the amount of take up between the different ends will be eventually equalised. If it is not, then considerable distortion of the cloth may occur. Periodic and temporary differences between neighbouring groups are permissible because the compensating system used is sufficiently flexible to allow for that. It may be observed from the illustration in *Figure 12.34* that tension variations which occur regularly in the crossing ends are controlled by the double-winged cam, C, mounted on the bottom shaft and the compensator bar, B. In the diagram the bar is shown at its highest point during the formation of the open shed; it will be at the same point on a crossed shed but between the sheds the major diameter of the cam will force the bar down taking up the slack which develops on those occasions. It must be appreciated, however, that the major function of the bar is to keep the crossing ends clear of the standard ends during crossed shed formation and that the real compensation for temporary yarn length differences, such as may occur between the crossed and the open shed or as are occasioned by structural differences within a repeat, is performed by the negative easing harness E. Each crossing end is threaded through an eye which is attached by a yarn loop to a lingo which depresses the end between two bars near the back rest of the loom thus creating a reserve length of yarn.

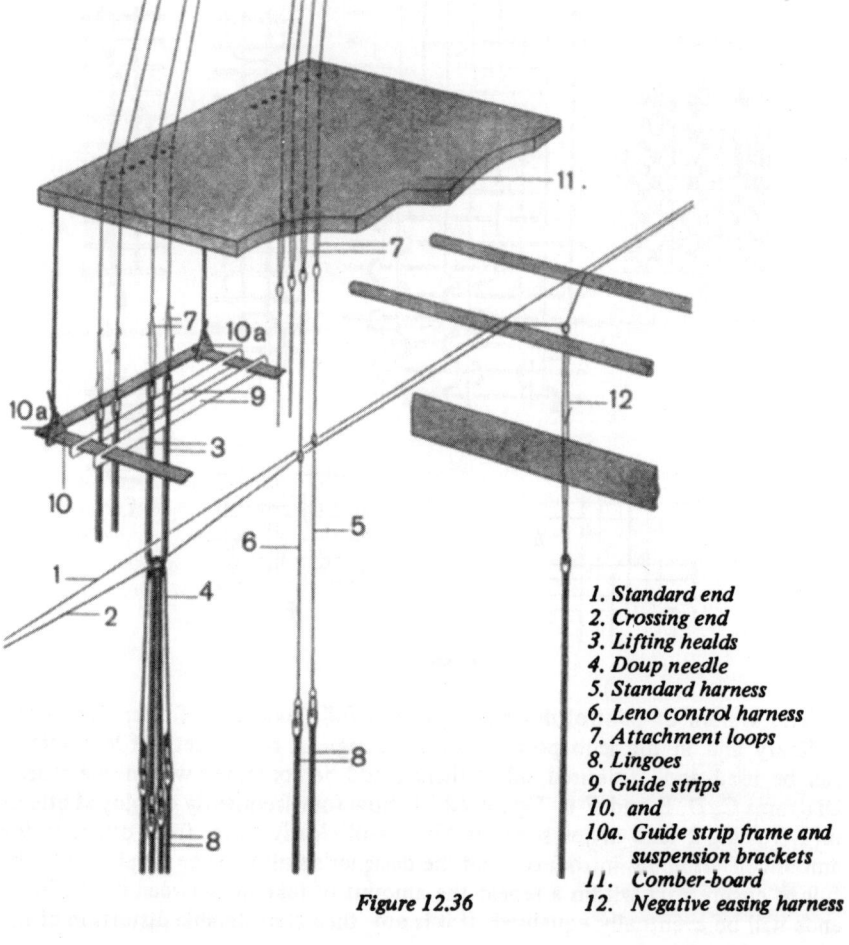

Figure 12.36

1. Standard end
2. Crossing end
3. Lifting healds
4. Doup needle
5. Standard harness
6. Leno control harness
7. Attachment loops
8. Lingoes
9. Guide strips
10. and
10a. Guide strip frame and suspension brackets
11. Comber-board
12. Negative easing harness

When the crossing end tightens up the reserve diminishes as the lingo is pulled up; when this end slackens off the reserve is restored again, the lingoes vibrating up and down in accordance with the variation in the demand for compensation by the ends to which they are attached.

The method of suspension of the individual jacquard doup assemblies is illustrated in *Figure 12.36* which shows a slightly different arrangement to the one described above. In this mounting the cam-controlled compensator bar is not used—instead the crossing ends are drawn additionally upon control harness which ensures that upon the formation of the crossed shed they are kept clear of the standard ends by lowering them behind the doup. To preserve the clarity of the diagram only one standard end is shown. As the key to the numbered parts is provided in *Figure 12.36* the arrangement is self explanatory, but attention is directed to the doup assemblies whose mounting involves special features not encountered in other jacquards. Each unit is controlled by two harness cords and two hooks and the lifting arms of the unit are individually weighted by separate lingoes. The return movement of the doup needle is ensured by two additional lingoes so that each assembly requires a total of four lingoes. To prevent the assemblies from turning sideways or twisting around during weaving each one is threaded through guide strips which fit between the double strips of the lifting arms without obstructing their shed forming movements. The guide strips themselves are held in a frame which is suspended from the comber-board. The positioning of the frame is such that the douping harness is kept the prescribed distance of 10 cm in front of the control and standard harness. The diagram in *Figure 12.36* also illustrates clearly the method of suspension of the negatively operating easing or compensating harness.

Two-crossing-two styles

The cloth usually produced in this style is less open than the previous ones with denser warp and weft settings. Consequently, to achieve a sufficient contrast in apparent density between the leno ground and the solid figure areas the ground is frequently produced as a 4-pick structure, i.e. the crossing is carried out over groups of four picks which being confined within the twists of the standard and crossing ends separate themselves from the neighbouring groups of picks thus creating open spaces. A number of ground weaves suitable for this purpose are given at A to E in *Figure 12.37*. A feature of all these structures is that on the crossed shed side the two crossing ends which form a leno row or group together with the two standard ends, must lift in pairs. On the open shed side, however, they can be operated individually. This is due to the type of mounting used which is basically similar to the one given in *Figure 12.36* and consists of individual leno assemblies but the doup is of the slotted type with two crossing ends per slot. Comber-board diagram A in *Figure 12.38* shows that all the ends are drawn through the figuring harness by which they can be operated individually. The crossing ends are additionally drawn through the slotted doup units in pairs the units being arranged in two rows in front of the figuring harness. The lifting arms of the leno assemblies are marked H1 and H2, as in all the previous examples, H1 being responsible for the crossed shed lifts and H2 for the open ones. When the crossed sheds are formed the figuring harness through which the corresponding crossing ends are drawn must remain down to permit the formation of a clear shed at the front but on open sheds the corresponding figuring harness may be operated in any way desired.

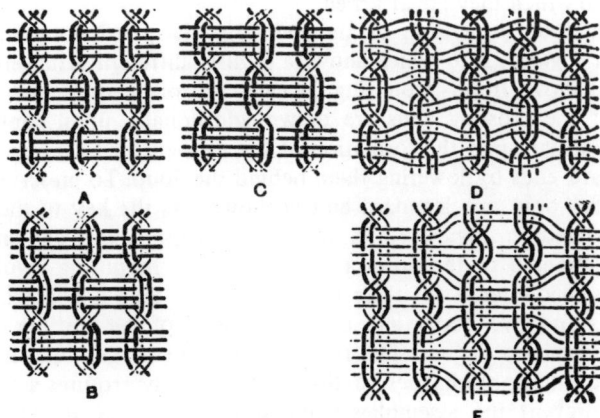

Figure 12.37

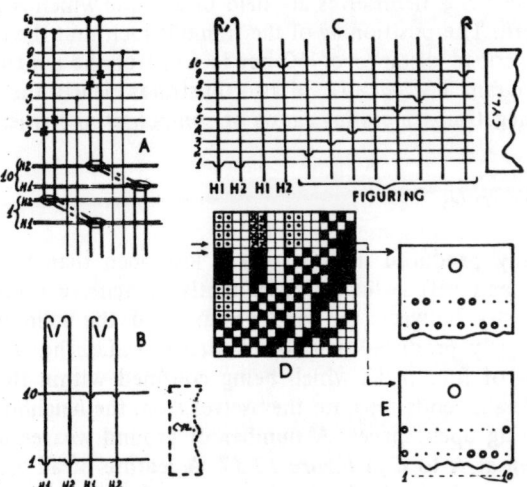

Figure 12.38

The lifting healds H1 and H2 are operated by separate hooks but each pair of hooks, if arranged on an inverted hook principle, can be controlled by a single needle because the two healds are never up or down together, i.e. when H1 is up, H2 is down and *vice versa*. The principle of inverted hook operation is shown at B in *Figure 12.38* in which the odd hook, controlling H1, is in the normal position facing the cylinder and over the knife when at rest; the even hook, controlling H2, is in the reversed position facing away from the cylinder and clear of the knife when at rest. A hole in the card opposite the needle will, therefore, cause a lift of the odd hook when the knives rise but will leave the even hook down as in the position indicated it cannot be engaged by the ascending knife. A blank in the card, on the other hand, pushes the odd hook clear and the even hook over the knife thus causing the lift of heald H2. Thus, unless the needles controlling the leno harness are cut, which results in the crossed sheds, it is the healds H2 which are lifted permitting the figuring harness to determine how the two crossing ends in the free slot are to be operated. The hook/needle connections in a full short row of the type of mounting described are given at C in *Figure 12.38*.

A small portion of design is shown at D and consists of a weft and warp float figure surrounded by plain weave the function of the latter being to spread the ends out after they have been pulled together by the leno weave in the ground areas. The solid marks used indicate warp up. The leno ground is represented in a conventional form in which the open sheds are indicated by solid marks and the crossed sheds by the dots and crosses and it will be noted that the ground weave selected corresponds to the structure A in *Figure 12.37*. From the diagram C in *Figure 12.38* it is seen that the figuring harness is operated by needles 2 to 9 and the douping harness by the needles 1 and 10. In order, therefore, to produce the structure depicted at D the following card-cutting instructions are formulated:

> Solid marks—cut for needles 2 to 9
> Dots—cut for needle 1
> Crosses—cut for needle 10

The result of the above cutting is indicated at E where the first two rows are shown in two cards which correspond with the horizontal rows of the design D marked by the arrows. To comprehend the above cutting fully is must be borne in mind that the hooks controlling lifting arms H2 of the leno assemblies are up on every pick unless their automatic lifts are cancelled by the cutting of holes for needles 1 or 10 which, of course, is equivalent to forming a crossed shed.

Figure 12.39

The edge of the figure is developed in steps of four to fit with the leno weave, but if it is considered that the outline so produced is too coarse then it can be modified together with the leno weave in the boundary areas. The easing or compensation is most conveniently carried out by the negative system as shown in *Figure 12.36*. A leno brocade constructed according to the method given above is illustrated in *Figure 12.39*.

MADRAS MUSLIN STRUCTURES

Structure of the cloth

In the Madras muslin texture a plain gauze foundation, which is formed continuously throughout the cloth, is ornamented with extra weft. *Figure 12.40* shows the appearance of a typical fabric, as viewed from opposite sides, while the diagram 1 in *Figure 12.41* illustrates the interlacing of the threads. The warp consists of crossing ends and standard ends arranged 1-crossing-1, and in the gauze structure, which is very light and open, there is one ground pick in each shed. The extra weft is softer spun and usually much thicker than the ground

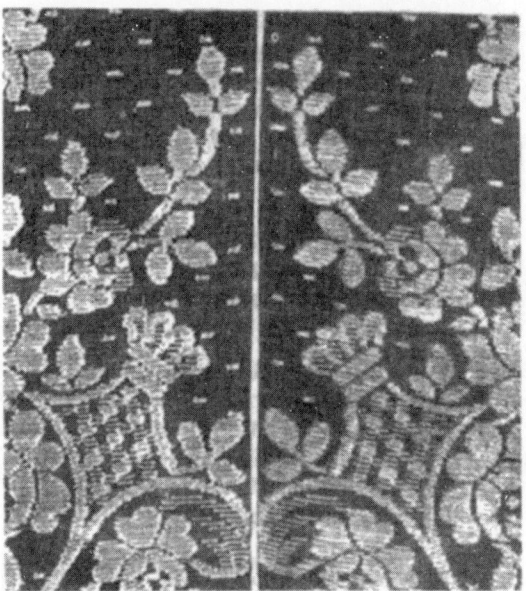

Figure 12.40

weft; it is interwoven with the warp where required, and floats loosely in the remaining parts of the design. The loose floats are afterwards cut away and a texture with an opaque figure on a transparent gauze ground, or *vice versa,* results, which is particularly serviceable for use as window curtains. The appearance of the cloth after the cutting operation is represented in *Figure 12.40,* which shows, on the left, the fabric as viewed from the cut side—i.e., the side on which the extra picks float loosely during the weaving of the cloth, and on the right, as viewed from the uncut side. The cloths are normally woven with the cut side uppermost. When used as window curtains either side of the cloth may be taken as the right side, but as shown in *Figure 12.40* the ornament has a bolder outline on the cut than on the uncut surface, hence for certain purposes the cut side is taken as the right side. The uncut surface, however, is neater, and for such cloths as dress fabrics, for which the structure is to some extent adapted, this is mostly taken as the right side.

The Madras loom

Special types of looms have been built for weaving the cloth. The following illustrations and description correspond with a system in use at present. In the diagram 1 in *Figure 12.41* the structure is represented from the same side as the cloth on the left of *Figure 12.40*, i.e. as viewed from the cut side. When the texture is woven with this side uppermost the crossing ends are raised and the standard ends are left down on every ground pick. On the figuring picks all the crossing ends are left down, but the standard ends are raised where the extra weft has to be interwoven, and the latter is, therefore, firmly bound in between the standard and crossing ends. The plan given at Y corresponds with the effect shown in the diagram 1; the full squares indicate the lifts of the standard ends and the crosses of the crossing ends. The crossing ends, however, are operated independently of the jacquard harness, while no standard ends are raised on the ground picks; therefore in constructing a card-cutting plan it is only necessary to indicate the lifts of the standard ends on the figuring picks, as shown at Z, which illustrates the usual method of designing a figure.

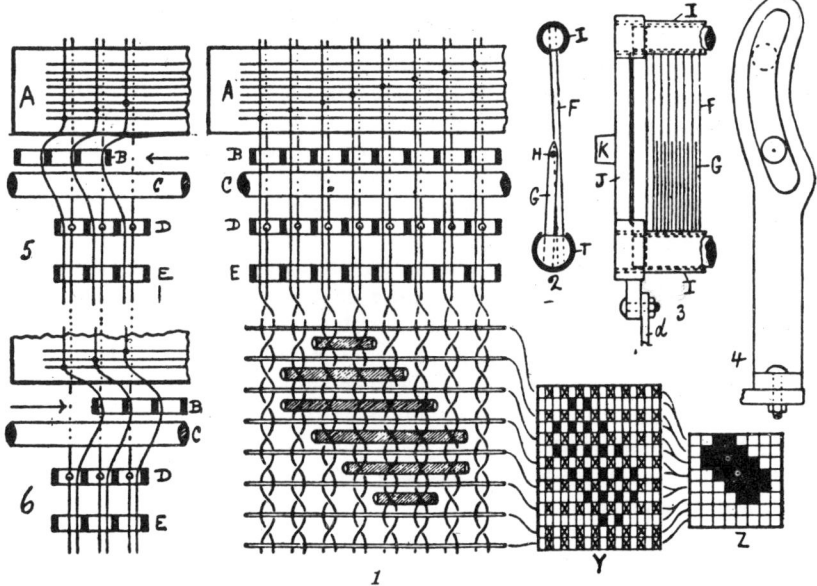

Figure 12.41

An ordinary form of double-lift jacquard and harness is generally employed, but in front of the harness A there are, in the following order, a tug reed B, and easing bar C, a gauze reed D, and an ordinary weaving reed E. These are represented in the diagrams given in *Figures 12.41* and *12.42*, while the draft of the ends is indicated in the upper portion of 1 in *Figure 12.41*. The standard ends are drawn through the harness mails, and then are passed through the reed B, above the bar C, and through the reeds D and E. The crossing ends are passed between the harness cords, under the tug reed B and the bar C, then through the eyes in half-dents in the gauze reed D, and afterwards each end is passed along with its accompanying standard end through a split in the weaving reed E.

The crossing of the ends is effected by means of the tug reed B and the gauze reed D. The tug reed is suspended in front of the harness and is moved a small distance to left and to right on succeeding ground picks. The form of the gauze reed is shown separately in the diagrams 2 and 3 in *Figure 12.41.* Between each pair of full dent wires F there is a short, pointed dent G which is slightly inclined, and is provided near the top with an eye H through which a crossing end is drawn. The baulks of the reed fit within metal cases I, and this arrangement makes the reed very strong and rigid. Each end piece J is provided with a stud K, which fits within a curved shot formed in a guide, as shown in diagram 4. The gauze reed is raised and lowered on each ground pick, during which the studs K slide within the curved slots in an arcuate movement. When the reed is down the points of the short dents G are below the lower line of the shed, as shown in diagram 7 in *Figure 12.42.* When it is raised all the crossing ends are lifted sufficiently high to form a shed for the shuttle to pass through in front of the weaving reed E, as shown in 8. Each time the gauze reed is lowered the tug reed is moved laterally, and the standard ends, which pass through the latter reed, are pressed by the wires and traversed to left or to right. This is illustrated in the diagrams 5 and 6 in *Figure 12.41,* in which small circles represent where the crossing ends pass through eyes in the short wires of the gauze reed. Diagram 5 shows the tug reed moved to the left and diagram 6 to the right, so that each standard end is moved above the point of a short wire of the gauze reed from one side to the other. Hence on succeeding ground picks, as the gauze reed rises it lifts the crossing ends first on the right and then on the left of the standard ends, in which positions they are bound in by the ground weft. The movement illustrated in the diagram 5, will take place previous to the insertion of the odd ground picks of the structure represented in diagram 1, and in the diagram 6, previous to the insertion of the even ground picks.

The easing bar C has an important function, as will be seen from a comparison of its position in the diagrams 7 and 8 in *Figure 12.42.* The bar rises and falls on each ground pick in coincidence with the movement of the gauze reed, and maintains a uniform tension on the crossing ends.

The Madras loom is usually made with four boxes at each side, but the two series of boxes are connected so that they rise and fall together. Not more than three figuring shuttles, as well as the ground shuttle, can, therefore, be employed at the same time, but in different parts of a design it is possible to employ the ground shuttle or to insert one, two, or three extra picks to each ground pick according to requirements. The box motion is governed by the jacquard cards through four needles and hooks that are set aside for the purpose. The ground weft shuttle is placed in the fourth or bottom box. The figuring wefts may be inserted in almost any sequence, but one figuring colour—usually that which is inserted most frequently and most regularly throughout the design—is taken as the leading colour, and is placed in the third box (the next box to the ground shuttle). An important feature is that when the gauze ground shuttle (the bottom box) is brought into operation the special motions of the loom are automatically put into action.

The rising and falling motion of the gauze reed on each ground pick is imparted by the backward and forward movement of the slay. The position of the parts, at the extremities of the movement, is represented in the diagrams 7 and 8 in *Figure 12.42.* The lower end of a rod L passes loosely through a slot in a bracket M which is bolted to the inner side of the box lever N, while the upper end passes loosely through the slot of a projection that is connected to a horizontal lever P. Two collars are set-screwed on the rod L, the lower one of which, according to the position of the box lever N, is either just above or rests on the bracket M, while the higher one retains a spring O with its upper end against the underside of the projection on the lever P.

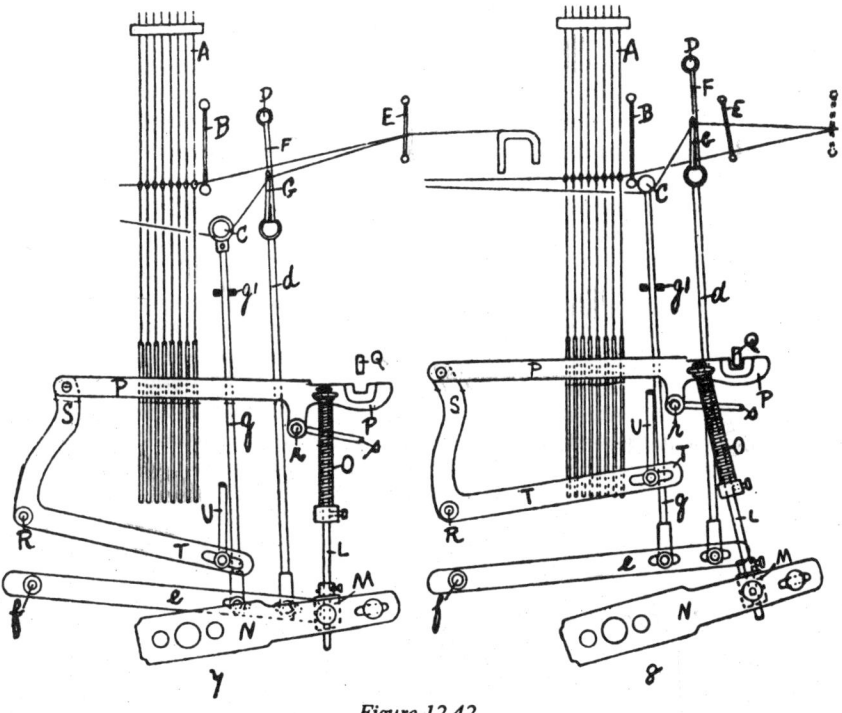

Figure 12.42

When the boxes are raised to their highest position so that the bottom box—which contains the ground weft shuttle—is brought into line with the race board, the upward movement of the box lever N lifts the rod L just high enough to cause the pressure of the spring O to raise the lever P at its forward extremity. This takes place at the time of beating up, when a bar Q, which is bolted to the back of a sword, is immediately above a recess formed in the upper side of the lever P, as shown in diagram 7, *Figure 12.42.* The lever P, when raised is brought into engagement with the bar Q, as shown in diagram 8, and as the latter moves backward and forward with the sword, a similar movement is imparted to the lever P. A bell-crank lever, which is fulcrumed on a stud R, has its upper arm S pivotally connected to the rear of the lever P, hence the movement of the latter causes the forward end of the lower arm T to rise and fall. This movement, by means of a rod U, and a series of levers causes the gauze reed and the easer, C, attached to e to rise and fall; and this takes place with each backward and forward movement of the sword so long as the gauze ground weft only is inserted.

On the insertion of a figuring pick, however, the lowering of the box lever N releases the pressure of the spring O against the projection on the lever P, the forward end of which then falls out of engagement with the bar Q, as shown in the diagram 7. The gauze reed then remains in its lowered position while the required number of figuring picks is inserted.

The lateral movement of the tug reed is obtained from the same source as the vertical movement of the gauze reed so that the two actions are fully synchronised. The movement transfers the standard end from one to the other side of the crossing end and the direction of the movement is the same as that of the preceding ground pick.

The picking motion is of the pick-at-will type and the direction of the pick is governed by the action of the shuttle box swell. At the side of the loom on which a shuttle is in line with the race, the pressing back of the swell, by suitable connections, causes the picking mechanism at the opposite side to be made inoperative, while the absence of a shuttle at one side brings into action the picking mechanism at the other side. In the event of anything occurring to bring a shuttle at each side in line with the race, the picking mechanism at both sides of the loom is put out of action.

The take-up motion.   In different parts of a design the number of extra picks to each ground may vary from none to three, but it is very necessary for the foundation texture to be uniform. For this reason the up-take motion is operated from the lever P so that the cloth is drawn forward only when a ground pick is inserted. A projection on the lower side of the lever P (see *Figure 12.42*) carries a stud *r* which is connected by a rod *s* to the lever that operates the take-up pawl. Each time the lever P is drawn forward the ratchet wheel of the up-take motion is turned one tooth. The arrangement ensures that the same number of ground picks per unit space will be put in however irregularly the figuring weft is inserted.

Madras designing

Very little variation can be made in the Madras structure on account of the special means employed in weaving the cloth, and because the shearing operation makes it necessary to avoid the formation of long figuring floats on the cut side of the texture. As viewed from the uncut side, all the crossing ends (which are operated simultaneously by the gauze reed) are above the figuring picks, so that on this side the floats cannot be longer than the space between two consecutive crossing ends. On the cut side the standard ends are above the figuring picks where the latter are interwoven, and so far as the weaving of the cloth is concerned these ends may be operated by the harness as desired. It is, however, generally recognised that in the figure the weft should pass over not more than one standard end at a place, otherwise the floats are liable to be cut away in the shearing process. Only the very simplest weave development is, therefore, possible, and the ornamentation of the texture is chiefly dependent upon the formation of different degrees of density, and upon colour. A large number of ends per cm cannot be employed because the fineness of the gauze reed is limited, and the setts usually vary from 12 to 18 ends per cm, the latter number being seldom exceeded. The ground picks range from 10 to 14 per cm, but the total picks vary according to the proportion of extra picks to each ground pick, and whether the extra weft is introduced regularly or intermittently. The ground yarns are mostly of very good quality and fine in counts—ranging from 7 tex to 11 tex cotton or polyester filament yarn—while the soft spun figuring weft varies from 60 tex to 32 tex cotton. The following are typical weaving particulars: Warp 84 dtex filament yarn, 18 ends per cm; figuring weft, 60 tex cotton; ground weft 84 dtex filament yarn; 12 ground picks per cm.

The most common modifications of the Madras structure, as viewed from the cut side of the cloth, are illustrated at A, E, H, L, O and R in *Figure 12.43*, in which the crossing ends are those which are over all the ground picks. The corresponding condensed plans on the right of the examples show how the effects are indicated on design paper for the card-cutting, while on the left the complete plans are given, in which the crosses represent the lifts of the crossing ends by the gauze reed, and the full squares and dots the lifts of the standard ends by the harness. A in *Figure 12.43* shows the ordinary opaque structure, and a modification in which only the alternate figuring picks are interwoven. The

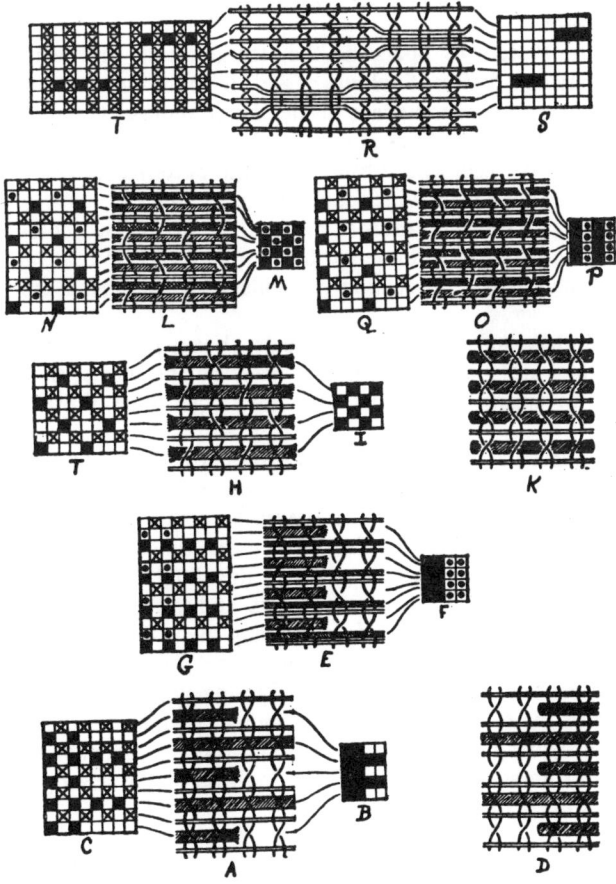

Figure 12.43

latter produces a semi-opaque effect which forms a pleasing contrast between the transparent ground and the opaque portions of a design, and may be used effectively in shading a figure. The marks in the plan B indicate where the standard ends are raised on the figuring picks when the cloth is woven with the cut side up. The appearance of A when turned over from left to right is represented at D; the texture is the same on both sides, except that the free ends of the figuring picks show more prominently on the cut side.

E illustrates another method of producing different degrees of density. In this case two figuring picks of different thicknesses are inserted to each ground pick; both wefts are interwoven in forming the opaque structure but in the semi-opaque effect only the finer weft is interwoven. Two figuring cards are cut from each horizontal space of the condensed plan F, the full squares and dots being cut on the first card, and the full squares only on the second.

In the modification shown at H the density is the same as in the ordinary structure, but the figuring weft is brought more prominently to the surface on the cut side of the cloth. The standard ends are raised alternately on the figuring picks, as shown by the plain order of marks in the plan I. On the uncut side, a view of which is given at K, the figuring weft shows less prominently than in the ordinary structure.

The alternate system of operating the standard ends can be effectively employed in mixing two different colours of weft in the manner represented at L in *Figure 12.43*. Each horizontal space of the plan M represents two figuring cards, in one of which the full squares are cut, and in the other the dots. The mixed colour effect may be used as a subsidiary to the patterns formed separately by the two colours of weft from which it also differs in density.

O in *Figure 12.43* shows another method of mixing two colours, in which each weft is bound in a straight line by the alternate ends, so that vertical lines of colour are formed. The card-cutting particulars for the plan P are the same as for M.

It will be understood that the small floats, shown in H, L, and O, remain in the cloth as they are too short to be cut away by the shears; the latter are set close enough to engage any floats that are longer than those shown in the examples.

The diagram given at R in *Figure 12.43* shows how the gauze ground structure may be modified by lifting certain standard ends at the same time as the crossing ends are raised by the gauze reed. This prevents a crossing taking place so that at these places the ends lie straight for three picks which are grouped together. The marks in the plan S, which show where the standard ends are raised on the ground picks, will be cut on the ground cards, and in this respect the example is different from the foregoing styles. In weaving the ordinary gauze foundation, ground picks are blank except where holes are cut for the purpose of operating the boxes and the selvedges. As regards the figuring picks, sometimes these are not interwoven with the selvedges, catch ends being provided for them at each side, so that on the figuring cards, in addition to cutting the marks of the design, it is necessary to cut holes to correspond with the lifts of the selvedges or the catch ends, as well as holes for operating the boxes.

Chintzed designs. An illustration is given in *Figure 12.44* of a Madras muslin texture which is termed a 'single cover', i.e. there is only one figuring pick to each ground pick. Three different figuring wefts are, however, employed, but these are chintzed, one following another in succeeding sections of the design. The ground is transparent, and in the figure, in addition to the ordinary opaque structure, effects are formed which correspond with those represented at A and

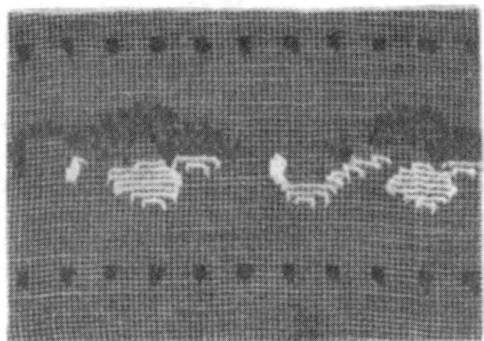

Figure 12.44

H in *Figure 12.43*. Where the ordinary opaque structure is formed the figure is simply painted in solid, but in producing the semi-opaque effect the alternate horizontal spaces only are filled, while in bringing the weft more prominently on to the cut side a plain order of marking is employed, as at A and H in *Figure 12.43* respectively. It is important to note that if the transparent gauze ground is required to show distinctly between two detached portions of figure formed by the same weft, the two portions must be separated horizontally by at least two blank spaces of the design paper. Otherwise the weft floats between the parts of the figure will not be long enough to be engaged by the shears, and the two portions of figure will appear to join up.

Chintzing is a technique resorted to very often because production of multi-coloured double and treble cover effects, i.e. designs in which there are two or three figuring picks to each ground pick, is very expensive and, therefore, rarely undertaken. The expense is due not only to the reduced rate of production but also to a very much increased amount of waste. It will be appreciated from the examination of the fabrics illustrated that even in single cover structures the percentage of the extra weft retained may be as little as·30 or 50 per cent. In treble cover cloths the extra weft float sheared off during finishing may amount to 75 per cent of the total length inserted. As the extra weft is usually dyed in fast colours it is expensive and efforts are made to reduce the waste by producing a gauze figure on opaque ground, thus retaining greater proportion of the extra weft in the cloth, or by constructing detached spot effects demanding only inter-mittent extra weft insertion.

Modified double cover structures

In addition to their use of lightweight window curtains Madras gauze fabrics are also made for heavier interior hangings in double or even treble cover structures. In such cases, to reduce the amount of waste, most of the extra weft is retained with very little of the gauze ground exposed. When three colours of weft are used one usually forms a foundation on which the figure is produced by the other two wefts, the gauze ground in this case being chiefly employed for outlining. It has previously been shown that very little variation can be introduced in the Madras structure, as with each figuring weft only the ordinary structure and a semi-opaque effect illustrated at A in *Figure 12.43,* and a 3-and-1 structure, shown at H in *Figure 12.43,* can be made. The use of two or more differently coloured wefts at the same time, however, enables considerable diversity of density and colour effect to be obtained. Thus, with two figuring wefts (a double-cover arrangement) six effects can be produced by using each weft separately, as described above and illustrated at A and H in *Figure 12.43,* while with the two wefts used together the seven effects may also be formed, which are represented at A to G in *Figure 12.45.* Thirteen ways are, therefore available for varying the appearance of the different parts of a double-cover design, and further emphasis may be given to a figure by suitably outlining the parts with the gauze ground.

The simplified designs A to G in *Figure 12.45* are each accompanied by an enlarged weave which indicates how the figuring picks are interwoven with the ends (the even ends are operated by the gauze reed and the odd ends by the harness). Two cards are cut from each horizontal space of the plans A to G, as follows: First card—cut the shaded squares, the solid marks, the crosses, and cut the diagonal strokes plain; second card—cut the shaded squares, the dots, the diagonal strokes, and cut the crosses plain.

In the enlarged weaves only the figuring picks are shown and the marks indicate weft on the surface of the cut side of the cloth; the solid marks represent the first weft colour and the dots the second weft colour. In the structure shown at A both wefts are bound in alike, as firmly as possible, and a finely intermingled colour effect is produced. In B and C both wefts form a 2-and-1 effect, but whereas in B the colours are intermingled, in C they form separate vertical lines. In D the first weft is firmly bound in, while the second weft forms a 2-and-1 effect, so that the second colour gives the predominating hue to the surface. In E the structure is opposite to that shown at D, the second weft

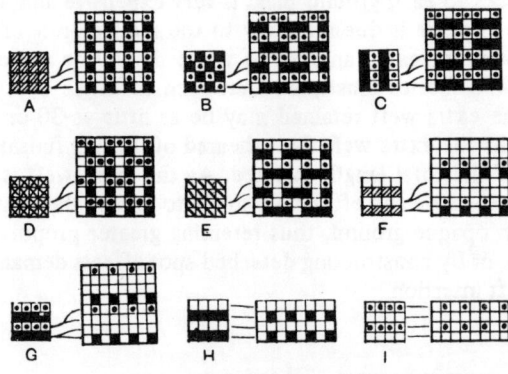

Figure 12.45

being firmly bound in while the first colour gives the predominating hue by interweaving in 3-and-1 order on the surface. In both F and G only half the picks of each colour are interwoven, but in F, which is a modification of A, the two colours are in the same shed and alternate with two gauze ground picks, whereas in G the two colours are in following sheds and are separated from each other by one gauze ground pick. F therefore has a more open structure than G, in which the figuring picks are evenly distributed.

An economical method of figuring with two colours of weft upon a gauze ground consists of arranging the weft in the order of—first colour, ground, second colour, ground, one figuring pick being inserted to each ground pick. Three effects which can be used in combination in a design are illustrated at G, H, and I in *Figure 12.45*, but only one card is cut from each horizontal space of the plans. The first colour forms the effect shown at H, and the second colour that shown at I, while a mixed colour effect and a more compact structure result from using both wefts together, as shown at G. Further variation of the structure shown at G can be obtained by interweaving one or both wefts in 3-and-1 order. Compared with a double-cover cloth the single-cloth texture should be woven with rather more ground picks per cm, and somewhat thicker figuring weft, in order to get satisfactory results.

Madras gauze with weft pile figure

Figure 12.46 illustrates a type of Madras gauze cloth which is not only suitable for use as curtains and hanging textures, but is also employed for evening dress fabrics. It has previously been shown that after the surplus weft has been sheared away from an ordinary Madras cloth the severed tips of the picks at the edges of the figure show very prominently on the cut side of the fabric. In the style under notice, where the figure is formed, the extra weft is floated in such a manner that after the cloth has been sheared the surface is covered by the severed tips of the picks, and the ornament is thus given the appearance of being composed of cut pile. The structure of the texture is illustrated in *Figure 12.47*, in which A corresponds with a portion of the design shown in *Figure 12.46*, while the weave which is indicated on the figure is shown separately at B. In order to produce the required fullness of pile two extra picks are inserted to

Figure 12.46

each ground pick, and two figuring cards are cut from each horizontal space, as follows: First card—cut the solid marks; second card—cut the dots. The method in which the figuring picks interlace with the ends (including the crossing ends) is shown at C, each pick interweaving in 7-and-1 order. The shearing is not done so closely as in an ordinary Madras cloth, but a portion of each float is cut away about the centre, and the severed tips of the picks then stand up from the surface, as represented at D. The two picks which are inserted to each ground pick may be alike or in different colours, a mixed colour effect being produced in the latter case.

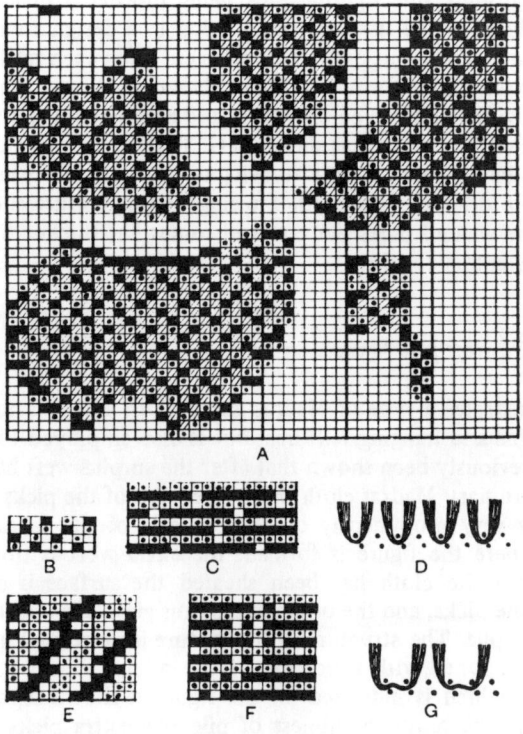

Figure 12.47

Another card-cutting plan for inserting on the figure is given at E in *Figure 12.47,* and a portion of the corresponding enlarged weave at F, while G represents how the severed tips of the picks stand up from the surface. In this case each weft floats over nine ends and is bound in by two consecutive harness ends, whereas in the plan B only one end at a place is used for binding. Binding with two consecutive harness ends makes the figuring weft less liable to fray out of the sheared cloth. (In ordinary Madras designing it is customary to bind on at least two consecutive horizontal spaces).

A further point, illustrated by the plan given at A in *Figure 12.47,* is that, in drafting a design to be developed in imitation pile, five spaces require to be left horizontally between the separate parts of the figure. This compares with a minimum of two spaces for medium sett, and three spaces for fine sett Madras cloths of the ordinary kind. Where in an ordinary cloth the severed ends of the weft increase the size of each portion of figure by about the width of half-a-split at each side, the less close shearing of the imitation pile cloth makes each detail of a design appear larger by the width of a split, or a split and a half at each side.

For this reason the ornamentation of the style requires to be on simple and massive lines.

LENO STRUCTURES PRODUCED IN A SLIDER FRAME AND NEEDLE DEVICE

Apart from the steel doup systems and the gauze and tug reed methods leno structures are also produced by a variety of devices which incorporate slider bars for the crossing action and eyed needles for the shed formation. Most of such devices together with the rotating cog-wheel mechanisms are suited mainly for the production of narrow bands of leno structure consisting of two to four rows of leno for the centre or side selvedges. One of the systems of this type, however, is also suitable for wider fabrics and can be employed for the production of very open leno fabrics in which high stability of construction is a desirable feature.

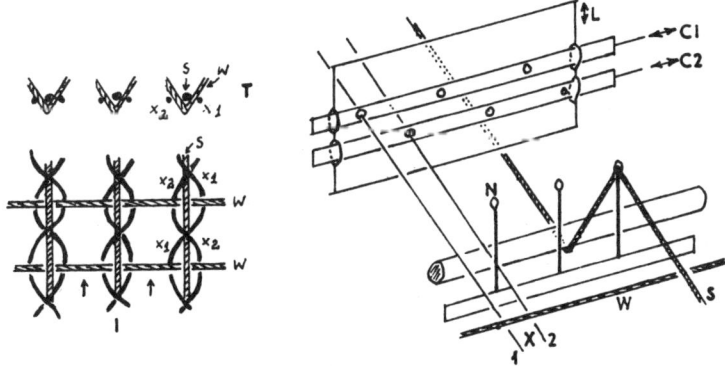

Figure 12.48

The system is illustrated by means of simple schematic diagrams in *Figure 12.48*. The standard ends, S, which are usually quite coarse, are threaded through the eyes of the needles, N, which are fixed in a stationary needle bar. Two crossing ends make a leno row in conjunction with the thick standard end and are designated X1 and X2. Each crossing end is threaded through an eye of its own slider bar, C1 and C2, which both fit into a lifting frame, L. When, between picks, the lifting frame is raised the crossing ends rise above the needles and the sliders, C1 and C2, are operated laterally, one to the left and the other to the right, so that when the frame descends X1 is on the left and X2 on the right of the standard end. After the insertion of weft, W, when the frame is raised again the sliders traverse in the opposite directions and the crossing ends assume reversed positions in respect of the standard end before the next pick is inserted.

This system has been originally developed for the weaving of chenille fur (tufted weft) for chenille Axminster carpets (see Appendix II). The tufted weft is produced by cutting the woven cloth along the lines indicated by the arrows at I and its cross-sectional appearance is represented at T in *Figure 12.48*. High tension in the crossing warp forces the severed weft picks into a 'V' formation which is particularly useful in carpets and similar one-sided pile structures. If the tension in all the warp elements, however, is about equal and the weft is not severed between the leno rows then a very stable cloth structure results in which the weft will not slide even in exceptionally open settings. The fabric construction obtained in this system is depicted at I in *Figure 12.48*.

13

Weft Pile Fabrics

In pile fabrics a proportion of the threads, either warp or weft, are made to project at right angles from a foundation texture and form a pile on the surface. The projecting threads may be cut or uncut thus resulting in tufted or looped pile. A different form of pile surface is produced by raising and cropping during fabric finishing operations but in this case the surface is formed of projecting fibres and not of projecting threads and the term nap rather than pile is more appropriate for cloths of this type. Weft pile fabrics are composed of one series of warp threads and two series of weft threads, the ground and the pile. The pile weft is cut in a separate operation after weaving resulting in a surface consisting of short and very dense tufts. A feature of weft pile structures, also termed velveteens, is very high density of shotting which in the finest fabrics may reach 200 picks per cm. In order to reach such weft densities the warp setts should be comparatively low and the warp yarn has to be kept very taut; also, the weaves must be so selected that successive picks can be beaten-up one on top of another. Due to the high warp tension positive shedding mechanisms are used and the highest qualities of cloth require specially constructed, heavy weaving machinery which cannot operate at high speeds and, therefore, aggravates further the already low production rates arising out of the high densities of shotting. For this reason the quantities of the top quality of velveteen produced at present are insignificant. On the other hand the low and the medium quality cloth in some constructions is very popular and can be produced on standard, high-speed automatic weaving machinery using reeds with special deep dent wires. The shottings at which such fabrics are produced range from 60 to 110 picks per cm.

The pile effect in the velveteens is not produced during weaving but is a result of a cutting operation during cloth finishing. The structure is so arranged that the surface of the cloth is covered by weft floats; these floats are severed by knife action and form the cut pile surface. The ground cloth, usually plain or twill, is unaffected by the knife action and forms a solid base from which the cut tufts project and in which they are anchored. The cutting method differs for the different classes of structures and is described together with the appropriate constructions. Before cutting the cloth is prepared for the operation by stiffening the surface float in order to define the cutting races more precisely and to ensure crisper cutting. The back of the cloth is also treated by an application of an adhesive, usually starch, to ensure that the tufts during cutting are not plucked out from the ground structure. The fabrics after cutting undergo a crosswise brushing operation and are then singed and dyed. If pastel shades are required the cloth may require to be bleached after singeing.

The yarns employed in these structures are mainly cotton although filament rayon pile velveteens are also sometimes produced. For furnishing purposes worsted or mohair pile yarns have also been occasionally used. Structurally, the velveteens may be classified as follows:

(1) All-over or plain velveteens in which the surface is uniformly covered by the pile.
(2) Weft plushes—similar to above but arranged to produce much longer tufts and used mainly for upholstery purposes.
(3) Corded velveteens—also known as corduroys and fustians in which the pile runs in orderly vertical cords of varying width.
(4) Figured velveteens in which pile figure is produced on bare ground.

All the above groups may be further sub-divided into plain back or twill back structures depending on the type of weave in which the ground picks interlace with the warp.

ALL-OVER OR PLAIN VELVETEENS

This class of velveteen has a perfectly uniform surface, the foundation texture being entirely covered by a short pile in which the projecting fibres are of equal length. In constructing designs for the fabrics the chief points to note are: (1) The weaves that are used for the ground and pile respectively; and (2) the ratio of pile picks to ground picks. These factors, together with the ends and picks per cm of the cloth, influence the length, density, and fastness of the pile.

The ground weaves mostly used are plain, 2-and-1 twill and 2-and-2 twill, the last weave being employed for very heavy structures. The interlacing of the pile is almost invariably based either on the plain weave, a simple twill, a sateen, or a sateen derivative. The pile and ground picks may be arranged in any reasonable proportion, but generally a particular ratio is most suitable for a given weave.

Plain-back velveteens

Examples A, B, C, D, and E in *Figure 13.1* are designs for standard velveteens, with the plain foundation weave. The latter is represented by the crosses, and the base weaves for the pile interlacings are shown at the left of the plans. In each design the number of pile picks to each ground pick is equal to the number of picks in the repeat of the pile base weave. This is a convenient ratio, but other proportions of pile to ground picks are quite easily arranged in the same weave.

A distinct feature to be noted in the designs is that the pile base weaves are indicated only on alternate ends; thus each plan is on twice as many ends as the base weave. Design A is arranged 2 pile picks to 1 ground pick, and the pile weave is based on the plain weave which yields a weft float of three. In a finely set cloth the pile from this design is short and poor, but at low warp settings a fairly good result is obtained.

In design B, the pile weave is based on the 1-and-2 twill which yields a weft float of five, and there are three pile picks to each ground pick. This design produces a fine and rich effect, and is extensively employed. Designs C and D are each arranged 4 pile picks to 1 ground pick, but whereas in design C the pile interlacing is based on the 1-and-3 twill, in design D it is based on the satinette weave. Both of these yield a float of seven, and produce identical results in the finished cloth. Design E is arranged 5 pile to 1 ground, and the base for the pile interlacing is 1-and-4 sateen, which gives a float of nine.

In order to produce a dense pile, a very large number of picks per cm are required to be inserted, the number varying from about 120 in 15 tex cotton weft for the design B in *Figure 13.1* to about 200 in 10 tex weft for the design E. There are two reasons why it is possible to insert such a large number of picks. First the warp is held under great tension, and the ends lie almost straight in the cloth, which causes the picks to do most of the bending. This results in the ground texture being formed on the weft rib principle, hence a comparatively large number of ground picks can be inserted. Second, the system in which the pile interlacing is arranged enables the pile picks to be beaten over one another, so that each group occupies not more than the space of one ground pick. Also, in the plain-back structures, all the pile picks go into the same shed as the first ground pick, but are in the opposite shed to the second ground pick. Therefore, so far as regards the space occupied by the picks, the structural effect of each design A to E in *Figure 13.1*, is somewhat as represented at F—i.e, the total number of picks in the repeat of each design go into the space of four picks, of which three are in the same shed.

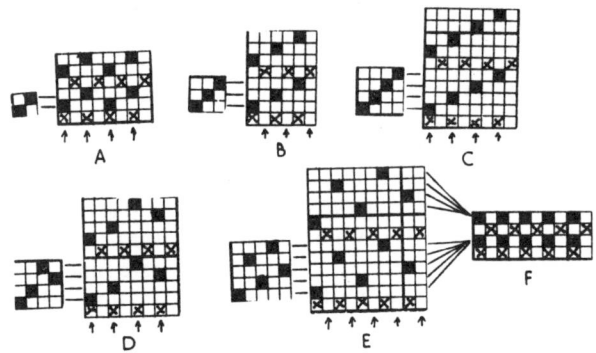

Figure 13.1

The diagrams given in *Figure 13.2*, in which design G is similar to the plan B in *Figure 13.1*, will enable various features of the velveteen structures to be noted. The flat view given at H, which corresponds with G, will serve to show somewhat how the pile picks crowd over each other in the cloth. This, however, is only a convenient representation of the structure, as in the actual fabric the ground picks are entirely concealed by the floating pile picks.

The purpose of binding in the pile picks only by the alternate ends (lettered A in H, *Figure 13.2*) is to enable the cutting to be more easily accomplished.

This will be understood from an examination of the cross-sectional drawing given at I in *Figure 13.2*, which represents how the picks 2,3,4, and 5 in the plan G interweave. Each pile float stands out furthest from the foundation cloth at

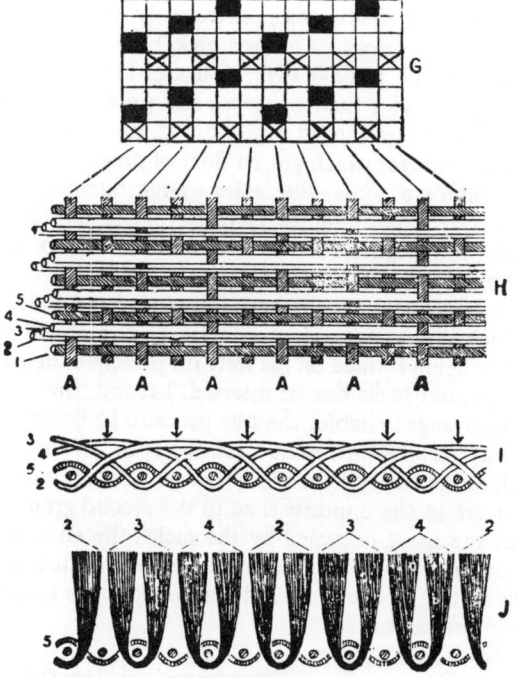

Figure 13.2

its centre, and the guide of the cutting knife is so adjusted that only those floats are engaged whose centres are in line with the longitudinal movement of the knife. The method of binding the pile picks causes the centres of the floats (indicated by the arrows above diagram I) to occur only on alternate ends, therefore only half as many longitudinal traverses of the knife are required as would be the case if the pile picks were bound in by every end.

An important feature, moreover, is that the alternate binding causes regular courses or races to be formed in the foundation texture, which are readily followed by the knife guide. Arrows are indicated below the designs in *Figure 13.1* to show where the cutting races occur.

After the cutting process, the twist runs out of the free ends of the weft threads which then project vertically from the foundation in the form of tufts of fibres, in the manner represented at J in *Figure 13.2*. Each repeat of the pile weaves produces one horizontal row of tufts, and in the plans A to E in *Figure 13.1* a complete row of tufts is formed to each ground pick.

Length of the pile

The length of the pile varies according to the ends per cm of the cloth and the number of ends over which the pile weft floats. An increased length of pile is obtained either by reducing the ends per cm or by increasing the number of ends over which the pile weft passes; and conversely, a decreased length results from increasing the ends per cm or from reducing the pile float. With the same number of ends per cm the designs A,B,C, or D, and E in *Figure 13.1* give successively an increased length of pile. For example, with 12 ends per cm in the cloth the approximate lengths are respectively 1·25, 2·1, 2·9 and 3·75 mm.

Density of the pile

The density of the pile varies according to the thickness of the weft, the length of the pile, and the number of tufts in a given space. An increase in the thickness of the weft tends to make the pile coarser, but other things being equal the density is increased. A long pile causes the surface of the cloth to be better covered, and thus gives a fuller handle than a short pile. The greater the length the pile is, however, the fewer are the number of tufts formed by each pile pick, and with the same number of pile picks per cm, an increase in density, due to increased length, will be counteracted by a reduction in the number of tufts. It is, therefore, customary for an increase in the length of the pile weft float to be accompanied by an increase in the number of pile picks per cm.

In each of the plans in *Figure 13.1*, the same number of tufts per cm² will result by employing the same number of ground picks per cm. Assuming that the warp is 20/2 tex cotton with 28 ends per cm, and the weft is 12 tex cotton, 32 ground picks per cm will be suitable, which will give the following number of pile picks and total picks per cm for the designs.

Design A.—64 pile picks and 96 total picks per cm
Design B.—96 pile picks and 128 total picks per cm
Designs C and D.—128 pile picks and 160 total picks per cm
Design E.—160 pile picks and 192 total picks per cm.

Comparisons of the number of tufts in different structures can be made by means of the following formula, which gives the number of tufts per cm²:

$$\frac{\text{Ends per cm} \times \text{pile picks per cm}}{\text{Ends in repeat of pile weave}}$$

For example, with the foregoing particulars, the design B will produce—

$$\frac{28 \times 96}{6} = 448 \text{ tufts per cm}^2.$$

It will be found in the same manner that the other designs with the particulars indicated will give exactly the same number of tufts per cm².

Changing the density of the pile

There are different way of changing the density of the pile, and in the same design and sett, alterations are frequently made simply by varying the number of picks per cm, or the thickness of the weft. Alternatively the design may be changed in order to obtain a different proportion of pile to ground picks. This is illustrated in *Figure 13.3* where the design K has the same base weave as B in *Figure 13.1*, but there are six pile picks, instead of three, to each ground pick. L and M are

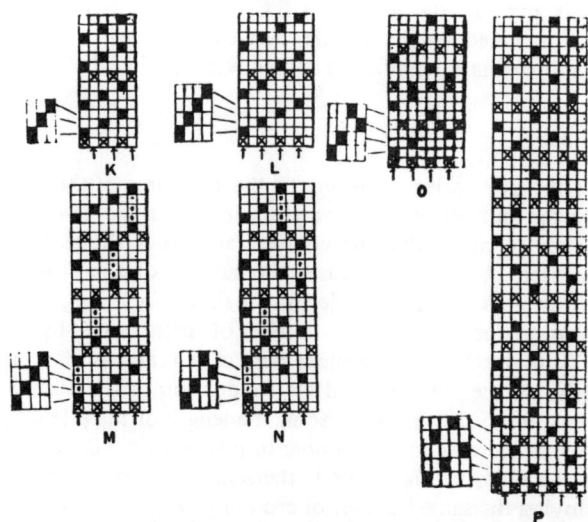

Figure 13.3

similar to C in *Figure 13.1* except that they have six and five pile picks respectively instead of four to each ground pick. In the same manner the designs N and O correspond with D, but have respectively five and three pile picks to each ground pick; while the design P is similar to E except that there are four instead of five pile picks to each ground pick.

The pile is most evenly distributed when, between each pair of ground picks, every binding end holds the same number of tufts, as shown at K in *Figure 13.3*. When the modified arrangement makes it necessary for the pile weave to be extended over two or more repeats of the ground weave, a sateen base is better for the pile interlacing than a twill base. The reason for this will be understood from a comparison of the plans M and N, both of which are arranged on a 4-thread base with 5 pile picks to 1 ground pick. Between each pair of pile picks two tufts occur on one end in the positions where the dots connect the full squares. In the plan M on account of the pile interlacing being based on a twill weave, these positions run in twill order, which may result in a slight twilled effect appearing in the finished cloth. In N, however, the positions occur in satinette order and the liability of twill lines being formed is avoided. Also, in the designs O and P, between each pair of ground picks, there is one end on which there are no tufts, and if these positions were to run in twill order there would be a similar liability of twill lines being formed in the cloth. This may be avoided by using a satinette base for the pile interlacing, as shown in the two examples.

In changing the proportion of pile picks to ground picks in a design the effect of the alteration should be considered in relation to the number of picks that it is proposed to insert under the new conditions. The alteration may be for the purpose of changing the density of the pile while retaining approximately

Table 2

	Design	Ratio of pile picks to ground picks	Ground picks per cm	Pile picks per cm	Total picks per cm	Tufts per cm	Remarks
Original structure	D	4 to 1	32	128	160	448	Original structure
To retain same total picks as original structure	N	5 to 1	27	135	162	473	Density of pile increased. Ground texture less firm.
	O	3 to 1	40	120	160	420	Density of pile reduced. Ground texture firmer
To retain same density of pile as original structure	N	5 to 1	26	130	156	455	Total picks reduced. Ground texture less firm.
	O	3 to 1	43	129	172	452	Total picks increased. Ground texture firmer.
To retain same ground texture as original structure	N	5 to 1	32	160	192	560	Density of pile increased. Total picks increased
	O	3 to 1	32	96	128	336	Density of pile reduced. Total picks reduced.

the same total number of picks per cm as the original structure; or of changing the density while retaining a similar ground structure; or the idea may be to obtain approximately the same density as before, but with a different ground structure. Table 2 shows the result which will occur under the different conditions named, assuming that the weave D in *Figure 13.1*—which has 4 pile picks to 1 ground pick—is changed to five and three pile picks respectively to each ground pick, as shown at N and O in *Figure 13.3*. The total picks of the original structure are taken as 160 per cm, giving 32 ground and 128 pile picks per cm; and the tufts per cm^2 are based on the cloth having 28 ends per cm.

Fast pile structures

A very important feature of these fabrics is the proper securing of the pile to the foundation cloth so that there will be no tendency of the tufts fraying out. In the examples given in *Figure 13.1* to *13.3*, the tufts are bound in by one end only at a place, and the fastness of the pile is chiefly dependent upon the pressure of the picks upon one another. It is therefore necessary, particularly in the longer piles for a very large number of picks to be inserted in order to keep the pile firm. If it is desired to introduce fewer picks per cm, or to make a very long pile, the necessary firmness can be secured by interweaving the pile picks more frequently and thus making what is termed a 'fast' pile. The examples Q,R, and S, given in *Figure 13.4*, respectively show how the plans C,D, and E in *Figure 13.1* may be made firmer. The section shown at T, illustrates how the tufts

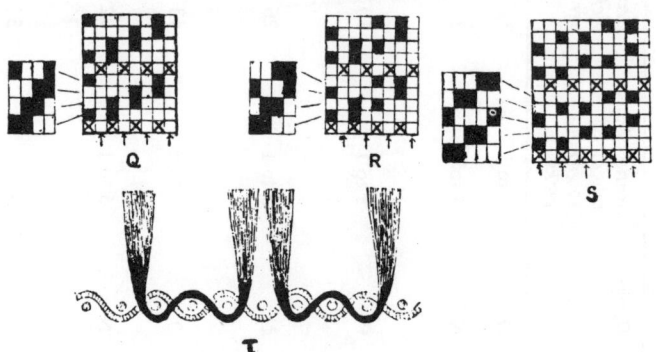

Figure 13.4

formed by the picks 5 and 6 of the design S, are bound in. By comparing the designs given in *Figures 13.1* and *13.4* it will be seen that with the same number of ends per cm Q and R will each produce the same length of pile as B, and S, as C or D. The firmer interweaving renders it more difficult to insert a larger amount of weft, and it is generally recognised that in a fast pile the richness of the cloth will suffer, but there is the advantage that the greater firmness gives the cloth better wearing qualities.

Twill back velveteens

Examples of velveteens with a twill foundation are given in *Figure 13.5* A, B, and C having a 1-and-2 twill or 'Genoa' back, while the ground weave of D and E is 2-and-2 twill. A twill foundation weave is looser than a plain, and therefore, not only permits, but, in order to maintain the same firmness of pile, requires a large number of ground picks to be inserted. Hence, with the same ratio of pile to ground picks, more pile picks can be put in and a denser pile formed. Also, a cloth with a twill ground is softer and more flexible than a similar cloth with plain ground; the latter ground, when very heavily wefted, tending to make the cloth handle somewhat hard and stiff.

In A,B, and C in *Figure 13.5* the pile weave is based on 1-and-2 twill, which, as before, is marked on alternate ends; the pile picks are arranged in the proportion respectively of two, three, and four, to each ground pick.

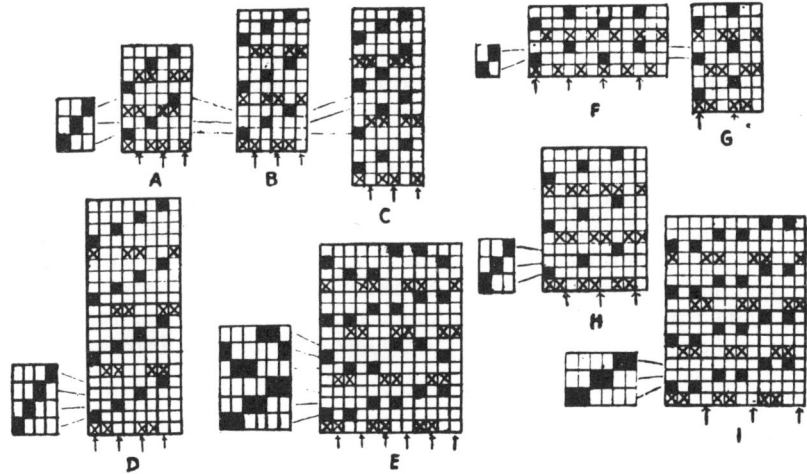

Figure 13.5

It may be noted that A in *Figure 13.5* is the standard design for the moleskin class of fabric, which is usually made in coarse cotton yarns. This is not a pile fabric, as the floating picks are not cut but remain in the condition they are after weaving. The cloth is very strong and leathery, and is, in fact, used as the 'leather' side in imitation sheepskin overcoats and car coats.

A 2-and-2 twill foundation weave enables a very large number of ground picks to be readily inserted, and is therefore used for the heaviest and densest velveteens. In the design D, *Figure 13.5*, the pile weave is based upon 1-and-3 twill, and there are four pile picks to each ground pick, while in E, a 6-thread sateen pile base weave is employed with the pile made fast, and there are three pile picks to each ground pick.

Designs which simplify the cutting operation

The designs F to I given in *Figure 13.5* illustrate a method of arranging the pile interlacing that is sometimes employed with the object of reducing the time occupied in the pile cutting. In F, G, and H, the pile base weaves are indicated only on every third end, therefore, only one-third as many longitudinal traverses of the cutting knife are required as there are ends in the width of the cloth. Compared with the examples in which the binding of the pile picks occurs on alternate ends, the number of cutting races is reduced by one-third. The distribution of the pile, however, is not so perfect, and the surface of the cloth has a coarser appearance. I in *Figure 13.5* shows a fast pile effect in which the pile interlacing is based on a 1-and-2 twill weave doubled; and as indicated by the arrows below the design, the cutting races occur only on every fourth end.

Cutting of all-over velveteens

The cutting of all-over or plain velveteens is a slow and costly process which adds considerably to the already high cost of production due to the great density of wefting. The finer qualities of velveteens can only be cut one cutting race at a time even with modern machinery. In a cloth of standard construction with 28 ends per cm there are 14 cutting races per cm which means that a length of fabric 60 cm wide requires 840 passages through the machine before it is fully cut.

The cloth, having been prepared for cutting in the manner described earlier, is stretched lengthwise and is guided with precision so that a knife guide enters a cutting race or 'tunnel' formed by the floats of the pile weft. The races are indicated by the arrows at I in *Figure 13.2*. The guide slides over the ground structure and expands slightly the pile weft floats which are above it. A razor-edge knife fits into a slot of the guide and as the cloth runs forwards the floats of the pile weft climb upon the inclined knife blade and are thus severed.

Qualities of all-over velveteens

Quality of any velveteen construction can be varied considerably by changing the warp and the weft yarn settings and counts. Reduction in the number of picks per cm is usually compensated for by the increase in the number of ends per cm if similar density of pile is required. As this results, however, in the increase in the number of cutting races the more favoured practice is to employ heavier yarns. This permits the reduction of the settings in both directions which reduces the costs of weaving and finishing whilst weight and density of pile cover is maintained. The height of pile can then be controlled by a suitable choice of the pile float length.

A feature to be noted in these fabrics is the considerable shrinkage in the width from the reed to the cloth which varies from 12½ per cent in the lighter velveteens to 20 per cent in the heavy ones. This explains why the pile floats are forced into a tunnel formation which permits the insertion of guides and makes cutting possible. The contraction in length is negligible and amounts to between 2½ to 4 per cent.

In the list which follows, ends per cm in the reed, and picks per cm in the loom are quoted. The settings represent a top quality all-cotton cloth in each of the selected weaves and yarn counts.

1. Weave B, *Figure 13.1*—Warp; 20/2 tex, 28 ends per cm; weft: 15 tex, 120 picks per cm; 420 tufts per cm^2; weft contraction 12½ per cent.
2. Weaves C and D, *Figure 13.1*—Warp: 17/2 tex, 28 ends per cm; weft: 10 tex, 176 picks per cm; 496 tufts per cm^2; weft contraction 15 per cent.
3. Weave A, *Figure 13.5*—Warp: 60/2 tex, 15 ends per cm; weft: 38 tex, 96 picks per cm; 160 tufts per cm^2; weft contraction 17 per cent.
4. Weave C, *Figure 13.5*—Warp: 17/2 tex, 28 ends per cm; weft: 10 tex, 208 picks per cm; 740 tufts per cm^2; weft contraction 17 per cent.

WEFT PLUSHES

These constructions are similar in principle to the ordinary all-over velveteens but are made with longer pile floats and in heavier weights, being chiefly employed as upholstery cloths. They are produced in insignificant quantities as most of the pile upholstery cloths are. at present made on the warp pile principles (see Chapter 15) in which similar effects can be woven faster and without the need for the separate costly cutting operation after weaving.

Due to the use of the cloth and the length of pile the pile weft is invariably anchored to the ground cloth on the fast pile principle. The pile consists usually of woollen, mohair, or acrylic yarns although other materials have also been used.

In the plain-back velveteen structures previously given, all the pile picks go into the same shed as one ground pick, and in the opposite shed to the other ground pick. This causes a slight irregularity in the lower picked cloths, which, however, is quite imperceptible in the finer fabrics when finished. In some of the long and coarse weft pile structures, the irregularity is got over in the manner illustrated by the design A in *Figure 13.6* in which the plain ground texture is modified, so that each group of two pile picks is in the same shed as the preceding ground pick, and in the opposite shed to the ground pick that follows. Suitable weaving particulars for the design are: Warp, 38/2 tex cotton, 20 ends per cm; weft, 1 pick 72 tex woollen, 2 picks 56 tex mohair, 60 picks per cm.

The design B in *Figure 13.6* is arranged 2 ground to 1 pile pick and the pile interlacing is based on an irregular 8-sateen weave. This structure is used for a heavy type of weft plush termed 'dogskin' in which a long mohair pile is developed with the following weaving particulars: Warp, 50/2 tex cotton, 16 ends per cm; weft, 2 picks, 60/2 tex cotton, 1 pick, 300 tex mohair, 24 picks per cm.

In the design C the pile interlacing is based on a 5-sateen weave, and to each ground pick there are two pile picks. One pile pick, however, has a longer float than the other, so that two different lengths of pile are formed in the cloth. Variety of effect can also be obtained by having the pile picks alternately in different colours or different materials. The following are suitable weaving particulars: Warp 30/2 tex cotton, 19 ends per cm; weft, 1 pick 24 tex cotton, 1 pick 32 tex acrylic yarn (shade 1), 1 pick 32 tex acrylic yarn (shade 2), 84 picks per cm.

The construction D illustrates an effective method of developing the pile in different materials or colours. The pile interlacing is based on an irregular 8-sateen weave, but the binding of the odd pile picks, shown by the full squares, occurs on one-half of the plan, while that of the even pile picks, represented by the dots, occurs on the other half. By arranging the pile picks alternately in different colours or materials, stripes of pile are formed on the surface.

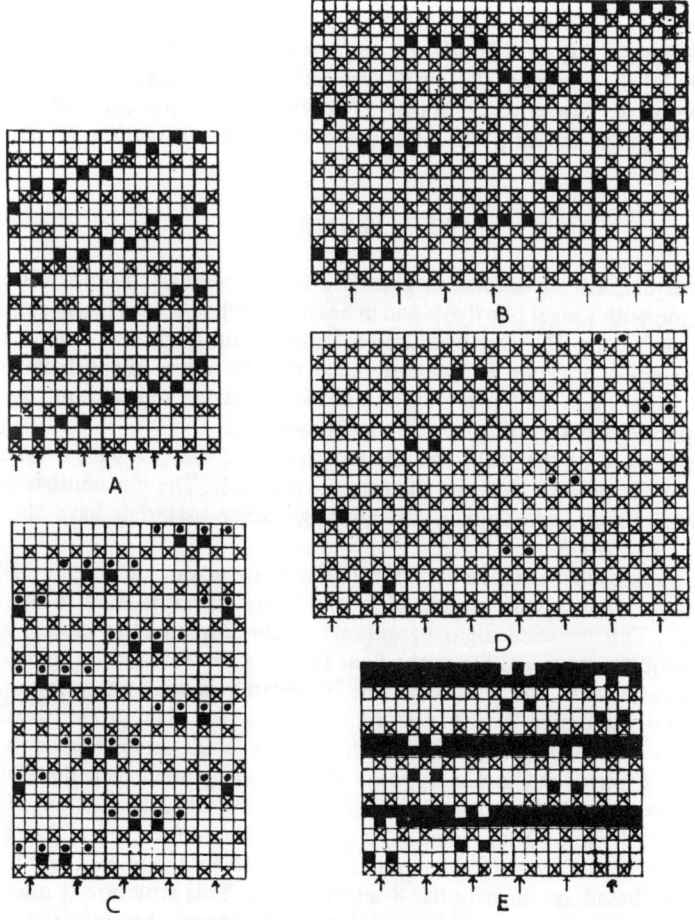

Figure 13.6

The design E in *Figure 13.6* shows the arrangement of a reversible weft plush, a structure that is sometimes used for luxurious rugs. The cloth has a plain ground, and the pile binding places are arranged in 6-sateen order on both sides. The following weaving particulars are suitable: Warp, 40/2 tex cotton, 19 ends per cm; weft 1 pick 66 tex woollen, 2 picks 74 tex mohair (face), 1 pick 66 tex woollen, 2 picks 74 tex mohair (back), 57 picks per cm. The pile on the back may be developed in a different colour from that on the face.

CORDED VELVETEENS

In these structures the pile picks are bound in, at intervals, in a straight line. The cuts are made right up the centre of the space between the pile binding points, with the result that the tufts of fibres project from the foundation in the form of cords or ribs running lengthwise of the fabric. An illustration of a cloth is given in *Figure 13.7*, which shows in the upper and lower portions respectively, the appearance of a corduroy before and after the operation of cutting.

Figure 13.7

The finer classes of cords, such as are used for dress fabrics, are largely made in fine yarns with a plain back. The corduroys used for men's clothing, are made sometimes heavier, in which case a twill ground weave is employed. In the heavier cloths thicker weft is used, and consequently fewer pile picks to each ground pick are necessary, usually not more than two being employed.

In the simplest cord designs, the pile picks are bound in plain order on two consecutive ends. J, K, L, and M in *Figure 13.8* are examples with a plain back which, in the same sett, yield successively an increased width of cord. Thus, with 24 ends per cm in the finished cloth the number of cords per 10 cm will be

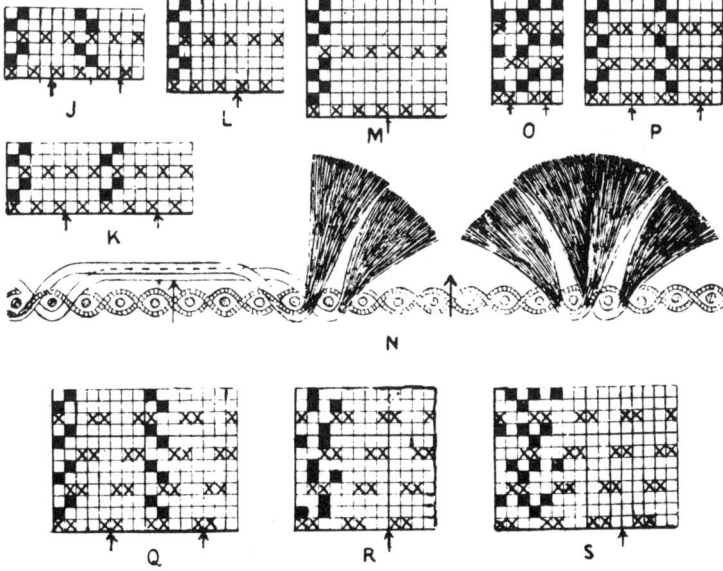

Figure 13.8

J-40, K-30, L-24, M-20. Designs may be constructed to produce other widths of cords simply by varying the space between the binding ends.

The plain binding weave of the pile picks may be reversed in alternate cords, as shown at J, in which case the design extends over the width of two cords, and each pile pick forms alternately a long and a short float. On the other hand, the pile binding may be the same in each cord, as shown at K, and in this case all the pile floats are equal. The result is practically the same whichever method of binding is adopted, because the floats are cut in the middle of the space between the pile binding points; consequently, in either case, one side of each tuft is longer than the other side. The difference in the lengths causes the ribs to have a rounded formation, as the long side of the tufts forms the centre, and the short side the outer parts of the cords. This is illustrated by the warp section given at N in *Figure 13.8*, which shows on the left how the picks of the plan K interlace, while on the right the appearance of the cord, after the cutting, is represented. The arrows indicate the position of the cutting races. Similar effects are produced by the designs L and M, but here there are three and four pile picks respectively to each ground pick.

Examples of cords with a 2-and-1 twill back are given at O and P in *Figure 13.8*, and with a 2-and-2 twill back at Q, R, and S. These are arranged two pile picks to each ground pick, and in producing very heavy structures they are woven with comparatively few ends per cm, the number varying from about 13 to 15.

The cords, produced by the design O in *Figure 13.8* are only three ends wide, and both sides of the tufts are of equal length, therefore, the ribs are not rounded, and a poor and bare structure results. P is similar to J except for the difference in the ground weave, the pile floats being of different sizes, and the complete design extending over the width of two cords. Q is similarly arranged, but in R and S (which are used for specially heavy and wide cords), alternate pile picks are interwoven more frequently with the object of producing greater variety in the length of the tufts, and so cause the rounded formation of the ribs to be more pronounced.

All the above constructions can be readily re-arranged to produce in the same cloth cords of different width which results in more interesting textures. Sometimes, alternate cords are left uncut so that a stripe of tufted cord alternates with a stripe of float construction. This modification is useful in fabrics intended for heavy wear as it results in an improved tuft anchorage.

Cutting of corded velveteens

Due to the distance between the cutting races corded velveteens can be cut in a single passage of the cloth through the cutting machine. All the cords are cut at the same time by means of circular knives, one to each cord, placed upon a revolving shaft. Each knife rotates within a slot formed in a guide, the pointed end of which is inserted under the pile floats in the centre of a cord. By means of tension rollers the cloth is drawn forward towards the knives, but at about the point of contact with the latter, it is taken downward over the edge of a transverse bar. The floating pile picks are brought by the guides into the path of the revolving knives and are cut, while the cloth passes downward and is either wound on a beam or is plaited down.

Qualities of corded velveteens

Quality of corduroy fabrics can be varied considerably by the changes in the density of weft shotting. Once the width of the cord has been determined the end setting has to remain unchanged but the scope for changes in the weft yarn counts and settings is sufficiently extensive to permit the construction of widely differing qualities of cloth within each structure. The examples which follow represent a good quality all-cotton fabric in each of the selected weaves.

1. Weave L, *Figure 13.8*—Warp: 20/2 tex, 28 ends per cm; weft: 12 tex, 140 picks per cm; 294 tufts per cm²; weft contraction 19 per cent.
2. Weave P, *Figure 13.8*—Warp: 60/2 tex, 12 ends per cm; weft: 32 tex, 168 picks per cm; 224 tufts per cm²; weft contraction 20 per cent.
3. Weave R, *Figure 13.8*—Warp: 74/2 tex, 13 ends per cm; weft: 32 tex, 172 picks per cm; 124 tufts per cm²; weft contraction 25 per cent.

FIGURED WEFT PILE FABRICS

These structures are not produced at present on account of very high costs of weaving and finishing and an example is given here merely for the sake of completeness. Similar effects can be produced much more economically on the principle of warp pile (see Chapters 15 and 16) and although, using the latter principle, the same density of pile population per area cannot be achieved the necessary hard wearing properties are obtained by the use of modern, strong, and resilient materials such as polyamides, acrylics or polypropylene.

Figured velveteens

In figured pile structures most of the surface is occupied by a massive pile figure this being the most ornamental part of the design, and the bare ground is exposed only to separate the parts of the ornament. Practically any velveteen weave can be used for the figure, but in the ground the structure is varied according to the method in which the pile weft is prevented from showing on the surface. There are two chief methods of disposing of the surplus weft in the ground: (1) It is bound in on the underside in the same manner as on the face. (2) It is floated loosely on the back of the foundation texture, and after the cutting operation is brushed away as waste.

The velveteen weaves that are chiefly used for the pile figure are given at A and C in *Figure 13.9*. If the pile picks are bound in on the back, corresponding methods of interlacing, as shown at B and D respectively are employed in the ground but the pile binding points are placed in different relative positions to the face binding points. The design is painted as indicated at E, i.e. by painting-in the ground which occupies far less space than the figure. Full scale is retained warp-wise but weft-wise the design is condensed, the degree of condensation depending on the weave used. It is usual to condense so that the number of horizontal rows in the design correspond to the number of ground picks in the repeat, thus in the case of weave A the condensation would be by four and in the case of C, by five. The detailed weaves which are cut for the blank and the painted portions are shown at F. From each horizontal row of the condensed design a number of cards is cut which equals the number of pile picks to each ground pick; in addition, a ground weave card is cut (plain or twill) and inserted

Figure 13.9

between each group of pile cards. The edge of the figure is modified to overcome a difficulty which may arise during cutting on account of the liability of the knife guide leaving the races between the separate parts of the figure. It has been found that the best results, as regards the pile cutting, are secured, first, by starting and finishing the outer edges of the ground portions on the bound or odd ends; and, second, by making each pile weft float pass over at least five ends at the edges of the figure. It will be seen in E, *Figure 13.9*, that the edges of the ground effect are indicated on the odd ends, i.e. the bound ends in the plan F, while the outline is worked in steps of two ends. The crosses in F illustrate the method of taking out all pile floats of less than five at the edges of the figure. The marks in the plan F are cut.

The method of indicating a design, shown at F, may be employed when the pile weft is not bound in the ground, except that no binding marks are indicated in the ground portion of the design. It has to be taken into account, however, that the removal of the surplus weft from the underside causes a half tuft formed by each pile pick to be drawn away on both sides of every portion of figure. The latter method, as compared with the former, therefore increases the ground space, and if care is not taken in indicating the cutting marks of a design, a narrow portion of figure may be eliminated. In order to preserve the full mass of the figure, instead of throwing all the small floats to the back, as shown by the crosses in F, the three floats may be extended to floats of five on the surface by taking out marks in the ground.

Figured cords

Any standard cord weave may be employed as the basis of the structure in producing a figured velveteen cord, but as it is necessary for the outline of the figure to fit with the vertical cord lines, more elaborate ornamentation can be produced in narrow than in broad cord effects. In any case the steppy character of the lines makes it necessary for the design to be simple and massive, and, as a general rule, the styles are limited to simple geometrical figures. The ground effect is produced by floating the pile weft on the back between the binding ends, but there is the exception that in check patterns the horizontal lines can be formed simply by discontinuing the pile weave and inserting the required number of ground picks consecutively.

14

Terry Pile Structures

The terry pile, also known as the Turkish towelling, is a class of warp pile structure in which certain warp ends are made to form loops on the surface of the cloth. Only one series of weft threads is used but the warp consists of two series of threads, the ground and the pile. The former produces with the weft the ground cloth from which the loops formed by the pile ends project. The loops may be formed on one side only or on both sides of the cloth thus producing single-sided and double-sided structures respectively. Any one pile thread may alternate between the face and the back of the cloth a possibility that is frequently utilised for the purpose of ornamentation. The schematic diagrams in *Figure 14.1* show at A the single-sided and at B the double-sided continuous terry structures. C conveys the idea of a pile thread alternating between the face and the back which permits the formation of pile figure on exposed ground whilst at D the ornamentation is carried further by having two differently coloured sets of threads which mutually alternate between the face and the back thus forming a figure in one colour on the background of another. All the structures, apart from A, are reversible.

Structure A has been used for the production of mats, curtainings, ladies' overcoats and dressing gowns. Structures B, C, and D represent typical towellings which form by far the most important outlet for these fabrics. The looped structure is eminently suitable for towelling purposes as the long, free floats of yarn, if made from absorbent materials, are capable of wicking-up readily large amounts of moisture. The material best suited for the purpose is cotton which not only absorbs moisture easily but also stands up well to frequent and severe launderings which the towelling fabrics have to undergo. Linen is used for the pile when, either, the slightly harsh feel is desired as in athletic towellings, or, an article capable of withstanding very hard wear is required as in public institutions, etc. Viscose rayon staple yarns are also employed and whilst they possess adequate moisture absorption capacity their ability to resist frequent laundering is poorer than that of cotton yarns.

Formation of the pile.

The formation of terry pile depends on the creation of a gap between the fell of the cloth and two succeeding picks of weft. The gap, the length of which depends on the height of pile required, results in the formation of uninterlaced warp floats. To form the gap two succeeding picks are beaten up short of the true cloth fell and produce a temporary false fell as indicated schematically at E in *Figure 14.1*. On the third pick of the group full beat up takes place the

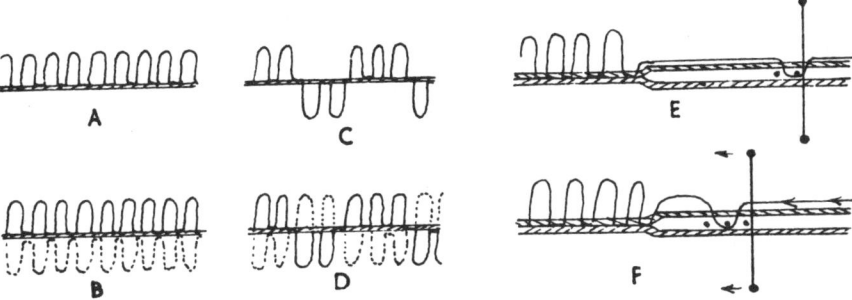

Figure 14.1

three picks being pushed forward together to the true fell position. During this action the three picks are capable of sliding between the ground ends, which are kept very taut, as depicted at F. However, they cannot slide similarly between the pile ends, firstly, because they are structurally locked with them and, secondly, because the pile warp at that moment is slack. Therefore, as they are pushed forward after the third pick they pull a length of pile warp from the beam and at the same time force the excess length of pile yarn in front of them into a loop. If the pile warp float is formed on the surface a loop is made on the face and if the float is on the back of the cloth a back loop results. From the description it will be obvious that in this construction two beams are necessary. The ground beam is very heavily tensioned whilst the pile beam is only under slight tension and in some systems it is, in fact, rotated forward positively during the full beat-up, i.e. after the insertion of the third pick of the group, to deliver exactly the length of yarn required for a loop.

The gap is created by a variety of devices which can be divided into two main classes, viz.: (1) Those in which the reed is drawn back the required distance before reaching the fell on the two picks in question (used in most of the conventional looms); and (2) Those in which the fell of the cloth itself is made to

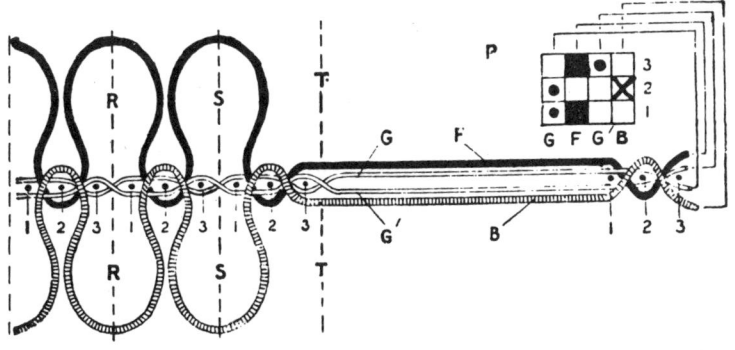

Figure 14.2

recede away from the on-coming reed during the insertion of the two succeeding picks (used in gripper and rapier machines).

The exact relation of the weft to the two warps and the principle of loop formation is depicted by means of the weft section in *Figure 14.2*. The broken vertical lines RR, SS, and TT divide the picks 1, 2, and 3 into repeating groups of three, line TT indicating the position of the fell of the cloth. On the right of the diagram, a group of three picks, which compose a repeat, is represented previous to being beaten up to the fell of the cloth. The ground threads G, G^1, and the face and back pile threads F and B are shown connected by lines with the respective spaces in the corresponding weave given at P. In weaving the cloth the group warp beam carrying the threads G and G^1, is heavily tensioned, as stated earlier so that these threads are held tight all the time. The picks 1 and 2 are first woven into the proper sheds, but are not beaten fully up to the fell of the cloth at the time of insertion in their sheds; but when the pick No. 3 is inserted the mechanisms are so operated that the three picks are driven together into the cloth at the fell TT. During the beating up of the third pick the pile warp threads F and B are either given in slack, or are placed under very slight tension.

The picks 1 and 2 are in the same shed made by the tight ground threads G and G^1, which, therefore, offer no obstruction to the two picks being driven forward at the same time with the third pick. The pile threads F and B, on the other hand, change from one side of the cloth to the other between the picks 1 and 2, and they are, therefore, gripped at the point of contact with the two picks. As the three picks are beaten up this point of contact is moved forward to the fell of the cloth, with the result that the slack pile warp threads are drawn forward and two horizontal rows of loops are formed one projecting from the upper and the other from the lower surface of the cloth in the manner represented in *Figure 14.2*.

In order to produce the loops on the three picks during the insertion of which the terry motion is in operation, the pile and ground threads must be interwoven with the weft in the exact order represented in *Figure 14.2*. The 3-pick terry structure is employed most extensively, but sometimes four, five, and even six picks are inserted in making each horizontal row of loops. The interweaving of the threads, on the subsequent picks, is, however, of little consequence so long as the cloth has the necessary firmness, and a natural connection is made with the weave of the three picks particularly referred to.

Terry Weaves

A number of standard weaves for producing the fabrics is given in *Figure 14.3*. These constructions have been grouped so that comparisons can be readily made. The dots in the designs represented the interlacings of the ground warp threads; the full squares show the interweaving of the face pile threads and the crosses, of the back pile threads. In A, B, C, D, and E the loops are formed uniformly on the face side of the cloth only, whereas the remaining structures are for producing a pile surface on both sides of the cloth. In A, B, C, D, and E, the warp threads are arranged 1 ground, 1 pile, and in F, G, H, I, J, and K, 1 ground, 1 face pile, 1 ground, 1 back· pile. The weaves L, M, N, O, P, and Q produce corresponding effects to the designs F to K respectively, but they are arranged 1 ground, 1 face pile, 1 back pile, 1 ground.

In each structure A to E in *Figure 14.3* there is a pile end on the surface to each ground end, but in the weaves F to Q, the proportion is one pile end on each side of the cloth in two ground ends. The single-sided pile cloths can, however, be made with 1 pile to 2 ground by leaving out the last thread in each of the constructions A to E. The plans A, F, and L are for producing one horizontal row of pile loops on three picks; B, G, and M on four picks; C, H, N, D, I, O, J, and P on five picks; and E, K, and Q on six picks.

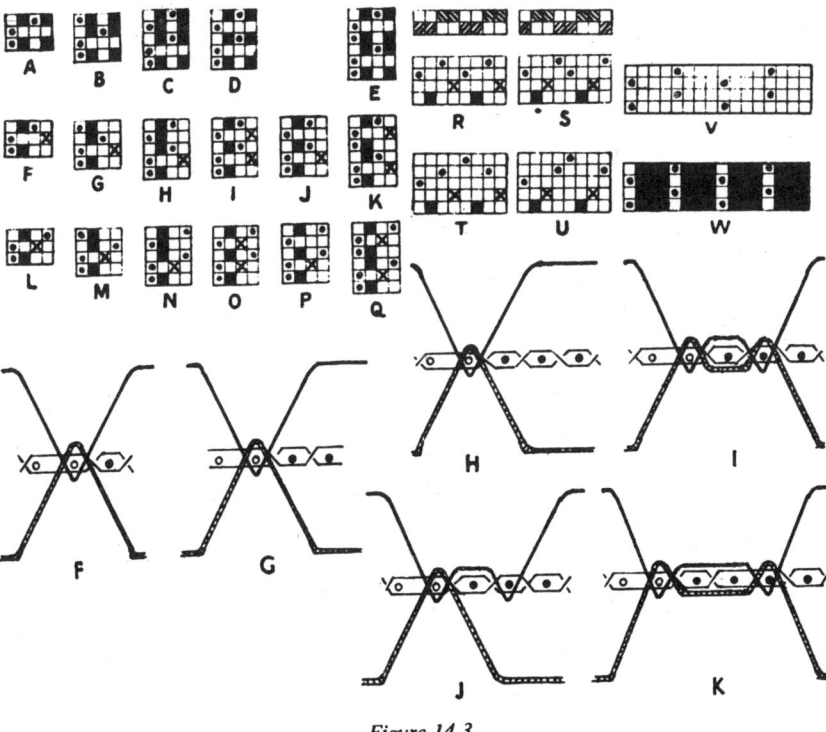

Figure 14.3

Every plan in *Figure 14.3* is constructed for the first and second picks to remain back from the edge of the cloth when they are first inserted, and for the full beat up to occur on the third and subsequent picks in the repeat. A comparison of the designs will show that in each case the interweaving of the respective threads is exactly the same on the picks 1, 2, and 3, and corresponds with the order of interlacing illustrated in *Figure 14.2*. Thus, on the picks 1 and 2, the odd ground threads are raised and the even ground threads depressed, while on the third pick they are in the reverse positions. The face pile threads are raised on the picks 1 and 3, and depressed on the second pick; the back pile threads being operated in the reverse order.

In the lower portion of *Figure 14.3* cross-sectional views of one loop unit of every structure F to K are given, and these are lettered to correspond with the squared paper plans. It will be noted that whilst in most of the constructions the loops are anchored around one pick of weft, I and K represent the fast pile anchorage in which the pile ends between each loop are interlaced in a tight weave order such as 1 up, 1 down, 1 up, or 1 up, 2 down, 1 up, or the reverse of it. This prevents the occurrence of one of the most undersirable faults in these fabrics which is loop 'sprouting' (i.e. formation, due to pulling, of an enormously elongated loop at the cost of the disappearance of some loops in front and behind it), and is used in fabrics expected to withstand particularly severe conditions of wear.

Most of the terry cloth is produced in the 3-pick structures, 4-pick weaves are used occasionally but the amount of 5-pick or 6-pick cloth made at present is very small being restricted by the high cost of production. It will be appreciated that in a 6-pick fabric six picks need to be inserted to make one horizontal row of loops as opposed to only three in a 3-pick fabric. Also, to produce the same pile coverage in a 6-pick, as in a 3-pick cloth twice as many picks per cm are required.

In drawing-in the two warps for weaving the pile threads may be on two healds at the front, and the ground threads (if looped formation is to be continuous) on two healds at the back, as shown at R in *Figure 14.3* for the 1 ground, 1 pile order of arrangement, and as indicated at S for the 2-and-2 order. When, however, the cloths are made in short lengths with a cross-border at each end, the drafts given at T and U are frequently employed. This arrangement enables a weave with a weft float over seven warp threads to be obtained by the alternate lifting of the third and fifth heald, the remaining healds being left down in forming the float on the face, and lifted in forming the float on the reverse side, as shown respectively at V and W. In dobby shedding specially crammed and coloured cross-over headings can be readily formed in the borders in this manner.

Usually two threads are placed in each split of the reed, and in the 1 ground, 1 pile order, one of each series is placed in the same split, as shown above R in *Figure 14.3*. In the 2 ground, 2 pile order, however, two ends of the same series are placed together, as shown above S. The two arrangements produce practically identical results, but the 2-and-2 arrangement has the advantage that by reeding as described, the threads in each split work opposite to each other, and at the same time the pile and ground threads, which on some picks work alike, are separated by the wires of the reed, so that a clear shed is more readily obtained.

The loom particulars of a good quality all cotton 3-pick terry cloth are as follows: Pile warp, two ends of 60/2 tex; ground warp, 66/2 tex; weft, 38 tex; ends per cm, 20; picks per cm, 22; 500m of pile warp and 120m of ground warp are required for producing 100m of terry. The shrinkage in width is about 12 per cent. In cheaper cloths the weft may be 30 tex and the picks 14 per cm and upwards; the pile warp 38 tex and the ground warp 42 tex; 300m and upwards of pile warp for 100m of cloth. The ground ends are usually slightly thicker than the pile ends. The feel of the texture varies according to the depth of the pile loops, a deep pile handling softer than a short pile. The depth of the pile is determined by the distance that the two picks are left away from the fell of the cloth, which is usually about 10 to 15mm. To assist the absorption of moisture the pile yarns are of low twist and, therefore, must be manufactured from reasonably long stapled fibres to prevent excessive fibre shedding during use.

Special mechanisms required in terry weaving

Most of the terry fabrics are produced in either dobby or jacquard machines. Cam shedding is used infrequently even when continuous self-coloured styles are manufactured although the weave repeats of such constructions fall well within the capacity range of cam motions. The reason for this is that most towels have a cross-border heading at each end which demands changes in the operation of various mechanisms at widely separated intervals which are impossible to achieve in cam-controlled shedding systems. At one time to avoid long pattern chains of lags, two, or three barrel cross-border dobbies were used with different portions of the towel pegged on different barrel chains. These are still sometimes employed but more commonly at present high-speed paper-roll dobbies are used in which the length of pattern does not create the same encumbrance as it does when the lags constitute the pattern chain.

For the production of the figured terries the inverted hook jacquard with a heald mounting was at one time highly favoured (see Appendix I), but at present the machine used most frequently is the fine pitch, large capacity jacquard capable of running at appreciably greater speeds than the coarse pitch machines.

The variable beat-up motions are an essential part of the terry pile weaving and they fall into two main categories. The function of these motions is, as mentioned earlier, to create a gap between the cloth fell and the first two picks of a pile forming a group of picks termed 'loose' picks, as opposed to the picks beaten up fully which are known as 'fast' picks. In the first category are those mechanisms in which the reed itself is drawn back on the loose picks thereby leaving them a small distance short of the cloth fell. A variety of devices exist to achieve this purpose in some of which the reed only and in others the sley itself may be controlled to provide a 'short' beat-up. On the following pick the reed or sley is locked fast so that the preceding loose picks and the fast pick are pushed together into the cloth fell proper. The two reed positions are shown at A and B respectively in *Figure 14.4*. In the second category of mechanisms, such as are used at present in some gripper and rapier machines, the reed is permanently fixed in position and has a constant stroke. To create the gap on the loose picks the cloth itself is drawn away from the advancing reed so that the two loose picks cannot reach the normal cloth fell position. On the third pick the cloth is brought forward again so that the three picks of a group join together with the previously woven cloth at the normal cloth fell point. These operations are illustrated schematically at C and D in *Figure 14.4*. All the above motions must be capable of precise adjustment to vary the size of the gap in order to produce shorter or longer pile. They are also normally controlled by a dobby or a jacquard to determine the occurrence of the loose and fast picks as the ratio of the loose to the fast picks varies in different structures. Thus, in a 3-pick structure two loose picks are followed by one fast one, in a 4-pick terry the order of operation is two loose, two fast, and so on. Also, when pile-less headings are produced the loose pick action is unnecessary and must be disengaged.

Another necessary adjunct in the formation of terry pile is the proper control of the pile yarn tension. The pile yarn must be slack when the first three picks of a group are being beaten up, i.e. at the point at which the loop is being formed, as otherwise the tension would pull the loops out. Imperfect tension control at this stage does, in fact, result in variable loop height. In some systems the necessary slackness is achieved by a very slight degree of tensioning of the pile yarn, the beam being only just sufficiently weighted to form a clear shed.

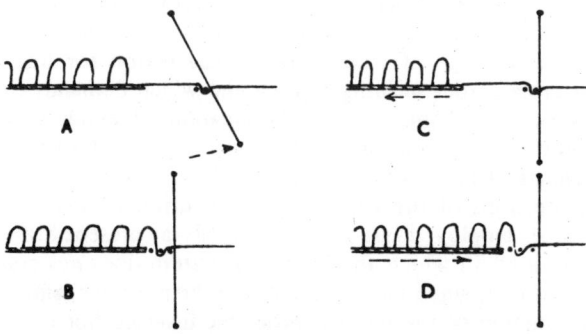

Figure 14.4

In other systems the pile warp is tensioned normally throughout but on the beat-up following the first fast pick in the group a length of the pile warp equal to the distance of the gap is delivered positively which results in more regular loop height along the length of a towel than can be usually achieved in the former method. When the plain, i.e. pile-less, headings are being produced extra weighting of the pile beam is required to ensure that the crimp of the pile ends in those portions of the towel is the same as that of the highly tensioned ground ends. In the second system during the weaving of the headings the positive delivery after every third pick must be stopped. For this reason the pile warp tension control motions must also be governed selectably through a dobby or a jacquard connection.

During the production of towel headings in which no pile is formed it is also usual to operate a cramming device to increase the density of shotting. This is often necessary because the number of picks per cm in the body of the towel is usually comparatively low. This does not give the appearance of a poorly set cloth because these areas are densely covered by the pile, a pile-less heading, however, would look unsubstantial at the same weft setting and to prevent this in such portions of the cloth the cramming device or the interrupted take-up device is brought into operation.

Some makers of terry weaving machines also include a fringing mechanism as one of the special devices. This is, in effect, an express speed take-up motion capable of pulling the cloth forward 30 to 60 mm in the space of time taken to insert two to four picks thus producing an almost uninterlaced gap between two towel lengths. This is employed when the towel is to be finished with a tasselled fringe instead of a hem.

TERRY ORNAMENTATION

When continuously repeating terry weaves are used, such as those given in *Figure 14.3*, the only possible form of ornamentation consists of introducing coloured pile threads to form stripes, but if the loops are formed on both sides of the cloth one side may be coloured independently of the other. In producing more elaborate ornamentation the pile yarns are caused to form loops first on one side and then on the other side of the cloth in the manner represented at C and D in *Figures 14.1* and *14.5*. This system may be employed either with one or two series of pile threads, and styles be produced ranging from simple checks to complex figures. The principle of ornamentation, in which certain pile threads form loops while others lie straight in the cloth, cannot be employed, since all the pile yarn is brought from one warp beam, and it is, therefore, necessary for all the threads either to form pile or to lie straight simultaneously.

When only one series of pile threads is used the pattern is due to the pile threads forming loops on the face and back in turn, so that alternate sections of pile and ground are produced on both sides of the cloth. In *Figure 14.5*, weft section A illustrates this method of interlacing, while the weave which forms the loops on the face is given at B, and on the back at C. In this system of ornamentation colours may be introduced either in the pile or in the ground threads.

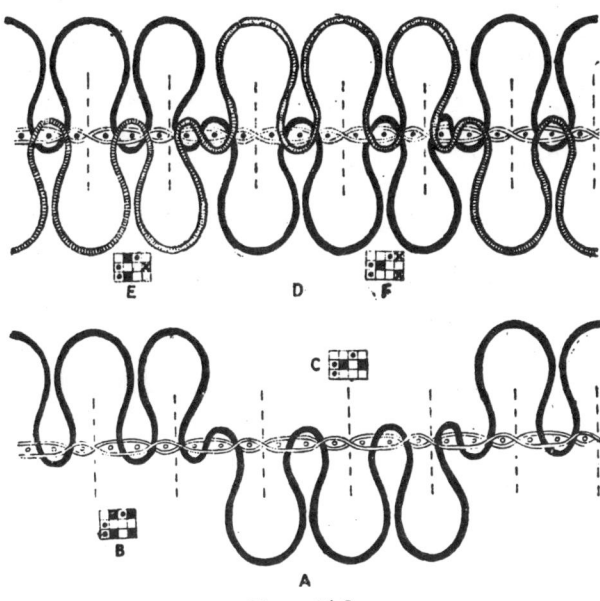

Figure 14.5

In *Figure 14.5* weft section D shows how the pile yarns interchange from face to back when two series are employed. In this case both sides of the cloth are covered by the loops, but one series of threads is differently coloured from the other series, so that alternate sections in different colours are formed. With the weave E, the loops are formed by the dark threads on the face and by the light threads on the back; and with the weave F they are formed by the light threads on the face and the dark threads on the back.

Stripe and check dobby patterns

With dobby shedding simple reversible designs are obtained usually of a check character, and an example is given in *Figure 14.6*, which shows an effect produced by a single series of pile threads, on the principle illustrated at A in *Figure 14.5*. In this case, however, the threads are arranged in the order of 2 ground, 2 pile. The design of the cloth is shown in *Figure 14.7*, in which sections G form

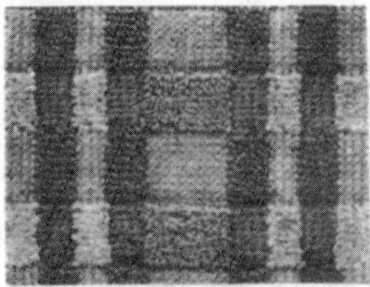

Figure 14.6

pile on the face and ground on the back, and sections H form ground on the face and pile on the back. The draft is shown at I and the lifting plan at J. In producing a given size of check, each section is repeated the required number of times.

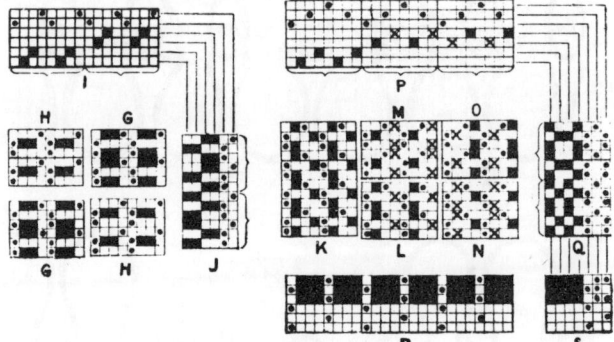

Figure 14.7

Figure 14.8 shows a check pattern produced in two series of pile yarns, on the principle illustrated at D in *Figure 14.5*. There is also a continuous stripe effect at each side of the check pattern. One series of pile threads is differently coloured from the other series, and in the corresponding design and draft given in *Figure 14.7*, the black squares represent red pile while the crosses indicate white pile. Section K shows the weave used in producing the continuous stripe, while sections L and O form red loops on the face and white loops on the back, and sections M and N form white loops on the face and red loops on the back. The change of effect between sections L and M, and also between N and O, is due to a change in the weave. In the sections L and N, however, and also in sections M and O, the weave is exactly the same, the change of effect in this case being due to a change in the order of colouring. Thus, as indicated by the black squares and crosses respectively, the pile yarns in sections L and M are arranged 1 red, 1 white, and in N and O, 1 white, 1 red, two white pile threads coming together in the centre. The drawing-in draft is shown at P—three healds being used for the ground threads, and the lifting plan at Q.

The lower portion of *Figure 14.8* shows a cross-border heading, the bulk of which is formed by continuing the centre weave with the terry motion out of action, but there is also a repp heading produced by floating thick picks over seven successive ends. The weave for the thick picks is shown at R in *Figure 14.7*, the lifting plan being indicated at S. Three picks float on the face and then three on the back, in order that the border will be reversible similar to the main body of the towel.

Figure 14.8

An interesting modification of the latter style of check pattern consists of separating the rectangular spaces from each other by narrow lines of ground, the longitudinal lines being formed by bringing six or eight ends consecutively from the ground warp beam, while the transverse lines are obtained by throwing the terry motion out of action for about six picks. This system of forming checks can also be employed when the pile threads are all of one colour, and when no interchange is made from one side of the cloth to the other.

Figured terry pile fabrics

A representation of a figured terry pile texture, taken from the corner of a towel, is given in *Figure 14.9*. The example is simply an extension of the principle

Figure 14.9

illustrated at D, E, and F in *Figure 14.5*, in which two series of differently coloured pile threads are interchanged. In the fabric represented a figure in white terry pile is formed on a blue ground on one side of the cloth, and a blue terry figure on a white terry ground on the other side. The warp threads are arranged in the cloth in the order of 1 ground, 1 white pile, 1 ground, 1 blue pile, and the structure is a 3-pick terry.

The design for the above fabric is given in *Figure 14.10* and shows the construction condensed by 4 warp-wise and by 3 weft-wise so that each square represents one loop on the face and one loop on the back. Filled squares represent white loops on the face and blue loops on the back whilst blank squares indicate blue loops on the face and white ones on the back. The detailed weaves for each colour are shown at A and B in *Figure 14.10*, in which the dots indicate the lifts of the ground ends, the solid marks, the blue pile ends and the crosses, the white pile ends. The cloth is produced with 20 ends and 21 picks per cm which results in five vertical and seven horizontal rows of loops of each colour per cm and this ratio determines the proper count of design paper to be used. In the case of the design in *Figure 14.10* the paper is 8 x 11 corresponding sufficiently closely to the ratio of 5 : 7.

The condensed design could be taken to represent any terry structure and, indeed, the same design could be used to produce towels in a 4-pick or a 5-pick quality if appropriate detailed weaves were substituted for the 3-pick structures given at A and B in *Figure 14.10*.

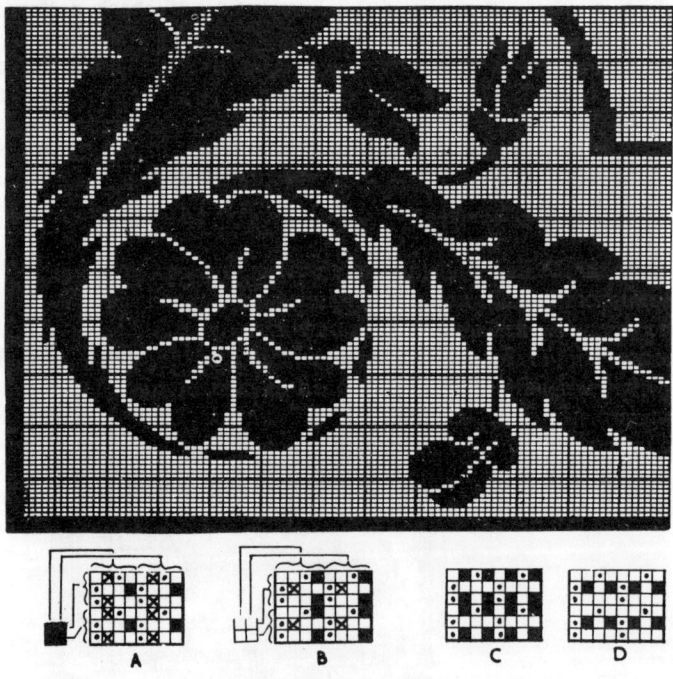

Figure 14.10

The system of designing, illustrated in *Figure 14.10*, is also suitable for the class of figured terry cloths in which there is only one series of pile threads.

In this case on one side a figure is formed in pile upon a ground of the foundation cloth, while on the other side the foundation forms in figure and the pile the ground. The principle is illustrated by the examples shown at A, B, and C in *Figure 14.5*. For the purpose of this structure one square of the condensed design represents a loop either on the face or on the back. If the paint is taken to indicate loops on the face, and the blank paper loops on the back, then the detailed weaves for the two different areas of the design will be, respectively, as shown at C and D in *Figure 14.10*.

Mixed colour effects

In a further development of the terry structure, which is applied to fancy towellings, beach wear, mats etc., white and two colours of pile warp are employed, and a design composed of four effects is produced. For instance, assuming that the pile threads are arranged 1 white, 1 pink, and 1 green, the ground may be formed in white pile loops and the figure by mixtures of pink and green, white and pink, and white and green loops in the different section of the design. There are really four series of pile threads in the cloth, two of which are on the surface and two on the back in every part.

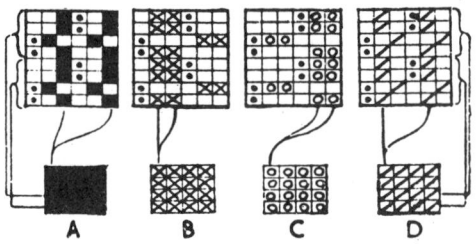

Figure 14.11

In the example given in *Figure 14.11* there are two pile ends to each ground end and the structure is a 4-pick terry in which the ground ends weave 2 up, 2 down. As there are four different effects the condensed design is painted in four colours represented by the different marks at A to D in *Figure 14.11*. The degree of condensation is by 6 warp-wise (4 pile and 2 ground ends) and by 4 weft-wise (2 loose, 2 fast picks) so that one square of the design equals two loops on the face and two on the back. The detailed weaves above the condensed design portions at A to D each correspond to one vertical and two horizontal rows of the designs.

The plan A represents the pink and green pile threads on the surface, and all the white pile threads on the back; B, white and pink on the surface and white and green on the back; C, white and green on the surface and white and pink on the back; and D, all the white on the surface and pink and green on the back.

Cut pile terry fabrics

Cut pile terry effects are sometimes produced by cropping, during a finishing operation, the tips of the loops in a terry cloth. Usually only one side of a

fabric is so treated the other retaining the normal loop formation. A very ornamental and rich appearance is thus created but the cloth is rendered less useful
for the purpose intended. Although the cropping does not reduce the absor-
bency it reduces the frictional characteristics of the fabric so that instead
of wiping the moisture off firmly the cloth tends to slide along a layer of
moisture without removing it efficiently.

15

Warp Pile Fabrics
Produced with the Aid of Wires

In this type of construction, frequently referred to as positive warp pile, only one kind of weft is required but at least two series of warp threads, separately beamed and tensioned, are essential, viz. ground ends and pile ends. The former produce with the weft the ground cloth from which the pile ends project and in which they are anchored.

To produce the pile a wire is inserted across the width of the warp into a shed formed only by the pile ends. When the pile ends are subsequently dropped into the bottom shed and interlaced with the weft they remain draped over the wires as shown at A and B in *Figure 15.1*. Thus, the cross-sectional dimensions of the wire determine the height of the pile. After the insertion of a number of picks (and wires) the wire furthest away from the cloth fell is withdrawn leaving the loops which were formed over its shank as a surface feature in the cloth as shown at C and D. The withdrawn wire is re-inserted at the front there being between 12 to 50 wires between the point of withdrawal and insertion. The special mechanism which controls the wire movement is designed to insert the wire rapidly, as fast as it takes to insert a pick of weft, and to withdraw it slowly. The large number of wires between the two points is necessary mainly to prevent the loops being pulled back by the tension on the pile yarn. The difference between the actual number of wires depends primarily on the weight of the fabric, fewer wires being required in lighter fabrics, and on the frictional characteristics of the pile warp.

The pile may be looped, if plain wires are used, or cut, if the wire has a cutting blade at its tip end. Both types are produced in exactly the same way, i.e. by draping the pile ends over the wires, and the difference in the nature of the pile is created only upon the withdrawal of the wire, the plain wire leaving in the cloth upon withdrawal the loops, whilst the cutting wire severs the loops formed upon its shank as it is withdrawn thus leaving the cut tufts in the cloth. This is shown schematically at E in *Figure 15.1*. The upper part of E and the cross-sectional shed diagram at F in *Figure 15.1* also show the normal shedding arrangements used in the manufacture of these fabrics. It will be noted that the wire is inserted into a special high shed formed by the pile yarn, simultaneously with the shuttle which inserts the weft into a low shed formed by the ground yarns. On occasions, to obtain special effects, it is necessary to insert the wire alone, but this is generally avoided as the wire insertion occupies the same length of time as the insertion of weft but by itself it does not add to the length of cloth woven because the take up is only operative during picking.

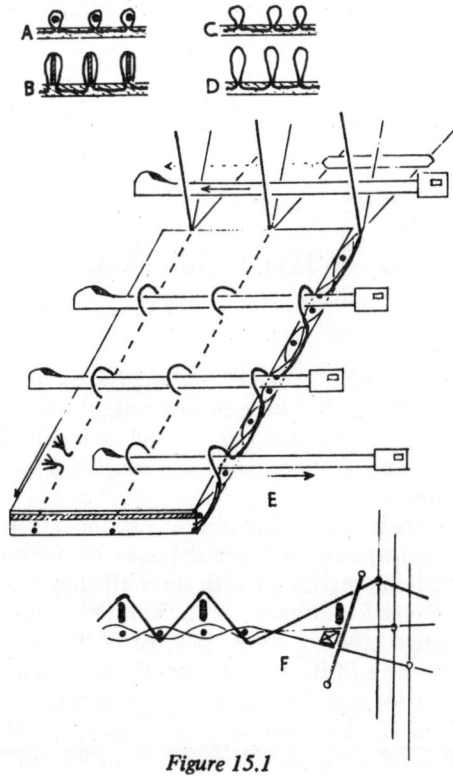

Figure 15.1

In dobby or jacquard shedding the high shed for wire insertion is obtained by special lifting arrangements. In cam shedding it can be achieved either by the use of tappets with a stroke bigger than that of the ground warp tappets or by suitable leverage connections to the heald shaft which controls the pile warp.

The wires vary in shape, as already mentioned, according to whether the pile is to be looped or cut. A and B in *Figure 15.2* represent the elevation and plan respectively of one form of a cutting wire whilst the cross-sectional appearance of it is given at C. It can be observed at B that to prevent damage to the reed the tip of the blade is turned slightly inwards. At present most of the cutting wires are made to operate with a disposable razor-type cutting edge which fits into a slot at the tip. This is done to save the time and the labour otherwise necessary for the frequent re-grinding and re-honing of the fixed blades. The appearance of a plain or looping wire is shown at D with two different cross-sections at E and F. The circular cross-section wire is only suitable for the production of short pile, long pile is produced on wires with a rectangular cross-section. The depth of the wires differs considerably and ranges from 1·5 mm for the short pile fabrics to as much as 25 mm for imitation fur fabrics and carpetings. As shown at A and D in *Figure 15.2* each wire has a shaped handle at its extremity by means of which the wire mechanism can insert and withdraw the wires. It will be appreciated that this mechanism requires at the side of the loom a space which is at least equal to the width of the cloth being woven. This adds considerably to the overall floor area required for each machine in the weaving of positive warp pile fabrics.

Figure 15.2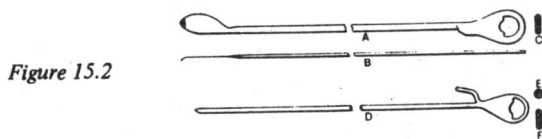

The loop pile fabrics in this system of weaving are produced mainly for upholstery purposes and are known as uncut moquettes, or for carpetings, which are termed Brussels, cord or boucle. The cut pile effects are used for apparel wear, curtainings and upholsteries and are known as velvets, plushes and cut moquettes, and also for carpets of the Wilton or velvet pile class. Very effective figured styles are also produced by combining in one cloth the loop with the cut pile. This form of figuring is used mainly in upholstery fabrics and in carpets.

The pile warp during weaving takes up much more rapidly than the ground warp, the difference in length varying according to the depth of the wires and the frequency in which the pile threads are raised over the wires. In an all-over pile structure the pile warp may require to be from five to twelve times the length of the ground warp. During weaving, in order that the pile face will not be injured, the temples act only on the selvedges, and in winding the cloth on to the cloth roller the underside is brought in contact with the friction roller. When the pile is long, however, the cloth is not wound on to a roller, but is passed directly into a box or other receptacle.

Apart from differentiating between the loop and cut pile and the different length of pile, the ratio of ground to pile ends and picks to wires may be varied to a considerable extent. Normally the ground weaves are very simple repeating on two, three or four picks but in some figured styles in which large amount of the ground is exposed more ornate ground constructions may be used. The wire to weft ratios are most commonly one wire to two, three or four picks. The fabrics can be grouped in three main classes depending on the surface effect formed:

(1) All-over or continuous pile effects.
(2) Figured effects with one series of pile threads which may consist of loop and cut pile figuring or pile and ground figuring.
(3) Figured constructions with up to five series of differentially coloured threads in which the ornament is chiefly due to colour.

ALL-OVER OR CONTINUOUS PILE STRUCTURES

The majority of cut pile effects produced for the apparel and upholstery fabrics in the all-over structures are at present made on the face-to-face principle (see Chapter 16); the constructions given in this section are representative of the effects still produced with the aid of wires.

All the pile over each wire

The term velvet is applied to the structures in which all the pile ends are raised over every wire as opposed to the term plush which refers to such pile effects in which alternate pile ends are raised over alternate wires, cut pile being produced in both instances.

A and B in *Figure 15.3* represent the weave and the weft cross-section respectively of a simple velvet structure with 2 picks to 1 wire and 2 ground, 1 pile ratio of the ends. At A the lifts of the ground warp are indicated by the dots and the lifts of the pile ends by the crosses where they are raised over the weft and by the solid marks where they are raised over the wire. It will be appreciated that when a pick of weft and a wire are introduced simultaneously, the former into a normal shed and the latter into a special high shed, as depicted at F in *Figure 15.1*, the high lift of the pile end results automatically in the end being over the pick of the weft. Nevertheless, a lift over the pick in question must be separately indicated on the design paper as otherwise the lifting instructions are liable to be misunderstood.

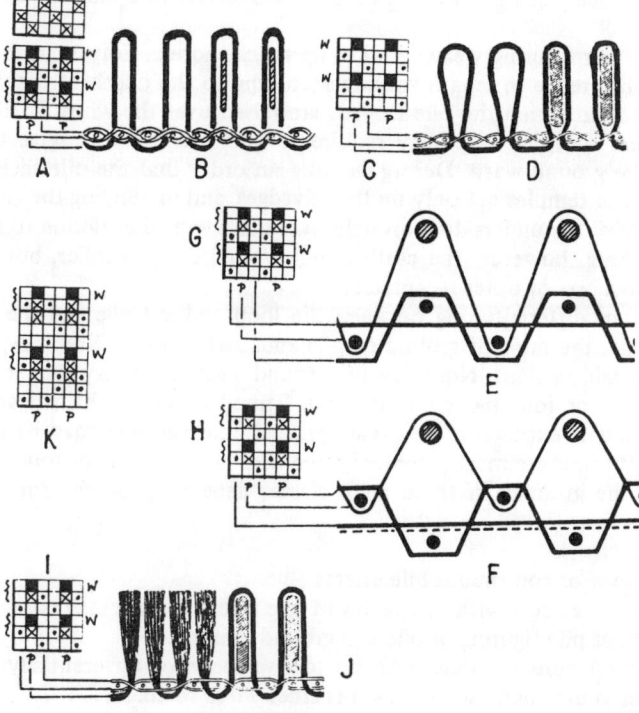

Figure 15.3

C and D in *Figure 15.3* show the construction which results when the wire is introduced on its own. Basically, there is very little difference between A and C; the ratio of picks to wires and ground to pile ends is identical in both cases, and so is the ground weave. Yet, the structure is quite different, as will be noted from the comparison of B with D. In the former the pile end is anchored under one pick of weft, in the latter it floats under two successive picks. Due to the longer back float construction D is inferior in respect of tuft anchorage there being a greater likelihood of tuft dislodgment by rubbing on the back of the cloth. As this construction is also slower to produce, the take up not being operative upon the insertion of wire alone, most plain ground velvets are produced as shown at A and B.

In velvets for furnishing fabrics the plain weave ground is also often used but the resultant construction is different from those depicted at B and D because the alternate ground ends which make the plain weave are taken from separate beams of which one is heavily tensioned and the other comparatively slack. A form of rib is thus produced in which the picks lie alternately above and below taut ends, as shown at E and F in *Figure 15.3,* whilst the slack warp is crimped to a considerable extent being made to operate over the top and under the bottom picks. In this manner a more secure pile anchorage is formed, especially in the construction E in which the pile yarn is woven through to the back. Apart from velvets both constructions are also used for uncut moquette self-colour upholstery fabrics for which purpose the cloths are often made heavier by additional warp stuffing yarns which form the same sheds as the taut ground yarns. The position of the stuffers in the structure is indicated by the dotted line at F which is more often employed for the uncut or loop pile effect than E. Although E results in a superior pile anchorage it is costlier to produce because greater length of pile yarn is required for the same surface depth of pile. The weaves for the two constructions are shown at G and H alongside the respective cross-sectional diagrams.

Weaves other than plain are sometimes also used for the ground and I and J show the design and the section respectively of a cloth with a 2-and-2 rib ground structure. In this case a higher warp sett can be used and the ends are often arranged in a 2 ground to 1 pile ratio. K in *Figure 15.3* shows a construction in which a 2-and-2 hopsack ground is employed with 2 ground: 1 pile ratio of ends and with 4 picks to 1 wire.

Fast pile anchorage.

The ordinary 'U' binding of the tuft shown in the previous constructions is adequate for most purposes especially when short, dense pile is produced. This type of binding may be further improved by using the alternate tight and slack ends as demonstrated at E and F in *Figure 15.3.* However, in some circumstances when the cloth is expected to be subjected to a degree of rubbing and particularly when long pile is produced a superior 'W' binding is used in which each pile is additionally interlaced with the weft between the wires. A plain or similar tight interlacing is used to anchor each tuft firmly in the ground structure so that it cannot be easily pulled out.

A typical fast pile structure based on a 2-and-1 rib ground weave is given at A and B in *Figure 15.4.* It will be noted from the cross-section B that between each wire shed the pile end is bound in the ground in a down-up-down order providing a very secure anchorage. Using the weave A the wire is inserted into a separate shed without simultaneous insertion of the weft, therefore, to simplify the weaving arrangements a 3-and-1 rib may be used as shown at C and D without in any way weakening the binding of the pile yarns. A is woven with three, and C with four picks to one wire, whilst the ratio of ground to pile ends is in both cases 2:1.

Alternate pile ends over alternate wires.

These constructions are particularly suitable for the upholstery plushes because fast pile binding can be easily arranged and the tufts, not being regimented in horizontal rows, are staggered and provide a more uniform pile cover. A construction of this type is shown at E and F in *Figure 15.4* which has a 2-and-2 rib ground arranged 1 ground, 1 pile. There are two picks to one wire and this, in effect, means that there are four picks between the wires over which each pile thread is raised. A fast pile bind is obtained as indicated clearly in the cross-section F. A similar fast pile plush effect can also be produced on a 2-and-1

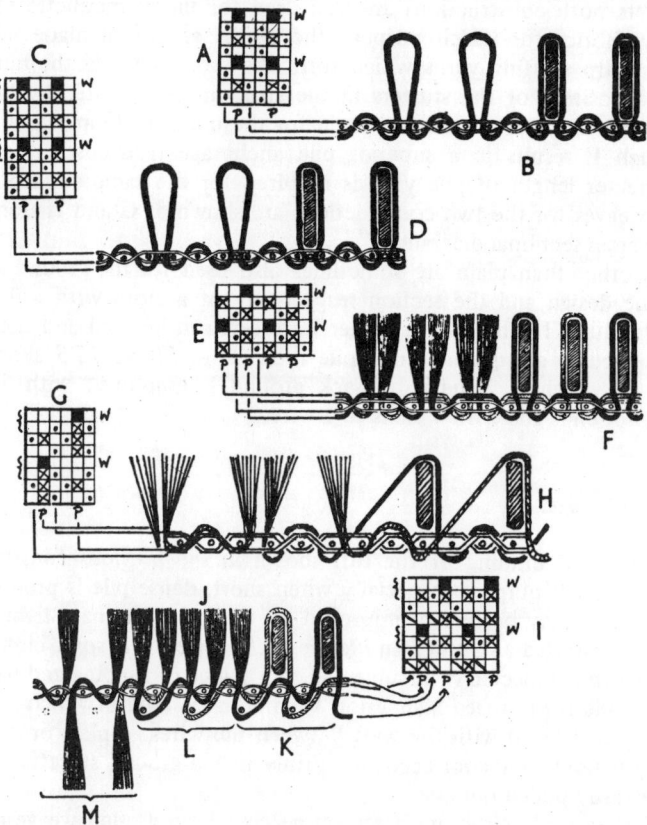

Figure 15.4

rib base with 3 picks to 1 wire and 2 ground to 1 pile arrangement of ends. This is shown at G and H and has the added merit of more convenient weaving arrangements due to the simultaneous insertion of wire and weft. In both structures the pile warp may be brought from the same beam, but the weaving is facilitated by using separate slackening of easer bars for the odd and even pile ends.

At one time plush structures were also produced upon warp-backed or double ground cloths. This was useful in overcoating fabrics where raised finish on the back was desired. The raising operation in such cloths could be carried out without any fear of pulling out the tufts from the back as the pile anchorage points were fully protected by another continuous yarn or cloth layer. Such constructions, however, are at present produced only very infrequently.

Reversible warp pile structures.

Warp pile cloths that are to be used as hangings are sometimes made with a cut pile on both sides. A method of accomplishing this is illustrated at I, in *Figure 15.4* and the corresponding section given at J, which shows how the warp threads interlace. Half the pile is over each wire, and in this case, after the insertion of a wire, an extra pick is introduced on which all the ground ends are raised (as shown by the diagonal marks in I), and alternate pile ends. In the section J the extra picks are indicated below the level of the plain ground threads, while one pile thread is shown shaded and the other solid in order that the system of interlacing may be readily seen. Exact repeats of the weave are indicated by the brackets, the portion lettered K representing the cloth previous to, and L following, the withdrawal of the wires. After the cloth is woven, the extra picks, which are usually thicker than the ground picks, are drawn away from the underside of the cloth; this causes one-half of each double tuft to pass to the reverse side, as shown in the section M, where the dotted circles indicate the positions which the extra picks previously occupied.

Settings of warp pile fabrics.

As mentioned earlier the density of thread spacing, the thickness of the yarns and height of pile in this group of structures can be varied to a considerable extent depending on the end uses of the cloths. In the following several typical qualities are given for a variety of purposes. Other settings are given in further parts of this chapter connected with certain specific uses of some of the structures dealt with.

(1) Dress velvets: Warp setting—10 to 16 pile ends per cm the number of ground ends per cm depending on the ratio of pile to ground ends; ground warp—20/2 tex to 32/2 tex cotton; pile warp—10 to 20 tex single or two-fold in a variety of materials such as mercerised cotton, filament rayon, synthetic yarns and occasionally spun silk; weft settings— 6 to 12 wires per cm the number of picks per cm depending on the ratio of picks to wires; weft yarn counts—20/2 tex to 32/2 tex cotton; pile height—1·5 to 3mm.

(2) Upholstery plushes: Warp setting—10 to 12 pile ends per cm; ground warp—60/2 tex cotton or staple viscose rayon; pile warp—50/2 tex to 72/2 tex worsted or equivalent counts in polyamide, acrylic or polyproplene yarns; weft settings—8 to 12 wires per cm; weft counts 60 tex two-fold or single cotton or staple viscose rayon; pile height 2 to 5 mm.

(3) Upholstery uncut moquette: Warp setting—13 pile and 13 ground ends per cm; ground and pile warp—74/2 tex cotton; weft settings—6 wires and 12 picks per cm; weft counts—120 tex cotton or staple viscose rayon; loop height—1·5 to 2 mm; construction as at F, *Figure 15.3* (without stuffer yarns).

Ornamentation of all-over warp pile fabrics.

Continuous warp pile structures can be ornamented in a variety of ways during manufacture or during finishing. Stripe effects can be readily achieved by introducing differently coloured pile threads. Horizontal bars, which add considerable interest to the appearance of the cloth, can be obtained by employing both cutting and looping wires which may be arranged in different sequences, such as eight of each, or ten cutting followed by six looping wires, and so on, as shown schematically at A and B in *Figure 15.5*. Stepped or waved pile effect can be

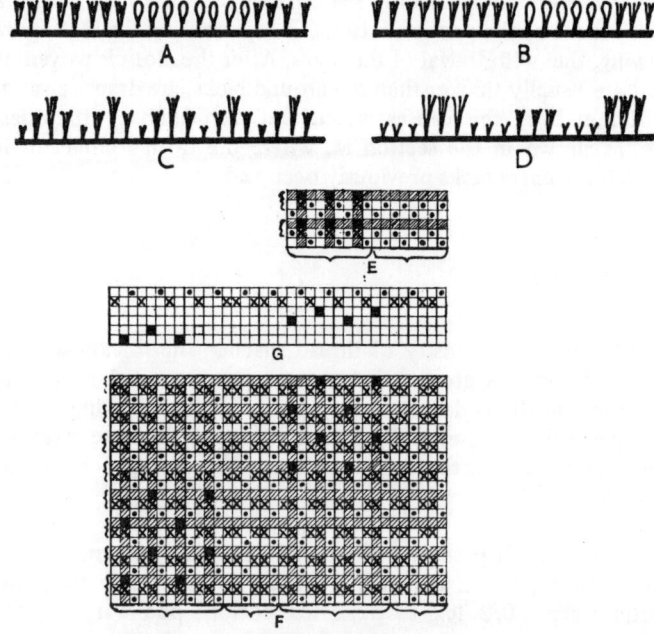

Figure 15.5

created by using different heights of wire in varying arrangements as indicated in the schematic diagrams at C and D in *Figure 15.5*. Other modifications may include a combination of small diameter looping wires with tall cutting wires and so on. Specially prepared and constructed pile yarns may be used which upon cutting are capable of curling tightly in various directions to produce an imitation Astrakhan fur surface.

Figure 15.6

Most of the ornamentation of the continuous pile fabric is at present carried out during finishing. Varied surface effects can be achieved by subjecting a cloth to the pressure of embossing rollers. Frequently, heat is utilised in conjunction with various thermoplastic yarns. Incorporation in a construction of synthetic yarns with different thermal properties followed by an application of heat results in many excellent effects due to the differential heat shrinkage. The cloth given in *Figure 15.6* represents one example of an imitation fur achieved by the above technique but there are numerous other possibilities of obtaining spectacular surface textures by utilising the knowledge of the cloth structure in combination with an appreciation of the behaviour of the different materials in varying conditions of heat, moisture or susceptibility to chemical agents.

Stripe and check effects.

A form of ornamentation obtained without having recourse to a jacquard machine consists of combining the pile structure with other forms of interlacing in stripe and check form. The design E in *Figure 15.5* is an illustration of a stripe composed of pile and 2-and-2 warp rib. The pile interlacing corresponds with I and J in *Figure 15.3*, while the rib stripe is simply a continuation of the weave of the ground threads, and can, therefore, be produced by the same healds. Different widths of stripes can be obtained by repeating the sections enclosed by brackets.

The design F illustrated the formation of alternate squares of pile on a warp rib ground, with longitudinal spaces of ground between, which may be coloured differently from the pile sections. There are two picks to each wire, and the warp is arranged 1 worsted ground, 1 worsted pile, and 1 cotton ground; the worsted threads work in pairs except on the wire sheds, in order to develop the rib formation. The rib structure is made more pronounced by weaving the cotton ground threads at greater tension than the worsted ground threads, and by wefting one pick fine, one pick coarse. Fine warp beams are necessary— viz. two for the pile threads and one each for the worsted and cotton ground threads, while six shafts are required, as shown in the draft indicated at G. The pile ends and the wires in both, E and F, are indicated by the shaded lines.

Continuous pile carpet structures.

Warp pile carpets produced with the aid of wires are classified in two main groups, viz. Brussels, in which a loop pile is produced, and Wilton in which the surface consists of cut pile. The ornate, multi-colour carpets are described in a subsequent section of this chapter but the technique of weaving the all-over or continuous pile carpets of this group is, apart from the weight of the cloth, similar to that already described in connection with the velvets and plushes.

Self-colour Brussels carpets, also known as cord or boucle, are woven in two-shot structures, i.e. 2 picks to 1 wire, using plain wires. Most of the self-colour Wilton or velvet carpets are also woven in identical two-shot structures only with cutting wires. Some of the better qualities of velvet pile carpets are, however, woven in three shot, i.e. 3 picks to 1 wire, structures which, although slower to produce, offer the advantage of superior pile anchorage.

A standard self-colour two-shot structure is shown at A and B in *Figure 15.7*. The weft cross-section at A shows clearly that the construction, in addition to the ground or chain ends and the pile ends, also embodies stuffer warp yarns, the stuffer being a heavy and stiff jute yarn which separates the weft into two layers. Each vertical row of pile encompassed within a reed dent contains one pile end, two chain (ground) ends and two or three stuffers. The weave repeats over four picks. Usually, the construction is woven in three heald shafts. The front shaft, which has a greater movement than the other ones, carries the pile and the stuffer ends in special mail eyes shown at C in *Figure 15.7*. On pick 1 of the structure, when only the weft is inserted, the front heald is down lowering the pile ends and the stuffers to the bottom shed line. The second chain heald shaft is also down so that the top shed line is formed only by the odd chain ends as depicted at D. On the following pick of the sequence the weft is inserted simultaneously with the wire. It will be noted from the diagram E in *Figure 15.7* that the pile ends are raised high to form an upper shed for the insertion of the wire by virtue of being controlled by the eye in the special mail whilst the stuffers being operated by the slot are raised to only half that height. They thus form together with the odd chain ends the bottom line of the upper shed and the top line of the lower shed into which the weft is inserted. The manipulations of the front heald repeat over two picks, therefore, the positions of this shaft on picks 3 and 4 are the same as on picks 1 and 2 respectively. The chain ends, however, operate in a 2 up, 2 down sequence and whilst on the first two picks the odd chains are up and the even ones are down on picks 3 and 4 these positions are reversed. The lifts of the healds can be easily traced in the full weave given at B in *Figure 15.7* in which the appropriate ends are connected by lines to their counterpart in the cross-section A. The high lift of the pile ends on the even picks is denoted by the solid squares at B whilst the low lift of the stuffer ends is indicated by the double vertical lines. The operation of the chains is shown by the circles in B. It will be appreciated that exactly the same procedure will apply when the two-shot cut or velvet pile carpeting is produced the only difference in the manufacture between the two being in the type of wire used. Occasionally the same type of structure as described above is produced on four healds, separate shafts being used for the pile and the stuffer ends.

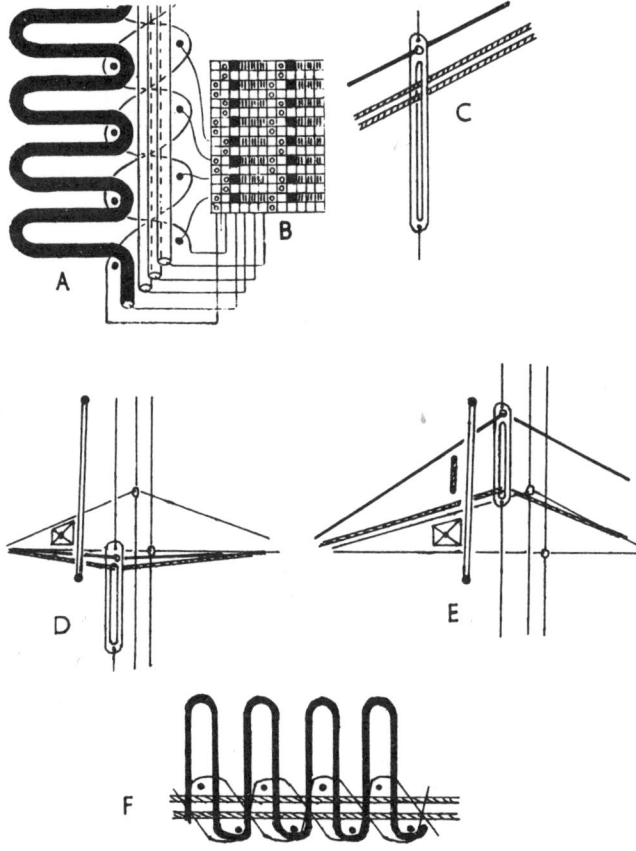

Figure 15.7

In areas of high wear such as hotels, public institutions, offices, etc., the standard self-colour Brussels structure provides insufficiently good pile anchorage. The loops are liable to be pulled and are said to 'sprout' when, due to pulling, a loop of an enormous length is created by robbing the neighbouring loops in the same row of all their length. To prevent this the loops are anchored around the back pick by weaving them through to the back as indicated at F in *Figure 15.7*. The construction is known as the 'wool back' construction because the pile yarn is clearly visible on the back of the carpet. The basic difference in the weaving is that the wire in this system is inserted together with pick 1, and not pick 2 as in the standard structure, and four healds must be used because the lifts of the stuffers do not synchronise with the lifts of the pile ends as they did previously. Due to the method of anchorage greater length of pile yarn is required to produce the same surface height of pile as in the standard structure.

Most carpets are produced in a standard pitch, i.e. with a set number of pile ends per unit space and the pitch is rarely changed, therefore, the quality is varied by changes of the pile yarn quality, the number of wires per unit space and the height of wires. A good quality self-colour loop pile carpet may be produced with 32 pile ends per 10 cm, using 350/2/3 tex pile yarn (80 per cent wool, 20 per cent polyamide fibre), 220/3 tex cotton chain and 480 tex jute stuffer with 36 wires per 10 cm and 280 tex jute or hemp weft; about 250 to 260 m of pile yarn being required to produce 100 m of carpet. Lower qualities may be produced with only 26 wires of lower height per 10 cm.

Many different materials are used for the pile—in addition to pure wool worsted or woollen spun yarns, wool is used in blended yarns, frequently with an admixture of polyamide fibres, and purely synthetic materials such as the acrylics, polyamides and polypropylene are also employed as well as staple viscose rayon yarns.

The two-shot Brussels construction was at one time ornamented by yarn printing to produce definite pattern effects. Carpets of that type were known as tapestry carpets and a section through this type of effect is shown at A in *Figure 15.8*. It will be noted that the construction of this type of carpet is identical with the standard structure given at A in *Figure 15.7* except that the pile yarn is differently coloured in small portions along its length. Each pile thread was coloured in a developing design sequence so that when they were all beamed together a complete colour design would result. Due to the vagaries of yarn movement during weaving these designs rarely achieved absolutely perfect registration and the outlines of the figures were usually somewhat hazy in the lengthwise direction. Also, each loop instead of being uniformly coloured would occasionally consist of half of one colour and half of another. Despite these faults the designs had their specific charm and were popular as cheap substitutes for the true multi-coloured Brussels constructions (see p. 307). However, the printing and beaming of these yarns was so time and labour consuming that the gains achieved by weaving what was virtually a continuous pile effect in preference to jacquard ornamented effects were eventually insufficient. For this reason the yarn printed tapestry carpets are no longer produced.

Although many cut pile effects are produced in the two-shot weave by the use of cutting wires in very similar settings to those employed for the loop pile carpets the three-shot structure is not equally interchangeable. It is used only for the finer qualities of Wilton or velvet pile carpets where improved tuft anchorage is desired. The construction is given at B and C in *Figure 15.8* from which it will be seen that there are 3 picks to 1 wire and that the chain warp ends operate in a 3 up, 3 down order. The stuffers lie in the middle of the construction separating the weft into two layers as before. The wire is introduced into the upper shed simultaneous with the second pick in each group of three the double shed technique being identical with the one used in the two-shot structure. Similar to the latter it can be produced with either three or four heald frames both drafts being shown above C. The first draft is, of course, only applicable when the front heald is equipped with the special eyed and slotted mails.

The velvet pile carpets are often woven with 32 to 38 pile ends per 10 cm using pile yarns ranging between the fine 350/2/3 tex worsted and the coarser

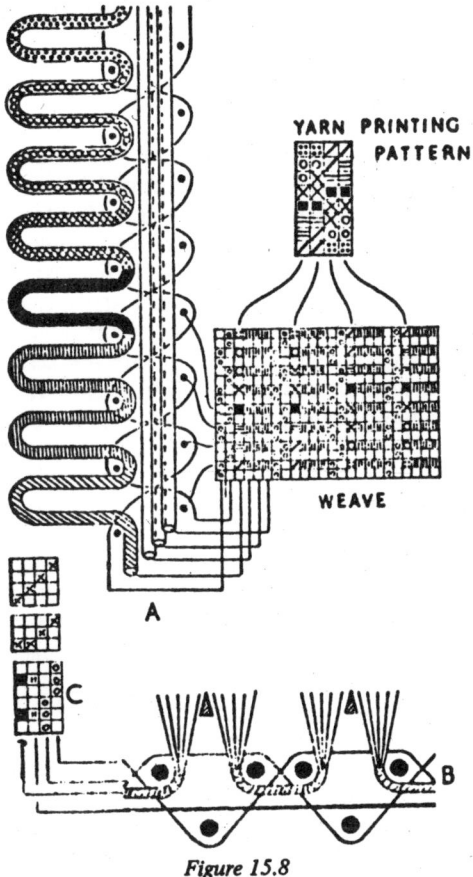

Figure 15.8

620/2 tex woollen spun qualities. The number of wires per 10 cm varies between 32 and 40. The pile height is generally greater than in the loop pile carpeting and may be as high as 12 or 15 mm but the commonest varieties are produced with the pile height of 5 mm.

The continuous pile Wiltons made in the two-shot construction are sometimes ornamented by stripe designs, occasionally described as 'candy stripe' Wiltons. These are frequently the result of utilising large and varied pile yarn remnants and are sold more cheaply than similar self-colour qualities being produced in comparatively short runs.

FIGURING WITH ONE SERIES OF PILE THREADS

Small quantities of high quality furnishing fabrics are produced in which the pile warp is jacquard controlled to produce figured effects. Although only one series of pile yarns is used several different groups of structures can be made which may be classified as follows:

(1) Pile figure on ordinary weave ground; several sub-types exist in this group as the pile may be cut or uncut and different ground weaves may be used.
(2) Loop pile figure on cut pile ground.
(3) Loop and cut pile figure on ordinary weave ground.

As the above fabrics are invariably used in furnishings the ground ends are normally taken from two differently tensioned beams to produce the alternate tight and slack end ground construction already described with reference to A and B in *Figure 15.4*. In most instances the ground ends are controlled by two heald shafts mounted at the back of the harness so that the jacquard controls only the pile ends. Often, the jacquard arrangement is such that the knives provide the high lift for the insertion of the wire whilst the comberboard is capable of lifting the pile ends which are not selected to go over the wires into the middle shed as indicated at B in *Figure 15.9*. Where loop and cut pile effects are produced looping and cutting wires are inserted alternately.

Pile figure on ordinary weave ground.

As shown in the fabric illustrated in *Figure 15.10* the pile threads in these styles are not forming pile continuously and the amount of take up of the pile ends may vary considerably. It is, therefore, necessary for each pile end in the repeat of a design to be 'beamed' on a separate bobbin; each bobbin may carry as many ends as there are repeats in the width of the fabric but in a large single repeat design each bobbin can carry only one end. The bobbins are mounted in a creel at the rear of the weaving machine and are individually tensioned.

The alternate ground ends are placed on different beams of which one is heavily tensioned and the other lightly weighted. The ground weave may be varied according to requirements but most frequently the ground ends are operated in the plain, or in the 2-and-2 rib weaves. The appearance of the ground cloth in these constructions is important because portions of it are exposed to view and often the ribbed effect is emphasised deliberately by using considerable density of warp setting and fine warp yarns as opposed to low shottings and coarse weft yarns. The healds which control the ground ends are usually operated from a tappet assembly and are placed at the back of the harness as shown by the comber-board diagram at C in *Figure 15.9* and by the operational diagram at A and B. In most constructions one wire is inserted to two picks.

The jacquard is normally of the single-lift type, it controls only the pile warp and is arranged to lift the selected pile ends into a high shed for wire insertion. The sequence of operation is shown at A and B in *Figure 15.9* and may be described as follows. During the insertion of pick 1 of the construction—diagram A—a card is presented to the needles and a selection takes place. On pick 2—diagram B—the knives lift the previously selected hooks and, therefore, pile ends to form the top line of the high shed into which a wire is inserted simultaneously with the pick of weft which is introduced into the lower shed. The unselected pile ends are taken on the same pick to the top of the lower shed by the working comber-board which is normally controlled by a positive tappet. The comber-board is capable of lifting the harness because the harness cords are knotted

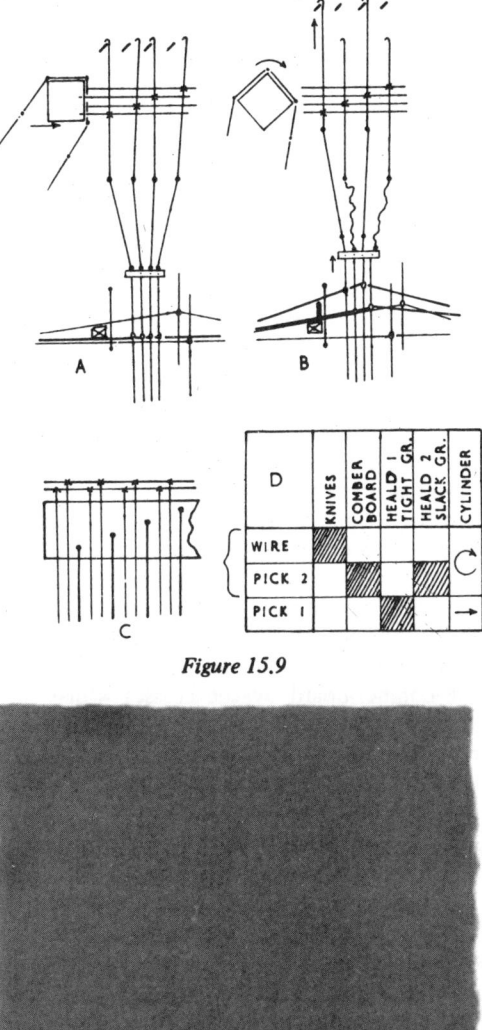

Figure 15.9

above the holes in the board and cannot slip through them when the lift takes place. The selected pile harness cords are not affected by the comber-board lift because they are lifted higher by the knives as shown at B. Some manufacturers prefer to use jacquards in which the function of the comber-board is performed by a bottom-board; i.e. the board upon which the hook bottoms rest in the jacquard engine. This has the advantage of obviating the slackening of the harness cords which is unavoidable when a working comber-board is used as clearly shown at B in *Figure 15.9*. The single-lift jacquard does not in this case represent a speed limit factor as it does in normal applications because, as may be noted from A and B, it operates at half the speed of the weaving machine. The selection takes place during the first pick of the sequence, the lifting action during the second pick at which time the cylinder is turned to present the new card for the following pick. At any rate, in all these structures it is usually the wire motion which represents the main impediment towards higher operating speeds. The lifting sequence of the various shed forming elements in this system is given at D in *Figure 15.9* in which the lifts are indicated by the shaded squares. The same diagram also relates the operation of the jacquard cylinder to the picking sequence showing the selection on pick 1 and the turning of the cylinder on pick 2/wire.

The resultant construction is given in the section at A in *Figure 15.11* in which the pile figure may be looped or cut depending on the type of wire used. It will be noted that when the pile ends are not selected to lift over the wires they operate by virtue of the working comber-board lifts in the same order as the slack ground ends thereby adding to the ribbed appearance of the ground already emphasised by a suitable warp to weft ratio and the alternate slack and tight ends. The method of painting is represented by the small portion of design shown at B which is related to the fabric in *Figure 15.10*. In the design each horizontal row represents a wire and each vertical row a pile end and, therefore, the marks in the design are cut as they stand as they represent the only lifts for which the jacquard itself is responsible. The full construction for the painted and the blank portions of B which results from the combined lifts of the jacquard hooks, the working comber-board and the healds is given respectively at C and D. The lifts of the selected hooks are denoted by the solid marks, those of the comber-board by the crosses whilst the operation of the healds is indicated by the dots. The ratios of 2 ground ends to 1 pile and 2 picks to 1 wire represent the usual arrangements in this type of structure. E and F in *Figure 15.11* correspond respectively to C and D and show the same design developed in the alternative 2-and-2 rib ground structure. The change in

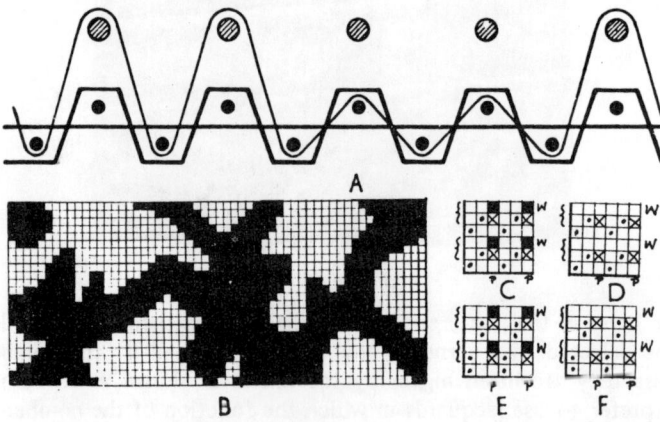

Figure 15.11

the ground structure does not affect the operation sequence for the jacquard and the comber-board, the only change occurs in the order of manipulation of the ground healds.

The form of heald and harness mounting described above and the system of jacquard operation are suitable only when simple ground weaves are used. Ornate, figured weaves in the exposed ground areas usually demand jacquard control of all the ends, ground as well as pile, and jacquard selection on every pick and not only on the alternate picks. Such arrangements are more often employed when more than one series of pile threads is used in the structure and are described in a subsequent section of this chapter.

Loop pile figure on cut pile ground.

This type of structure, illustrated in *Figure 15.12*, is used mainly for upholstery purposes. The ground structure may be plain or 2-and-2 rib with alternate tight and slack ground ends controlled, as previously, by two healds placed at the back of the jacquard harness. The jacquard arrangement may be exactly the same as described with reference to *Figure 15.9*. The construction is achieved by the insertion of alternate looping and cutting wires so that one structural unit in the weft-wise direction consists of four picks and two wires. Within this unit a given pile end will be raised only once—either on the looping or on the cutting wire, but not on both. As every pile end is raised on one of the two wires which are all the same in height the take-up of all pile ends is identical and they can be beamed on the same beam without the need for bobbins and creels necessary in the construction illustrated in *Figure 15.10*.

Figure 15.12

The cross-sectional appearance of the construction is given at A in *Figure 15.13* with each structural unit being separated from its neighbour by the dotted lines. The looping wires, represented by the oval shapes, are inserted together with pick 2 of each unit whilst the cutting wires, indicated by the triangular shapes, are inserted simultaneously with pick 4 of each structural unit. The working comber-board is also raised on picks 2 and 4, therefore, any pile ends not selected to lift over a given wire will be nevertheless raised on the pick which accompanies that wire. This is clearly shown in the section. Cut pile, being the richer in appearance, occupies the greater proportion of the repeat area. In preparing the design it is usual to paint the loop pile and at B a small portion of design is shown in which the loop pile figure is represented by the solid marks the blanks indicating the cut pile. Each horizontal row, therefore, represents a looping and a cutting wire and each vertical row corresponds to one pile end. Two cards are cut from each horizontal row and the cutting instructions can be formulated as follows:

Looping wire   —   cut all marks
Cutting wire   —   cut all blanks

The two bottom horizontal rows are expanded at C in *Figure 15.13* to show the full construction which results from the co-ordinated action of the jacquard selection mechanism, the comber-board, and the two healds. The lifts of the hooks on the looping wires are indicated by the circles, on the cutting wires by the solid marks whilst the comber-board lifts are represented by the crosses and the heald lifts by the dots.

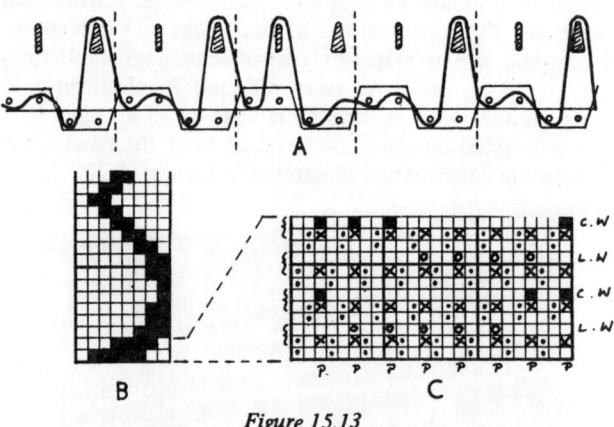

Figure 15.13

The quality of these fabrics varies considerably and whilst the ground yarns are usually cotton or staple viscose rayon the pile yarns may be worsted, acrylic, polyamide, or polypropylene. A good quality upholstery fabric may be constructed as follows: Pile warp—6 double ends (two ends weaving as one in decked harness mails) of 60/2 tex worsted per cm; ground warp—12 ends (6 slack, 6 tight) of 74/2 tex cotton per cm; weft—24 picks of 40 tex cotton per cm; 12 wires (6 looping, 6 cutting) per cm, each 3mm high; density—36 double loops or tufts per cm^2.

Loop and cut pile effects on ordinary weave ground.

This construction, which is used for upholstery purposes, may be regarded as a combination of the previous two effects. The shedding and the wire insertion order may be exactly as described for the loop and cut pile figuring but in addition selected pile ends are occasionally left down on both wires to produce sunk ground effects. The cross-sectional view of the structure is represented at A in *Figure 15.14*. At B a small portion of design is shown in which the blank spaces correspond to cut pile, the solid squares to loop pile, and the crosses to sunk ground. As in the design B in *Figure 15.13* each horizontal row corresponds to two wires and each vertical row to one pile end. The cut pile occupies usually the greatest area in the repeat being the richest in appearance and to save time and labour it is used to paint the loop pile and the sunk ground areas leaving the cut pile portions blank. The card-cutting instructions, using the harness mounting previously described, are as follows:

> Looping wire — cut all solid marks
> Cutting wire — cut all blanks

The two bottom horizontal rows are expanded at C to show the full construction in which the lifts of the pile ends on the cutting wires are indicated by the solid marks, on the looping wires by the circles whilst the comber-board lifts are designated by the crosses and the ground heald lifts by the dots.

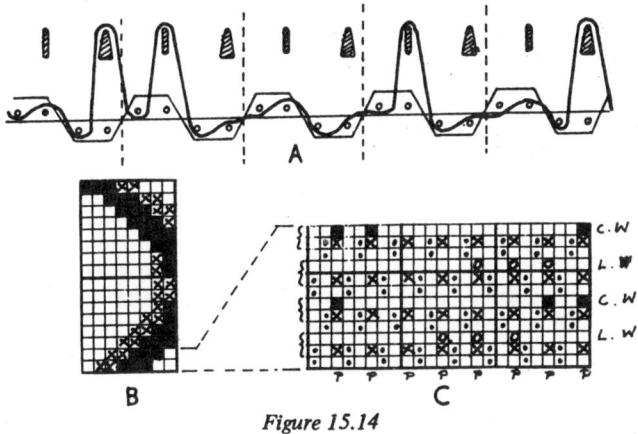

Figure 15.14

The operational system described above represents one of a number of methods in use. Other systems with different shedding arrangements also exist but all are designed to result in similar structural effects. Fancier constructions with two different heights of pile and greater structural variety are also sometimes produced but usually the more elaborate effects are made with more than one series of pile threads and are, therefore, described in the following section of this chapter.

FIGURING WITH SEVERAL SERIES OF PILE THREADS

In this form of figuring the effect is due primarily to colour. As opposed to previously described constructions in which one longitudinal row of loops or tufts consisted of one pile end, in multi-coloured arrangements one row of pile contains from two to five differently coloured pile ends. Only one end is raised over any given wire whilst the remaining pile ends at that point lie 'dead', simply awaiting the occasion when they will be required to form the surface effect in further portions of the design. Each vertical row of pile thus consists of a group of threads which are technically referred to as 'frames' and the cloths are classified as two, three, four or five frame constructions depending on the number of pile threads per group. The term frame is applied because each pile is on a separate package—bobbin or cheese—placed in a creel or 'frame' at the back of the loom. Separate packages are necessary because each pile thread may have a vastly different take up from any other thread within a group and between the groups. Each group will contain one thread from each creel or frame and the total number of groups used is equal to the number of loops or tufts in a horizontal row of pile in the cloth width.

The greater the number of pile threads in a group the greater is the consumption of the pile yarns and, therefore, the cost of the fabric. The benefits of the increased number of pile ends per group are in improved design and colour scope and better resiliency combined with higher bulk of the cloth. The quantity of pile yarn used is not in direct proportion to the number of the frames because no matter how many frames, i.e. pile ends per group, are employed only one thread of a group is raised over a wire. For example, assuming that 300 m of continuous pile are required to produce 100 m of cloth and 104 m of stuffer yarn (the dead pile assumes the same configuration as the stuffer being raised and lowered together with it) each group of threads in a 5-frame structure will require $(1 \times 300) + (4 \times 104) = 716$ m of pile yarn, whereas a 3-frame pattern will require $(1 \times 300) + (2 \times 104) = 508$ m of yarn.

The multi-frame constructions are employed in the manufacture of loop pile effects for upholstery work, usually referred to as uncut moquettes, and for the Brussels carpets. Cut pile effects are also produced for both the above purposes, the carpets being known as Wilton carpets and the upholstery fabrics as moquettes. In addition, mixed effects are also made in which cut and loop pile may be combined with sunk ground areas and the structural diversity may be further extended by producing effects with high and low pile due to different heights of wire. The upholstery cloths, obviously, differ from the carpet structures in the weight, degree of rigidity or pliability, and the pile height, the difference being mainly due to the thickness of constituent yarns. Some structural differences between the two groups of constructions also exist and these are detailed subsequently but the method of designing is common to both. It must not be understood that the design styles or sizes are similar between the carpet and the upholstery structures. It is only the procedures of design preparation which are the same. The jacquard mountings are also similar in principle although the machines may vary considerably in weight or size. Multiframe carpets are produced in a variety of settings the finest containing as many as 40 groups of threads and 52 wires per 10 cm, and the coarsest, 20 groups of pile threads and 28 wires per 10 cm. The upholstery cloths are made in finer settings ranging from 72 to 56 groups of threads and wires per 10 cm.

Standard multi-frame structures.

A two-shot structure with four differently coloured pile ends in a longitudinal row (4-frame) suitable for Brussels or Wilton carpets is depicted at C in *Figure 15.15*. The section shows one vertical row of pile and the yarns are appropriately connected by lines to the corresponding rows of the fully worked-out weave at B in which the circles denote the lifts of the chains (or ground ends), the double vertical strokes the stuffers, the solid marks the high jacquard lifts of the selected

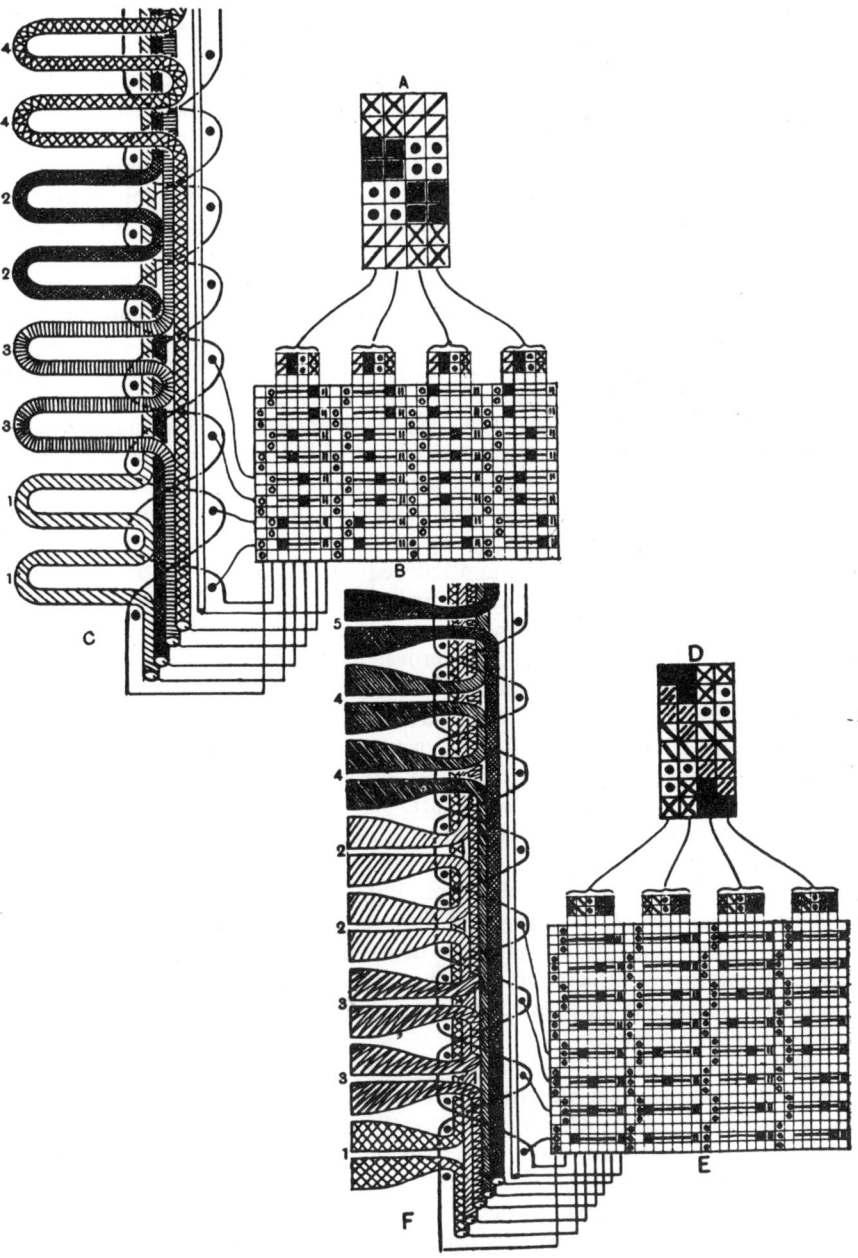

Figure 15.15

pile over the wires, and horizontal lines the comber-board lifts of the unselected or dead pile yarns. The wire is inserted simultaneously with pick 2. The simplified design A indicates where a given colour of pile is raised upon a wire and is condensed by 7 warp-wise (4 pile ends, 2 chains, 1 stuffer) and by 2 weft-wise. Each group of pile ends with the attendant chains and stuffer occupies one dent in the reed. Suitable particulars for a high quality loop pile carpet in the two-shot structure as shown at C are as follows: Pile warp 340/2/3 tex worsted, chain warp 220/3 tex cotton, stuffer warp 420 tex jute, weft 280 tex linen; 36 groups of pile threads and 36 wires per 10 cm resulting in 13 loops per cm². A similar high quality cut pile carpet in a two-shot structure could be similarly set although slightly finer pile yarns combined with a slightly greater density of settings and a higher pile are also frequently made. Lower quality Wiltons with coarser woollen spun yarns are produced often with as few as 10 tufts per cm².

A three-shot Wilton carpet construction using five pile ends in each vertical row is given at F in *Figure 15.15*. The simplified design at D and the fully worked-out weave at E are similarly arranged to A and B only in this case the degree of condensation in the design is by 8 warp-wise and by 3 weft-wise and each vertical row of pile consists of five differently coloured threads eight ends in all occupying one reed dent. The chain ends operate in a 3 up, 3 down order and the wire is introduced simultaneously with the second pick in each group of three. The three-shot structure is used only for cut pile constructions and offers superior pile anchorage the tuft being held under two picks of weft each of which belongs to a different group of three as can be seen at F. It is an expensive cloth to produce due to a reduced rate of production one horizontal row being made in three picks as opposed to one row in only two picks in the two-shot structure. Consequently, it is only made for situations in which the higher degree of wear demands the improved binding of the pile as, for example, in stair carpeting, and for hotels, stores and other public institutions. A good quality three-shot Wilton may be produced in accordance with the following specifications: Pile warp 300/2/3 tex worsted, chain warp 220/3 tex cotton, stuffer warp 560 tex jute, weft 280 tex linen; 40 groups of pile threads and 40 wires per 10 cm resulting in 16 tufts per cm². On occasions 2-frame, or 3-frame Wilton structures are produced which contain two or three permanently dead pile yarns per vertical row in addition to the two or three which actually participate in the formation of figure. For this purpose pile yarn remnants are used which would otherwise be disposed of as waste. Although the permanently dead pile threads cannot be used for figuring as they may consist of an odd assortment of colours they improve considerably the bulk and the resiliency of carpets thus effectively raising the low frame construction into a higher quality bracket.

Multi-frame upholstery moquettes or uncut moquettes are produced in two-shot structures as a rule but with a different ground weave than the one shown at C in *Figure 15.15*. Usually the ground weave is plain with the alternate tight and slack ground ends. Two structures suitable for upholstery cloths are shown at C and F in *Figure 15.16*. Both are 3-frame constructions and can be used for loop and cut pile effects. Customarily, however, the structure C is more often used for the uncut pile effects, whilst the structure F, in which pile is woven through to the back, is employed for the cut pile fabrics. The construction F, as already observed in connection with self-coloured effects, results in better pile anchorage than C but requires greater length of yarn for the same height of pile on the surface. A good quality loop pile upholstery cloth can be produced as follows: Pile warp 98/2 tex worsted, ground warp 74/2 tex cotton, stuffer warp 98/2 tex cotton or staple viscose rayon, weft 98 tex cotton or staple viscose rayon; 64 groups of pile threads and 64 wires per 10 cm, 41 loops per cm². Cut pile cloths are sometimes produced in slightly denser settings than loop pile structures and polypropylene and acrylic yarns are often used instead of worsted for the pile with excellent results. The following represent the particulars of a high quality cut pile upholstery fabric: Pile warp 125/2 tex worsted, ground warp 118/2 tex cotton, weft 98 tex cotton or staple viscose rayon; 68 groups of pile threads and 64 wires per 10 cm, 44 tufts per cm². It will be noted that no stuffer yarn is employed in this structure. The heavy pile and ground yarns provide sufficient rigidity and as the taut ground ends assume the same position in the cloth as the stuffer the need for the latter is obviated. The simplified and the fully worked-out designs at A and B, and at D and E refer respectively to the sections C and F. The marks used in the fully worked-out weaves correspond with those employed in *Figure 15.15* for the carpet structures.

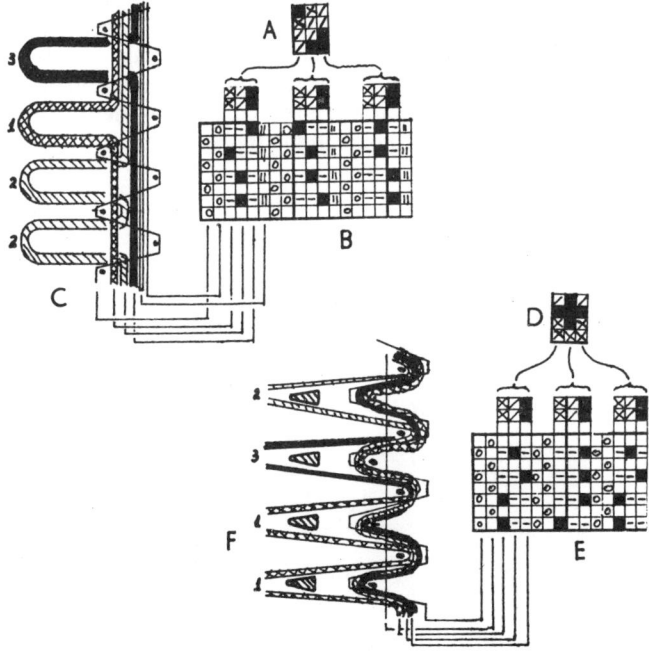

Figure 15.16

All the four constructions described in the foregoing can be embellished further by the exposure of the ground structure, or, in other words, by the creation of sunk places. This is achieved by deliberately failing to lift any pile ends in a group on some wires in selected areas. The sunk portions add effectively to the ornamentation and are useful for emphasising certain design features as shown in the fabric given in *Figure 15.10*. Used in excess, however, they reduce the quality of the cloth in respect of its resiliency and wear resistance by reducing the total number of pile points per area. As the use of sunk portions reduces the total length of pile warp required it makes it possible to produce constructions of pleasing appearance more cheaply.

Planting

The number of colours in the width of a fabric is not limited to the number of frames employed; the threads in different groups may be differently coloured, in which case one, two or more of the frames each contains more than one colour of pile. Thus, in a 5-frame structure one portion of a design may require the colours, 1, 2, 3, 4 and 5, and another portion the colours 1, 2, 3, 6 and 7, and yet another portion the colours 1, 2, 3, 8 and 7; the colours 1, 2 and 3 being constant, while the colours 4 and 5 are replaced by the colours 6 and 7, and then the colour 6 by the colour 8. The substitution of one colour for another is termed 'planting' and if this is judiciously performed a design may be produced in a 4-frame or 5-frame cloth which contains as many as, say 20 colours. In the same quality the higher the number of frames the more costly is the cloth on account of the greater quantity of pile yarn required and frequently a cloth, by successful planting, is given the appearance of being produced with a higher number of frames, and therefore, appears more costly than is actually the case. The chief point to note in planting is to avoid the formation of stripes in the woven design, and for this reason a planted colour is sometimes graduated at both sides towards the adjacent colours in the frame.

Method of Designing

In originating a large design a sketch of the figure is usually first made in pencil to a reduced scale on plain paper, and the proper colours are then indicated more or less roughly on the different portions. In transferring the design to squared paper it is customary to use paper that is ruled according to the pitch of the cloth, so that in drawing and painting the figure it is shown exactly the

Figure 15.17

size it will appear when woven. Also, it is usual to paint-in the several parts of the ornament in the exact colours that it is intended to employ in the cloth, although subsequently the colours of the woven design may be changed by substituting other threads in the loom. Each vertical space of the design paper represents a group of pile threads, and each horizontal space a wire, hence each small square of the paper represents a loop or tuft. An illustration of a 5-frame cut pile structure is given in *Figure 15.17*, in which the same five colours

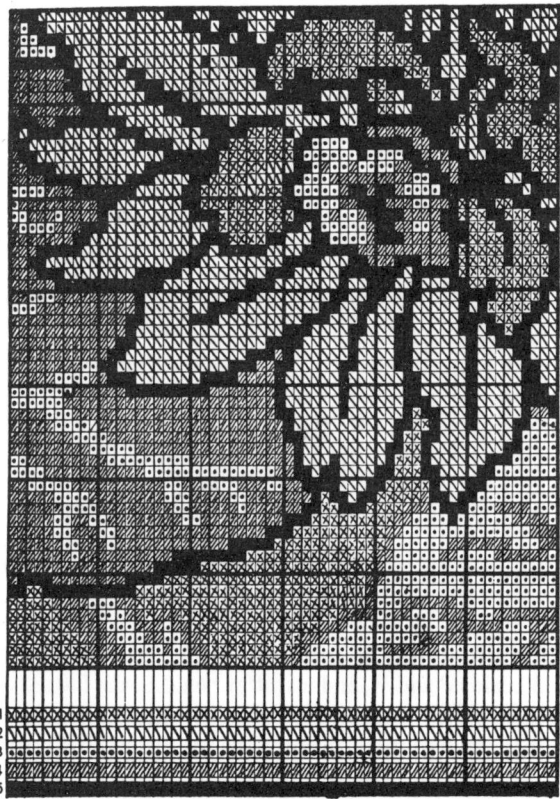

Figure 15.18

are employed throughout. In *Figure 15.18*, which corresponds with a portion of the design given in *Figure 15.17*, the five colours are represented by different kinds of marks, as shown in the 'gamut' below the plan; each mark in the plan indicated a pile tuft formed in the corresponding colour.

System of loom mounting

A form of harness and heald mounting is shown at A in *Figure 15.19* which may be used in weaving the textures. In each short row of the jacquard there are 10 hooks and needles which are connected in the same manner as in an ordinary single-lift machine. The arrangement of 10 per short row is convenient for 5-frame designs, and any smaller number of frames can be woven by casting out in long rows. The harness is knotted over the comber-board, and the comber-board M is supported at each side by a flat bar N to which a vertical movement is given by means of a cam, all the harness being thus capable of being raised by

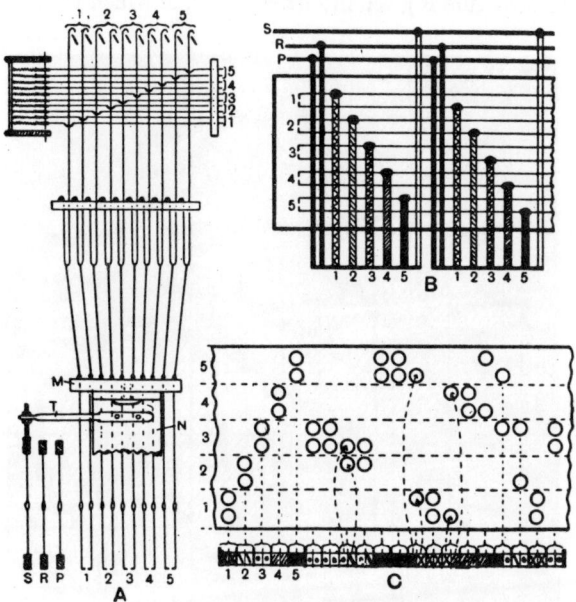

Figure 15.19

the comber-board M at regular intervals. Behind the harness there are two ground (or chin) healds P and R, and a stuffer heald S, the latter being connected at each side to a bar N by means of a rod T, so that the stuffer ends are lifted at the same time as the harness ends are raised by the comber-board. The ground healds P and R are operated in reverse order by means of positive tappets.

In the diagram A in *Figure 15.19* the hooks, needles, and harness cords are shown bracketed together in pairs, and numbered to indicate the numbers of the frames—i.e., the several colours of the pile warp that the respective parts control. At B, which represents how the warp threads are drawn on the healds and harness, the pile threads are correspondingly numbered; and the order of denting is indicated by the horizontal lines which connect the lower ends of the threads—two ground (or chain) threads, five pile threads, and one stuffer thread being passed through each split of the reed. In one split the five pile threads are drawn on the odd rows of the harness, and in the next split on the even rows, each colour being thus allocated to two consecutive rows of the harness, as shown by the numbers at the side of the harness draft. A comparison of the harness draft with the arrangement of the hooks and needles will show that the numbers coincide, and that each short row of the jacquard controls two pile threads of each colour.

Each vertical space of the design given in *Figure 15.18* represents one pile thread of each colour, so that two vertical spaces are equivalent to one row of needles and hooks, and one row of a card which is 10 holes deep. The size of the jacquard depends on the number of vertical rows of pile per repeat. Thus, if the repeat is 1 m wide and there are 40 rows of pile per 10 cm the required size of jacquard is 40 x 10 x 5 (frames) = 2000. For very wide looms several jacquards working in tandem may be necessary to cope with large design repeats. In upholstery fabrics single repeat designs are not normally made so that although the density of setting is higher (60 to 80 groups of pile threads per 10 cm) smaller jacquards may be adequate because the width of the repeat does not usually exceed 30 cm.

Card cutting

The system of card cutting which corresponds with the draft B is illustrated at C in *Figure 15.19*, where a portion of card is represented as having been cut to coincide with one horizontal space of a 5-frame design in which the same marks are used in *Figure 15.18*. A card may be considered to be in five longitudinal sections of two rows each, each section corresponding with a distinct colour of pile warp (a frame), as indicated by the numbers at the side of the example shown. The spaces in the card-cutting plan are bracketed together in pairs to coincide with the rows on the card, and two holes are cut in each row, the several colours or marks of the design being cut on the corresponding sections of the card. On the left of C the marks of the plan are arranged in the order of the frames, and numbered from one to five in order that they may be readily compared with the position of the corresponding holes in the card. Dotted lines also connect certain marks with the corresponding holes, and it will be seen that the first mark of a pair is cut on an odd row of the card, and the second mark on an even row. One card corresponds to one wire and the selected pile ends form the top line of the high shed underneath which the wires are inserted.

The effect of casting out in long rows on the card cutting when fewer frames than the maximum of five are used, and the method of dealing with planted designs, are shown in *Figure 15.20*. The second, third, and fourth frames are each in the same colour throughout but the first frame is planted in several colours, as shown in the gamut below the design at A. The system of designing is the same as in the previous example, each section of the design being painted out in the proper colour; and the foregoing system of jacquard mounting may be employed with the two unwanted long rows of the harness cast out. The system of card cutting is also the same, but all the colours of the planted frame are cut on the longitudinal rows of the card which belong to frame 1. As the card cutter reads the design he checks with the gamut to find out to which frame any colour has been allocated. Thus, in planted designs the gamut is an essential reference chart as there may be, say, 10 colours planted upon different frames and without the gamut the card cutter would be unable to tell what colours to cut on which longitudinal rows of the card.

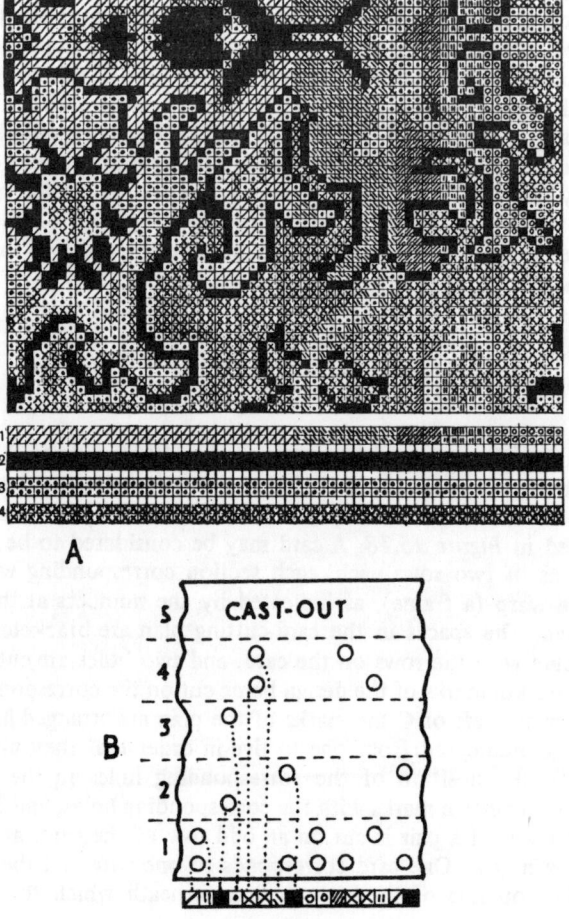

Figure 15.20

The diagram B in *Figure 15.20* shows a small portion of one row of a planted 4-frame design, marked as at A, and its relation to a corresponding portion of the card.

The cordage jacquard

In the carpet industry many manufacturers, instead of using the common jacquard system with wire hooks and lifting knives, favour the cordage machine. In this version of the jacquard the principle of selection is the same as in any other jacquard and depends on cards and needles. The method of lifting is, however, different. The needles act upon cords knotted above a lifting board which performs the same function as the knives. The lifting board, the plan of which is shown at A in *Figure 15.21*, is drilled with apertures in the shape of keyholes. When the lifting board is raised it lifts the knotted cords which are placed over the slits of the keyholes but the knots which remain over the round portions of the keyholes slip through and are not lifted. The lifting board provides the high shed line for wire insertion, x, whilst the working comber-board takes the unselected or dead pile ends to the middle shed line, y, as shown at B in *Figure 15.21*. The tilted lift of the lifting board ensures correct shed angle for the pile ends between the front and the back harness cord rows.

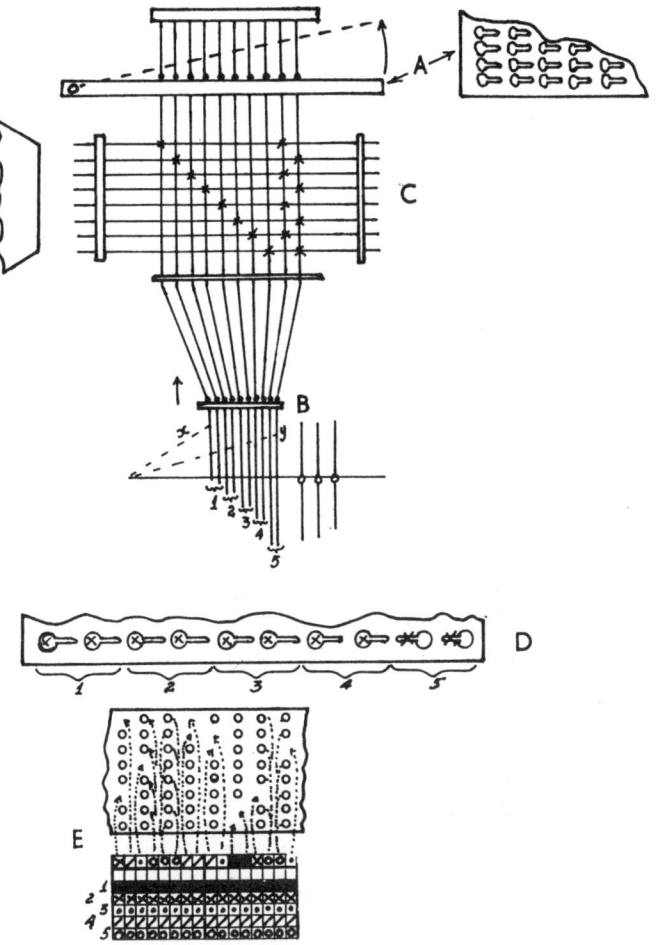

Figure 15.21

For five-frame work an eight-row card is used by arranging the lifting board on an inverted keyhole principle with suitable needle to cord controls. One short row of eight needles and ten cords (or bands) is given at C in *Figure 15.21* from which it can be seen that each needle acts against two cords, one of the front eight which control frames 1 to 4 and one of the back two which control frame 5. Odd needles control odd rows of pile threads and even needles even rows. A short row of keyholes in the lifting board is shown at D. In the first eight the slits are facing away from the cylinder and in the last two towards the cylinder. The normal resting position of the knots in the lifting cords is over the round portion of the keyhole for the first eight and over the slit in the last two as indicated by the crosses. It will be clear that if any one cord in the first eight is to be raised its knot must be pushed over the slit by the blank in a card, however, if one of the back two cords is to be raised its knot must remain over the slit which is achieved by leaving the needle immobile by presenting to it a hole in a card. Any two blanks in a short row, one against an odd and the other against an even needle, ensure that no lift of the back two cords will take place by pushing their knots over the round portions of the keyholes. If it is desired to lift the two back cords the short row of the card must be fully cut.

E in *Figure 15.21* shows a 5-frame gamut whilst at F a portion of one horizontal row of a five-colour design is given against a corresponding portion of a card. Studying the card in conjunction with the gamut it will be seen that in the first short row in the card a frame 2 end is lifted in the odd row of pile and frame 4 end in the even row. In the second short row there is only one blank and, therefore, the lifts are—odd row end from frame 3 and even row end from frame 5; in the third short row no blanks exist which means that both odd and even row ends from frame 5 are raised; and so on. For ease of identification the marks in the design are connected by fine lines to the operative portions of the card.

It should be appreciated that the above system of producing a 5-frame design from an eight-row instead of a ten-row card could be equally well adapted to a standard jacquard with hooks and lifting knives by using the inverted hook principle of operation instead of the inverted keyholes.

Methods of achieving textural variety in multi-colour pile fabrics

The design in the multi-frame pile fabrics is due mainly to colour. However, in addition to colour effects it is possible to create textural effects either with the aid of the wire insertion mechanism alone or in combination with special selection or shedding devices. The use of the exposed ground in figuring has already been explained earlier. In *Figure 15.22* a 3-frame fabric is shown in

Figure 15.22

which the effect is due to colour, to the use of sunk places and to the sequence of wire insertion which consists of two cutting, followed by one plain wire. Another 3-frame structure is shown in *Figure 15.23* in which two short and two tall looping wires are inserted alternately thus resulting in a. distinctly ridged appearance of the cloth. Such effects are produced by arranging the wire insertion according to a specific order disregarding the colour patterning which goes on notwithstanding the wiring sequence.

Figure 15.23

Other forms of embellishment consist of figuring in selected colours with cut and uncut pile so that in a 3-frame construction a three-coloured figure in loop pile could be produced on a three-coloured ornamental background of cut pile with sunk portions incorporated in the design as well. A cross-sectional view of the above type of design is given at A in *Figure 15.24* in which two frames are employed. The ground weave is a 2-and-2 rib with alternate tight and slack ends operated by healds. If stuffers are required they may be run-in together with the tight ground ends—in the example shown stuffer ends are not used. The jacquard operates on every pick and provides a high lift for the selected ends. The working comber-board may be employed but is not often used in this particular structure. A group of four picks forms a structural unit and a looping wire is inserted together with pick 2, and a cutting wire with pick 4 of each unit. Holes in the card cause a lift of the selected pile ends. The pile ends, to achieve the construction shown, are operated as follows:

Pick 1   — all pile ends down (a blank card)
Pick 2 and looping wire — ends selected to form loop and cut pile are raised
Pick 3   — all pile ends up (a fully perforated card)
Pick 4 and cutting wire — only ends selected to form cut pile are raised.

The working comber-board could be utilised to produce a wholesale lift of all pile ends on pick 3 of the sequence but in a system in which the jacquard operates on every pick the use of it is of insufficient benefit to offset the dis-advantage of reduced versatility which the fixed movement of a shedding element invariably implies.

As a result of the manner of operation given in the foregoing description the pile ends selected to form loops are raised only over the looping wire (and pick 3), but the pile ends selected to form cut pile are raised over both the wires and the intervening pick 3 of each group which makes the cut pile slightly longer than the loop pile. In the section A four structural groups or units are

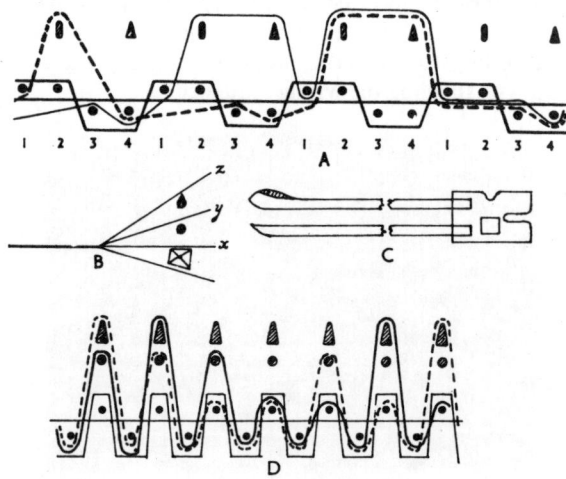

Figure 15.24

shown—in the first one an end from frame 1 makes a loop pile, in the second one an end from frame 2 makes a cut pile, in the third one both the pile ends make a cut pile tuft simultaneously whilst in the fourth unit both the ends are down on both wires which creates a sunk place in the design. The situation depicted in the third unit is used sometimes to produce mixed colour areas which are at the same time more heavily tufted than the single colour areas as two adjacent ends make the tuft instead of one. Such effects are described as moresque effects and are more often used in 3-frame designs in which any two out of three ends may be raised together which extends the colour range to six, i.e. three mixed and three pure colour areas. Other moresque effects were at one time also made in which the mixed and the pure colour areas were of the same density but these are not greatly favoured at present as the results are frequently somewhat indeterminate. Altogether figuring in colour with looping and cutting wires using the system described is not undertaken very often due to the expense involved in requiring four picks to produce one horizontal row of design. However, other systems which overcome this difficulty have been developed. Such systems depend on simultaneous insertion of two wires into separate sheds formed by a jacquard capable of lifting selected pile ends to two levels—high and very high—resulting in treble shed formation.

A treble shed is shown at B in *Figure 15.24*—the wire inserted into the middle shed is a plain or looping wire but the wire inserted into the top shed may be either looping or cutting. Thus, the possibility exists of figuring with low and high loop pile in selected colours or with low loop pile and high cut pile. In one system of operation the wire insertion mechanism is duplicated, i.e. a wire is inserted from the right-hand side into the middle shed whilst simultaneously another wire enters the top shed from the left. This is economically not a very desirable proposition because it effectively trebles the width of the loom. A more modern concept is to use a double-tier wire as shown at C in *Figure 15.24* inserted and withdrawn by a right-hand side mechanism. The construction which may be achieved with the high pile either looped or cut, is shown at D. The double wire is inserted together with pick 2 of the plain ground and the cross-section D shows all the possible combinations which may occur in a 2-frame construction. These consist of colour 1 high pile with colour 2 forming the low pile background and the reverse situation—these two possibilities are not often used due to a somewhat indeterminate colour resolution; the usual effects are—sunk places, either colour 1 or colour 2 over low wire, and either colour 1 or colour 2 over high wire. Excellent sculptured effects are produced with the high pile figure in one colour on the low pile ground in another colour. The construction is achieved with the aid of a jacquard with a working comber-board or bottom board the lifts of which raise the unselected pile on pick 2 over the weft only—level x at B in *Figure 15.24*. For lifts to the level y at B the hook is engaged upon a lower knife by one needle of a pair which control it and for lifts to the level z the hook is engaged by an upper knife.

Methods of weft insertion

Although the shuttle still represents the most common mechanism for the insertion of weft in warp pile fabrics other devices are also coming into use. Carpet looms have been developed in which a single rigid rapier is employed inserting a double shot of weft from a stationary weft supply package. In the weaving of upholstery fabrics other systems such as double rigid or flexible rapiers are also used. The carrier or gripper shuttle method has also been adopted for both the carpet and the upholstery cloth production. Apart from higher speed of operation the great merit of the new systems lies in the fact that weft is in continuous supply from magazined stationary packages. This is particularly attractive in fabrics in which coarse weft yarns are used as it obviates the need for frequent replenishment which exists when conventional shuttles are used and for the costly operation of winding the weft on to numerous and quickly exhausted packages.

16

Warp Pile Fabrics
Produced on the Face-to-Face Principle

Face-to-face weaving represents an alternative method of manufacture of the cut warp pile fabrics in which two cloths are woven simultaneously and the pile is produced without the aid of wires. By comparison with the wire insertion system there is greatly increased production, while the mechanism required for cutting the pile threads does not necessitate any increase in the normal width of the loom, so that there is great economy of floor space. For the above reasons this system is preferred to the wire method for the production of most cut pile fabrics at present.

Two separate ground fabrics with a space between them, each with its own warp and weft, are woven on the unstitched double-cloth principle, while the pile warp threads interlace alternately with the picks of both fabrics and thus are common to both. The distance between the ground fabrics is regulated according to the required length of pile and as the textures pass forward the pile threads extending between them are cut by means of a transversely reciprocating knife during the weaving process. Two cloths are thus formed—the bottom cloth with the pile facing up, and the top cloth with a similar pile facing down. The cloths pass in contact with separate take-up rollers and are wound on two cloth rollers as shown schematically at C in *Figure 16.1*. The double texture may be woven either on the single-shuttle or the double-shuttle principle, both of which are illustrated in *Figure 16.1*. The single-shuttle method is represented at A in which the wefting is shown arranged in the order of two picks top fabric, two picks bottom fabric, and one shed is formed at a time while the weft is inserted in the ordinary manner. Only one shuttle-box is necessary at each side of the sley, but a box motion may be employed when, for special reasons, two different kinds of weft are used or weft mixing is required. If the picks alternate in even numbers, as shown at A in *Figure 16.1*, the weft joins the two fabrics together at the extreme edge at one side only, but the picks for the respective fabrics may alternate in odd numbers, in which case the selvedges are joined at both sides. A knife situated between the top and bottom selvedges comes in contact with and cuts the joining picks as the weaving proceeds. The picks in each fabric are shown separated in pairs at A in *Figure 16.1* for convenience of illustrating the single-shuttle method, but in actual weaving they are evenly spaced.

In the double-shuttle method represented at B in *Figure 16.1*, two sheds are formed, one above the other, and two shuttles are thrown across simultaneously,

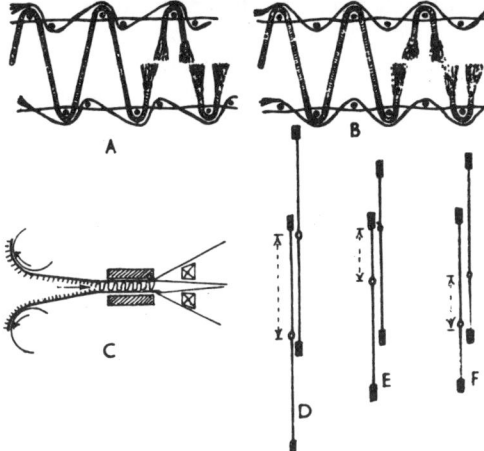

Figure 16.1

so that a pick is inserted in both the top and the bottom fabric at the same time. A shuttle-box is provided for each shuttle at both sides of the sley and the two shuttles are propelled by the same picking stick at each side. The lower shuttle runs on the warp on the race-board in the ordinary way, while the upper shuttle runs on the lower line of the top shed, which is usually higher than the upper line of the bottom shed. As each ground fabric is woven from its own shuttle there is no joining of the selvedges at either side. Double-shuttle weaving is much more productive than the single-shuttle system and is chiefly employed, but the latter method is sometimes preferred for certain makes of cloth as its use renders the production of defective cloth less liable. Most of the modern machines used for this class of structure instead of shuttles employ double or single rigid rapiers in a twinned (two-tier) arrangement, one set inserting weft at the top, and the other at the bottom cloth level. Apart from permitting higher speed of operation this system also offers the advantages of continuous weft supply from large stationary packages at the side of the machine and reduces frictional contact between the warp and the weft inserting element.

For single-shuttle work the healds are of the usual type. The pile shafts are placed centrally but the healds which operate the group warps are adjusted so that the ends for the cloth are rather higher than those for the bottom cloth in order to assist the shedding and to reduce friction. For double-shuttle weaving the mails of the pile healds are midway between the shafts, as the pile threads have to move between the top line of the upper shed and the bottom line of the lower shed or the full distance of both ground sheds as indicated at D in *Figure 16.1*. The ground warp threads have to move only half this distance and the healds are constructed with the mail eyes one-third of the distance between the top and bottom shaft. Thus, as shown at E and F in *Figure 16.1*, placed in position E the ground heald controls the ground ends of the top cloth, and placed in position F it controls the ground ends of the bottom cloth, an equal movement of the shedding mechanism being sufficient to move the ground ends of the respective fabrics to form the corresponding top and bottom sheds.

Special mechanisms required in face-to-face weaving

In weaving face-to-face pile fabrics by the single-shuttle method the only devices different from the normal are those connected with the cutting action, the take-up and the let-off of the pile warp. As the function of these mechanisms is identical with that of the corresponding devices in double-shuttle weaving the descriptions given in connection with the latter are equally applicable to the former.

There are several weaving machine makers both in this country and on the continent of Europe who specialise in the production of looms for the face-to-face double-shuttle weaving and although each version differs from another in some detail they are basically similar in principle.

Two shuttle boxes are provided at each side of the sley from which two shuttles are picked across simultaneously. On account of the shape of the two sheds the shuttles are specially bevelled and the boxes shaped accordingly. Fast reed warp protector motions are used and a weft fork is provided at each side of the loom, one acting for the top and the other for the bottom shuttle. In some instances the crankarm is constructed in two parts to provide sley dwell giving sufficient time in wide looms for the shuttles to cross before the forward movement commences.

In weaving continuous pile fabrics in double-shuttle machines a tappet shedding motion is usually employed and the tappets are most frequently of the positive type. Utilising the form of healds previously described the ground ends are moved the usual distance between the top and bottom of their respective sheds but the pile threads move about twice that distance oscillating between the top of the upper shed and the bottom of the lower shed and consequently require tappets with a bigger stroke or suitable leverage adjustment if normal stroke tappets are employed. In weaving certain construction three-position tappets are required the pile ends moving not only through the distance of the two sheds but also occasionally to the centre position. For figured pile fabrics three-position dobbies and jacquards are available.

Two cloths have to be taken up simultaneously and exactly at the same rate. The double take-up is of the positive type usually with pinion and worm gearing. Following the cutting of the pile threads the top cloth passes upward and the bottom cloth downward, each in contact with a pinned take-up roller. The two rollers are equal in size and turn in opposite directions, the cloths are suitably guided and are wound on to two cloth rollers in a stand in front of the loom.

A special stand for the ground and pile warp beams is provided at the rear of the loom. If all the ground ends in both fabrics are at the same tension one ground beam may be used which usually is placed low down and is controlled positively, but sometimes, for convenience, the ground ends of each fabric are placed on a separate beam. In weaving such fabrics as moquette, of which the odd and even ends are at different tensions, the tighter ends of both fabrics are run off one beam which is positively controlled, while the beam which contains the slacker ends of both fabrics is usually placed above the warp level and is sometimes provided with only a negative let-off. By means of dividing rollers the ends from the ground beam or beams are divided into two series to correspond with the top and bottom fabrics, the top series of ends passing over the upper roller and the bottom series under the lower roller, so that the ends of each fabric are retained at the proper level.

The delivery of the pile yarn ranges from five to ten times or more the length of the ground yarn, and the pile warp is therefore specially controlled in order to ensure that the proper length is let off according to the depth of pile and number of tufts, etc., required. A train of wheels from the low shaft drives a yarn delivery roller A, *Figure 16.2*, on which presses a free roller B, and both rollers are coated with suitable material to prevent slipping of the pile yarn. The threads

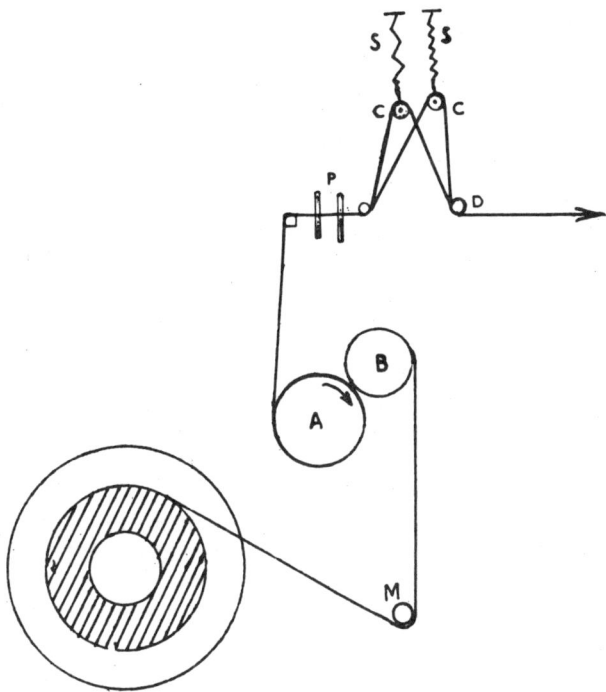

Figure 16.2

pass from the pile beam under a guide-roller M, over roller B, between the two rollers and under A, then over compensating rods C and under guide-rod D, and forward to the healds. The threads are gripped firmly by the rollers A and B and are given in at a predetermined rate at each pick, and this rate can be changed according to requirements by altering a change-wheel by which the surface movement of roller A is modified. The arrangement enables the proper length of pile to be fed-in to obtain the required depth of pile in both fabrics at the same time that no undue tension is put on the pile warp.

With certain weaves, although all the pile threads have the same degree of take up, they are in series which pass from one fabric to the other at different times. A separate pile beam and pile delivery roller may be employed for each series, but the use of compensating rods enables all the threads to be operated from the same beam by means of one pile let-off motion. The threads pass together from the pile yarn delivery roller A, *Figure 16.2*, then the different series are distributed separately over the compensating rods, as shown at C. Each rod passes across the width of the loom above the top ground warp, and is suspended at each end by a spring, S. A considerable number of rods can be used if necessary in the space available. The pile warp passes from the pile roller at a constant rate, and any length that is given in of a series that is not immediately taken up by tufting is taken up by the corresponding compensation rod through the recovery action of the springs. From the rods the pile threads pass downwards through the top ground warp, then under guide-roller D, *Figure 16.2*, situated about midway between the top and bottom ground ends, and in a straight line to the centre of the healds. If a warp stop-motion is employed the pile threads pass through it before they reach the compensating rods C, as shown at P.

The pile threads are firmly interwoven in both ground fabrics which are held some distance apart, while between them the pile threads extend until they come into contact with the cutting knife which passes to and fro across the loom. The distance between the ground fabrics, i.e. the depth of the pile, can be regulated by means of two distance plates which extend across the width of the loom between the fell of the cloth and the path of the knife. Machined regulating screws are provided at each side by means of which the distance between the plates can be modified according to requirements. Once set for a given cloth the distance is maintained at a constant value by the direction of tension in the two sets of warp threads and in the cloths after they are separated. The tension in the top set of ends and in the top cloth acts in the upward direction thus forcing the upper fabric to sit tight against the top bar. The lower cloth is tensioned in the opposite direction and sits tight against the bottom plate.

The cutting motion is a principal feature of face-to-face weaving and considerable attention must be paid to the accuracy of running and the sharpness of the knife. The knife runs in a carriage upon a rail placed between the two fabrics near to the point at which they are separated. The plane of the rail can be adjusted according to requirements once the distance between the two cloths has been determined. Normally the knife is set midway between the two fabrics to produce equal length of pile in each, but if so desired, it could be adjusted to produce low pile in one cloth and high pile in the other. After each traverse the cutting edge comes into contact with a pair of hones and is honed twice before commencing the reverse traverse. The knife is also racked so that a different portion of the blade is presented to the threads for each run. The knife rail is very accurately planed and has a comparatively broad span, and the carriage and rail are so shaped that vibration is reduced to the minimum at the same time that a free sliding movement is secured. It is upon the steady movement of the carriage, the frequency of the cutting and the sharpness of the knife that the production of a level pile of uniform depth which requires little cropping largely depends.

Apart from increased production and reduced space requirements the face-to-face system of weaving permits considerable savings in the consumption of pile yarn in multi-frame structures as will be shown later. It also eliminates some of the defects of wire-produced pile fabrics such as wire marks due to bad cutting or wire over-heating and the distinct diagonal alignment of pile in the direction of wire withdrawal.

ALL-OVER OR CONTINUOUS PILE STRUCTURES

Moquettes

The structure illustrated in *Figure 16.3* is used for firm and hard wearing cloths of the upholstery type known as moquette. There is one series of pile threads which passes from the top to the bottom fabric and back again, and the ground ends in each fabric, which interweave in plain order with the weft, are often arranged 1 three-fold single, 1 two-fold three-ply. Two ground warp beams are employed at different tensions, the single ends being held very tight in order that they will lie almost straight in the cloth, while the three-ply ends are lightly tensioned so that they will bend and impart a ribbed appearance to the back of each ground fabric. The pile threads are bound on alternate picks, and the plain shedding of the ground ends is so arranged that the binding position of each pile thread is covered by the slack three-ply ends. The bend or knuckle of the pile is thus protected, and the liability of the tufts 'rolling' or moving out of position is avoided. Usually all the weft is alike, but in some fine qualities of moquette thick and fine wefts are inserted alternately in each fabric, the cloth then being woven in a loom provided with a cross-pick arrangement by means of which the top and bottom shuttles are interchanged. In the pick-and-pick structure the pile threads are bound by the fine picks, the thick picks forming the ground.

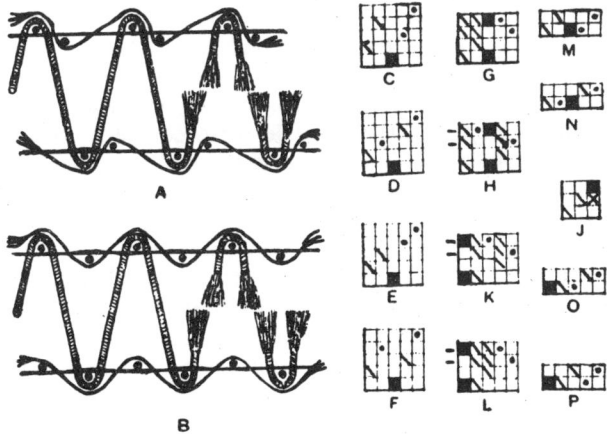

Figure 16.3

In face-to-face weaving the pile threads are invariably drawn on the heald or healds at the front, but considerable latitude is permissible as regards the arrangement of the ground ends and healds. Thus, four drafts that are suitable for the single-shuttle and double-shuttle structures illustrated at A and B in *Figure 16.3* are shown at C, D, E and F in which the solid marks represent the pile threads, the diagonal strokes the warp ends of the top fabric, and the dots the ends of the bottom fabric. In drafts C and D the ground healds are arranged top and bottom alternately, but in E and F the healds for the top fabric are in front of those for the bottom fabric. The ends for the respective fabrics are arranged in 2-and-2 order in drafts C and E, and in alternate order in drafts D and F. Each group of ends is dented in one split of the reed, and in all the drafts the pile thread is shown in the centre, but it may also be arranged to precede or to follow the ground ends in each split. Draft F shows the principle of arrangement largely used in this country.

G shows the weave of the structure represented at A in *Figure 16.3* with the ground ends arranged 2 top and 2 bottom alternately, as indicated in drafts C and E, while H shows the weave if the ground ends are arranged top and bottom alternately, as shown in drafts D and F. For comparison the weave which will produce the same structure when the pile is produced by means of wires is given at J, the solid mark representing the lift of the pile thread on the wire shed. The lifting plan for diagram A in *Figure 16.3*, to correspond with drafts C and D, is given at K, and to correspond with drafts E and F, at L. The marks at the side of plans G, H, K and L indicate the picks of the bottom fabric.

For the double-shuttle structure shown at B in *Figure 16.3* the same drafts as for single-shuttle weaving can be employed, but as two sheds are formed and two picks are inserted at the same time each horizontal space of the weaves shown at M and N (to correspond with the single-shuttle weaves given at G and H) represents a pick of each fabric, and each weave therefore repeats on two horizontal spaces. The lifting plan for the double-shuttle structure, to correspond with drafts C and D, is given at O, and to correspond with drafts E and F at P. The solid marks show the lifts of the pile threads from the lower line of the bottom shed to the upper line of the top shed, the diagonal strokes the lifts of the ground ends of the top fabric from the centre to the top line of the upper shed, and the dots the lifts of the ground ends of the bottom fabric from the bottom line to the top line of the lower shed.

Yarns in the moquettes and similar fabrics vary considerably according to the purpose and quality of the cloth. Pile yarns consist of cotton, staple or filament rayon, mohair, worsted, and various synthetic materials. Ground yarns are mainly cotton or staple viscose rayon. Silk, either net or spun, which at one time was very widely used for pile in dress goods is at present employed very infrequently. Mohair, worsted, and polyamide pile yarns are largely used in the manufacture of different kinds of imitation fur, while acrylic and poly-propylene yarns are particularly suitable for upholstery plushes because of their lustre, springy nature and resistance to wear. Cotton pile is liable to flatten under pressure and, therefore, is not very appropriate for upholstery cloths, but for curtainings and similar purposes and for dress goods cotton pile yarns, either mercerised or ordinary, produce very suitable and attractive textures. Combed yarns are desirable in order that there will be a minimum of short fibres which are liable to leave the surface of the cloth. Broken ground ends cause defects to show in the pile surface because the adjacent tufts lack support, and strong two-fold or three-fold good quality cotton yarn is there-fore largely used for the ground warps, while a medium quality of cotton is employed for the weft, but with rather more twist than is usual for weft.

Pile yarns are sometimes sized in order to facilitate clean cutting which improves the lustre of the pile. Curled mohair and worsted pile yarns are used for mats and for the manufacture of imitation. Astrakhan fur, etc. As a rule pile yarns are rather soft twisted as there is no great strain on the threads, and with soft twist the tufts more readily open out. The pile generally has to be dense enough to conceal the ground fabric, and to secure this condition fine setting is required if the pile is short, while in lower setting coarser yarns with longer tufts are necessary. In low set cloths the cover can be improved in the finishing process by laying the pile in the direction of the length, or sideways.

The following are average particulars of the moquette structure illustrated in *Figure 16.3*, for 100 m of cloth: pile warp, 74/2 tex lustre worsted, 700 to 1000 m; tight ground warp 98/3 tex cotton, 110 m; slack three-ply ground warp, 38/2 tex cotton, 160 m; weft 74 tex cotton, 14 picks per cm in each cloth, 72 splits per 10 cm with one pile, and two tight and two slack three-ply ground ends in each split. Tufts are formed in each cloth on alternate picks, hence there are approximately 7 tufts per cm in width and length giving 49 tufts per cm² (not allowing for contraction).

Velvet structures

In the moquette structure only one series of pile threads is used and the tufting in each cloth is complete on two picks, on one of which all the pile threads are interwoven, so that the tufts of pile are distributed in horizontal lines or ridges. In the examples shown in *Figure 16.4*, the tufting is again complete on two picks in each cloth, but in this case two series of pile threads are employed, one of which is interwoven on the odd picks and the other on the even picks. The alternate binding of the pile thread yields a more uniform distribution of the pile tufts so that a well-covered surface is produced and this type of structure has been found very useful for upholstery and hangings. The ground weave is plain but the foundation texture is different from that of the moquette structure as all the ground ends are alike and only one ground beam is essential. However, the ends for each ground fabric are brought from separate beams while the use of two compensating rods enables one pile beam and one pile let-off motion to be employed for both series of pile threads.

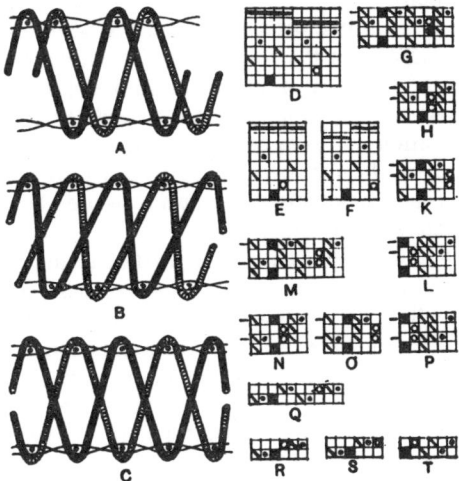

Figure 16.4

The single-shuttle system is illustrated at A in *Figure 16.4*, with the picks arranged two top fabric, two bottom fabric alternately, and three drafts with the orders of denting indicated above, are given at D, E and F. In drafts D and E two top and two bottom ground ends are dented in each split, but D has one and E has two pile threads in each split. In the same setting draft D would give the same number of tufts per cm^2 as the moquette structure illustrated in *Figure 16.3*, whereas draft E would yield twice as many tufts per cm^2. Draft F has the same proportion of pile threads to ground threads as E, but it is arranged for the threads to be dented three per split instead of six, giving one thread of each in each split. The weaves for the structure represented at A for the drafts D, E and F are given respectively at G, H and K, while the lifting plan shown at L is applicable to the three arrangements.

B in *Figure 16.4* is similar to diagram A except that for the purpose of illustration the structure is shown wefted one pick top fabric, one pick bottom fabric instead of 2-and-2. The corresponding weaves for drafts D, E and F are given at M, N and O respectively, while P shows the lifting plan. In the examples solid marks and circles are used to distinguish the two series of pile threads; diagonal marks and dots respectively represent the ends of the top and bottom fabrics, while the marks at the side of the plans indicate the picks of the bottom fabric.

C in *Figure 16.4* illustrates the double-shuttle structures to correspond with diagrams A and B. The weaves, which repeat on two picks in each fabric and two horizontal spaces, are given at Q, R and S for the respective drafts D, E and F, and T shows the lifting plan.

Average particulars for 100 m of cotton velvet curtaining fabric are as follows: Pile warp, 38/2 to 42/4 tex combed, gassed and mercerised cotton, 600 to 800 m; ground warp, 24/2 to 30/2 tex cotton, 107 m; weft 26 to 30 tex cotton, 26 to 32 picks per cm in each fabric, 14 to 18 ground ends per cm in each fabric, 14 to 18 pile ends per cm (for draft E or F). Each pile thread forms a tuft on alternate picks, or from 13 to 16 tufts per cm, giving from 182 to 288 tufts per cm^2 (not allowing for contraction).

The pile surface of a cloth is usually satisfactory when the pile is dense and the tufts stand vertically from the foundation. To obtain these conditions in a structure such as that represented in *Figure 16.4* the ground ends and picks require to be set close enough to nip the pile threads and hold the knuckles of the tufts firmly in position. In lower quality cloths, however, in which the density of the pile is deficient, sufficient cover can be obtained by laying the pile over in the finishing process, and the required firmness is secured by treating the ground fabric with resin or latex on the underside.

In *Figure 16.5*, A and B represent the single-shuttle and double-shuttle structures respectively of a style in which the ground weave is 2-and-2 warp rib; there are two series of pile threads which are bound in alternately in each cloth. The tufting in each cloth is complete on four picks, and each tuft is bound by one pick. The examples illustrate also the introduction of extra ends which work in 1-and-3 order alternately and are given in rather slacker than the ground ends from a separate beam. The object of the extra ends is to form a backing to each fabric and, by covering the knuckles of the tufts on the back, to reduce the possibility of the pile fraying out or becoming displaced.

The draft for A and B in *Figure 16.5* is given at C, and except for the backing ends is similar to draft F in *Figure 16.4*. In the same setting, however, the latter draft will produce twice as many tufts per area as draft C in *Figure 16.5* because each pile thread in *Figure 16.4* forms a tuft every two picks, and in *Figure 16.5* only every four picks.

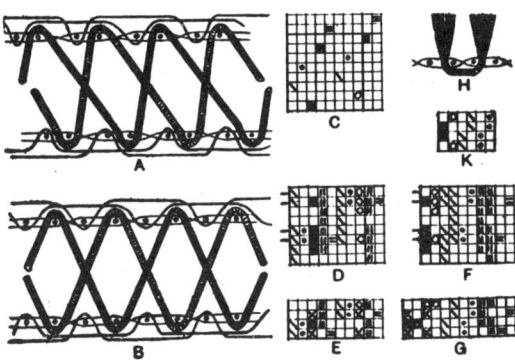

Figure 16.5

D and E, *Figure 16.5*, are the respective weaves for A and B, the first repeating on eight picks which are inserted singly, and the second on four double picks. The corresponding lifting plans are given at F and G. The diagonal marks and dots respectively represent the top and bottom ground ends, the vertical and horizontal marks the top and bottom backing ends, and the solid marks and circles the respective pile threads. In addition, in plans E and G crosses are shown which represent the lifts of the pile threads to the upper line of the bottom shed on the picks on which they have to be interwoven in neither fabric. Thus, in plan G which is arranged for a three-position dobby, two solid marks or two circles are indicated alongside each other to show the lifting of two jack levers in order that a pile heald will be raised the full depth of both sheds, while a cross represents that one jack lever will lift a pile heald the depth only of the bottom shed.

The warp-backed structure is largely woven with acrylic or lustre worsted pile threads, and in proper setting a very upright pile is formed which will resist pressure, so that the material is very suitable for upholstery work. The same order of interlacing of pile and ground threads, but without the backing ends, is also used with mercerised cotton for the pile, and the cloth has a softer handle and is suitable for table covers and curtaining fabrics. The following are suitable particulars for producing 100 m of upholstery velvet in the structure illustrated in *Figure 16.5*: Pile warp, 60/2 tex acrylic yarn or lustre worsted, 800 to 1000 m; ground and backing warps, 42/2 tex cotton, 106 m and 120 m respectively; weft, 60 to 74 tex cotton, 22 picks per cm in each fabric, 13 ground and 13 backing ends per cm in each fabric, 13 pile threads per cm. Each pile thread forms a tuft every four picks giving 72 tufts per cm^2 (not allowing for contraction).

The form of tuft previously illustrated, which is bound by one pick, is often referred to as 'V' pile, but in the 2-and-2 rib ground weave shown in *Figure 16.5* the 'U' form of tuft is sometimes formed, the shape of which is illustrated at H.

The pile thread is bound by two picks and each side of the knuckle of the tuft is gripped between two picks which are in the same shed, so that improved pile anchorage can be secured by wefting the cloth suitably. The lifting plan for diagram H on draft C for double-shuttle weaving is given at K, assuming that no backing ends are employed.

Fast pile structures

A system of pile interlacing is illustrated in *Figure 16.6* which gives very firm binding of the pile and the term 'W' is applied to the tufts on account of their shape. It is extensively used for light textures such as dress fabrics, hat trimmings, etc., which may be composed entirely of silk or rayon or of rayon pile and cotton ground warp and weft. On account of the fastness of the pile textures can be made in which the ground fabric is light and somewhat open, while a short pile may be produced above a fancy effect in the ground which shows more or less clearly through the fibrous surface. This order of pile interlacing is also very commonly used in figured velvet in which a pile figure is formed on plain voile crepe-de-chine, or georgette ground. The 'W' pile represented in *Figure 16.6* is similar to the fast pile structure shown at D in *Figure 15.4*.

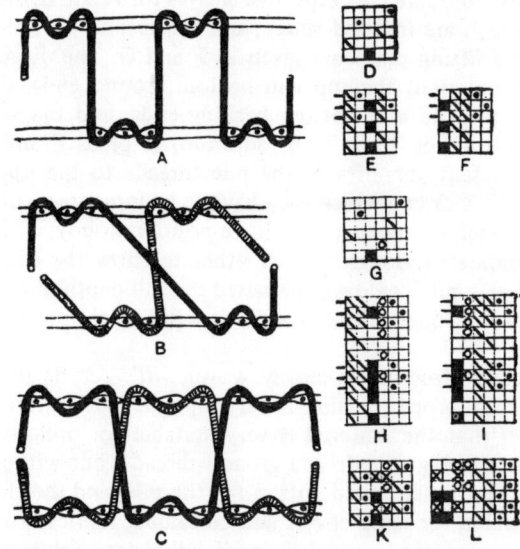

Figure 16.6

In the single-shuttle structure shown at A in *Figure 16.6*, the order of wefting is 3 picks top fabric, 3 picks bottom fabric, to correspond with the tuft formation. One series of pile threads is employed and the ground weave of each fabric is 2-and-1 warp rib, the last pick of each group of three picks, being in the same shed as the first pick of the next group of three picks, the weave thus repeating on three picks in each fabric. The arrangement is very suitable for the 'W' form of tuft as the two picks which are in the same shed readily approach each other, so that the tendency of the picks to ground in three's is reduced at the same time that both sides of the 'W' tufts are gripped by these picks. Sometimes, in order to accentuate the 2-and-1 ribbed effect in the ground, odd and even ends in each fabric are woven with about 10 per cent difference in the let-off from separate warp beams.

The draft for A in *Figure 16.6* is given at D, the arrangement consisting of 1 pile thread to 2 ground ends in each fabric. The corresponding single-shuttle weave is given at E and the lifting plan at F. B in *Figure 16.1* represents a single-shuttle structure which is similar to that shown at A, as regards the ground weave and the shape of the tufts, but the example illustrates the use of two series of pile threads which interlace alternately in each fabric. In the draft G there are 1 pile thread, 2 top ground ends and 2 bottom ground ends in each group which, in the same setting, will give the same number of tufts per cm^2 as A and draft D. The corresponding weave is given at H and the lifting plan at I, the repeat in each case being on 12 picks or six picks in each fabric.

C in *Figure 16.6* represents the double-shuttle structure to correspond with A, but in this case the ground weave is plain and two series of pile threads are used so that the weave in each fabric repeats on six picks. Draft G is suitable for C, while K is the corresponding weave and L the lifting plan which repeat on six double picks. The double solid marks and circles in plan L show the full lifts and the crosses the half lifts of the pile threads.

In correctly set cloths an erect pile can be obtained in the structure, represented in *Figure 16.6*, but on account of the formation of horizontal lines of pile every three picks it is difficult to obtain good cover, and the pile may therefore be laid over in the finishing process in order to secure a better covered surface.

The examples illustrated in *Figure 16.7* also show the production of the 'W' form of fast pile tuft, but they are designed to get over the 3-pick grouping and to secure a more even distribution of the pile. A represents a single-shuttle structure which is wefted four picks top and four picks bottom fabric, and two series of pile threads are used, one of which is interwoven on the first three and the other on the last three of each group of four picks. The ground weave is plain, and the corresponding draft is given at C in *Figure 16.7*, the weave at D and the lifting plan at E.

The plain interlacing of the pile threads in the 'W' structure represented at A in *Figure 16.7*, cuts with the plain weave of the ground ends on one side only. By arranging the draft, however, so that a ground end on both sides of a pile thread is drawn on the same heald, as shown at F, the ground ends will form plain weave with the pile interlacing with the result that the tufts are more firmly secured, as they are held in position by the cutting of the ground ends at both sides. The style is used for light supple cloths composed of silk or rayon. The corresponding weave for diagram A and draft F for single-shuttle weaving is given at G in *Figure 16.7*, while H is the lifting plan.

The double-shuttle structure illustrated at B in *Figure 16.7* contains six series of pile threads, three of which work opposite to the other three, the idea being to distribute the pile to the greatest possible extent and thus secure a cloth with a surface that is most effectively covered. Three pile warp beams may be employed, but by making use of three compensating rods all the pile threads can be brought from one beam. The draft for diagram B is given at K, the corresponding weave at L and the lifting plan at M. Different marks are used to distinguish the various pile threads, the half-lifts of which, in plans L and M, are indicated by the crosses.

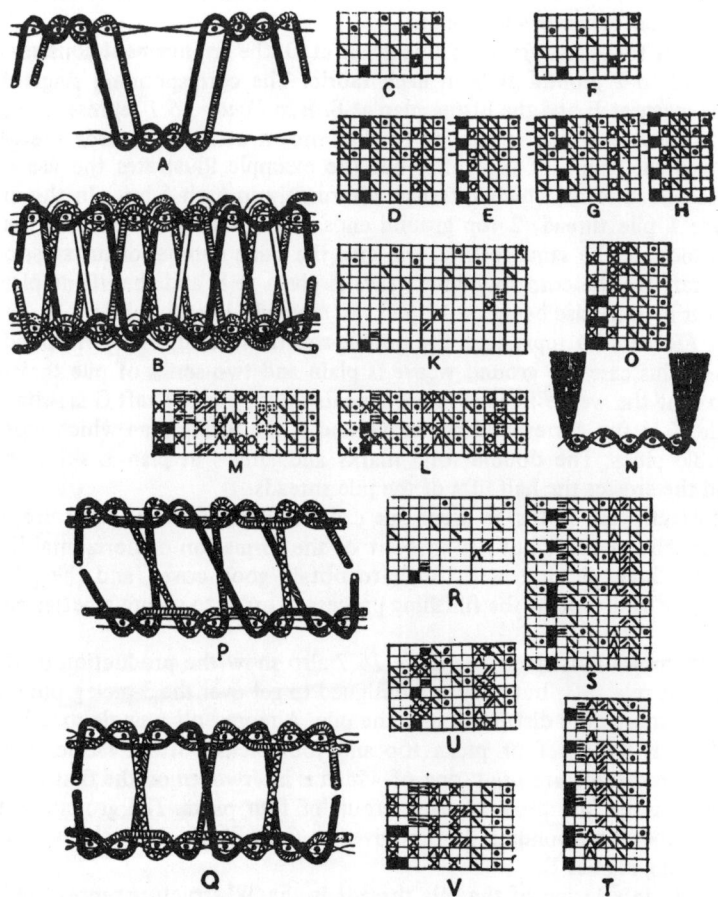

Figure 16.7

Particulars for 100 m of dress fabric suitable for draft G, *Figure 16.6*, or draft K, *Figure 16.7* are: Pile warp, 17 tex filament rayon, 400 m; ground warp, 20/2 tex cotton, 110m; weft, 20 tex cotton, 29 picks and 29 ground ends per cm in each fabric, 29 pile threads per cm. Each pile thread forms a tuft every six picks giving 140 tufts per cm^2 (not allowing for contraction).

A modification of the 'W' form of tuft is shown at N in *Figure 16.7*, which is used to obtain great firmness when a very long pile is formed or when the cloth is liable to be subjected to hard wear, as in the case of pile rugs and mats. In addition, the draft of the ground ends can be arranged, as shown at F, so that the plain ground weave cuts with the interlacing of the pile threads and the tufts are held in position at both sides. The lifting plan for double-shuttle weaving is shown at O.

The order of interlacing, illustrated in *Figure 16.7* at P and Q, produces very firm pile anchorage and is particularly appropriate for 2-and-2 warp rib and hopsack ground weaves, which, themselves, are specially suitable for the production of heavy pile fabrics. Each tuft is interwoven on four picks, and in order to distribute the pile four series of pile threads are used, two of which work opposite to the other two, so that they can be operated from two warp beams or from one beam if compensating rods are used. This pile structure may be compared with that shown at F in *Figure 15.4*.

P in *Figure 16.7* represents the single-shuttle structure wefted 2 picks top fabric, 2 picks bottom fabric. The draft is given at R, the complete weave at S, and the lifting plan at T. The double-shuttle structure, shown at Q corresponds to P except that the 2-and-2 warp rib ground weave is in a different relative position to the interlacing of the pile threads. The draft R is suitable and U and V show the weave and lifting plan respectively, the crosses representing the half-lifts of the pile threads.

The following are suitable particulars for an imitation fur woven in the structure, represented at P and Q. Pile warp, 64/2 tex mohair, 800 m or more of warp for 100 m of cloth; ground warp, 40/2 tex cotton 108 m; weft, 50 tex cotton, 28 picks per cm in each fabric, 14 ground ends per cm in each fabric, and 14 pile threads per cm. Each pile thread forms a tuft every 8 picks giving 49 tufts per cm^2 (not allowing for contraction).

Carpet structures

Double shuttle face-to-face weaving represents the principal method of producing self-colour, all-over cut pile carpets which after separation by the knife become structurally identical with those produced by the wire insertion method. Any constructions described in connection with the latter method can be used although the two-shot structure is mainly employed, as depicted in *Figure 16.8*. All the healds are tappet controlled; the chain or ground ends are operated alternately 2 up, 2 down in each cloth, the stuffer yarns 1 down, 1 up, whilst the pile threads oscillate between the two cloths and are anchored around alternate picks in each fabric.

Figure 16.8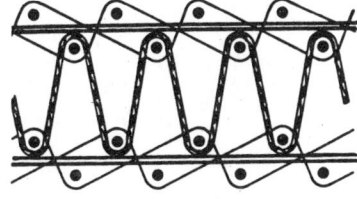

Similar qualities can be made to those described with regard to carpets produced by the wire insertion method. Frequently high twist yarns are employed for the pile which have the merit of yielding carpets which are not only hard wearing and resilient but also less prone to show furniture marks, etc.

Production rate of the face-to-face system compares very favourably with the wire loom. Weaving identical quality body width carpeting in a two-shot structure the face-to-face loom will produce 105 horizontal rows of tufts per min as opposed to 55 by the wire loom. Similar proportional advantage is retained in broadloom weaving.

FIGURED PILE STRUCTURES

Face-to-face moquette textures, which are used for upholstery and similar purposes, are ornamented in diverse ways by means of colour and design. In the ordinary structure in which only one series of pile threads is used stripe patterns and marl effects are formed by combining yarns that are different in colour, material or dyeing property, etc., and figured styles are obtained by printing. Bulky threads similar to the pile yarn are substituted where required for the slack cotton ground ends (corresponding empty mails being left on the pile healds) so that stripes of pile are separated by sunken repp lines as shown in the fabric at A in *Figure 16.9*. The special surface effects obtained in finishing and described in connection with all-over pile structures produced with the aid of wires are, of course, equally applicable to similar face-to-face fabrics.

In dobby shedding more elaborate effects, ranging from combinations of hopsacks and horizontal cords to simple figured patterns, are produced by making use of the system of compensation and, instead of using only one series of pile threads, by employing two series which are differently coloured.

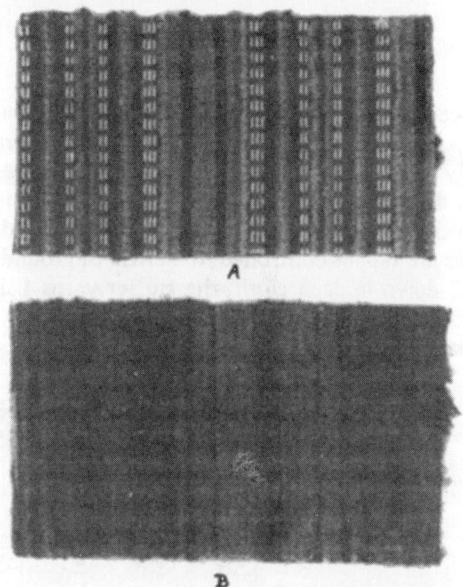

A

B

Figure 16.9

By planning a design so that an equal number of tufts is formed by each series of pile threads only one pile warp beam and one pile let-off motion are necessary, as one series is a duplicate of the other. A separate pile warp beam and pile let-off motion may, however, be used when greater freedom of design is desired as the pile ends used for forming the figure can then be considered independently of those forming the ground pattern. Figured effects which require a jacquard machine for their production are also made, but for these separate bobbins are necessary for the pile threads. Multi-frame jacquard designs are produced mainly in upholstery moquettes and in carpet structures.

Production of loop and cut pile effects is possible with the aid of frieze wires and different height of pile can be formed by using special stitching weft which is removed after weaving but neither of the above two structures is utilised to a great extent. The results are not as good as in similar structures produced with wires and in employing the special devices the face-to-face system tends to jeopardise its main advantage over the wire method which is the high rate of production.

The use of duplicate series of pile threads

The principle by which two duplicate series of pile threads are used in forming a fancy effect is illustrated in *Figure 16.10*. A motif is shown on one vertical space at A in which the solid marks represent a dark pile thread and the circles a light pile thread, the plan thus indicating that in a longitudinal line three tufts of dark pile alternate with three tufts of light pile. The corresponding diagram, given at B, shows that when one series of threads is forming pile the other series is interweaving in the bottom fabric in the same way as the slack ground ends. The pile threads, when not forming pile, might be interwoven similarly in the top fabric, but the method illustrated is generally preferred, and to allow for the addition of the pile threads to the ground only one slack ground end instead of the usual three may be used in the bottom fabric. This causes the top and bottom fabrics to differ from each other in composition and handle, but the difference is not considered objectionable as long as the two fabrics are not combined in the same article.

As only one pile beam is used, the length of pile warp must be the same for both series of threads, and the yarn is let off regularly at the average rate required for tufting. The threads which are not forming pile, however, are taken up only at the same rate as the ground ends so that there is an accumulation of length of this series during the period that the other series is tufting. The accumulated slack, which each series forms in turn is taken up by means of compensating rods, the operation of which by means of stretched springs has been described previously. The length that can be taken up by the rods is limited, but as each series of pile threads forms only its own proportion of the tufts, in this case one-half, the rate of pay-off is very much less than when only one series of pile threads is employed.

The draft for the structure, represented at B in *Figure 16.10*, is shown at C; six healds are employed and each group of threads consists of 2 top and 2 bottom ground ends and 2 pile threads in each split of the reed. The complete lifting plan is given at D for double-shuttle weaving, the weave of each fabric repeating on 12 picks. It is assumed that a three-position dobby is used, hence two solid marks alongside each other, or two circles followed by blanks, represent the full movement of the pile threads for tufting, while the single crosses show the half-lifts of the pile threads during the period they are being interwoven in the bottom fabric. One vertical space is used for each ground heald and the lifts of the top and bottom fabrics are represented by diagonal marks and dots respectively.

Design E in *Figure 16.10* shows an elaboration of the motif A in the form of a stripe, but the structure throughout corresponds to that represented at B, while G and D show the draft and lifting plan respectively. The change of pattern is

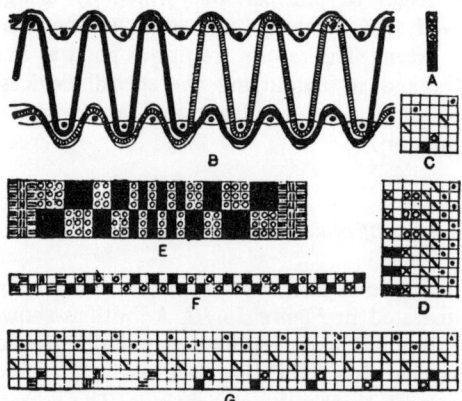

Figure 16.10

due to the arrangement of the pile threads as to colour. Each vertical space of design E represents two pile threads, while F shows how the colours of the first 19 vertical spaces of E are arranged and how the pile threads are drawn on the two pile healds. Where continuous lines of the same colour are formed, as represented by the horizontal and vertical marks in E, both pile threads of a pair are of the same colour. Where the pattern changes colour longitudinally in 3-and-3 order, as represented by the solid marks and circles, the two pile threads of each pair are differently coloured, and to alter the pattern in a horizontal direction two threads of the same colour are brought together at the change of effect. Thus, assuming that in design E the horizontal marks represent black, the vertical marks green, the solid marks blue, and the circles gold, the warp arrangement of the continuous lines is two black, two green, two black; in the 3-and-3 colour effect it is one blue, one gold three times, then one gold, one blue three times; in the 2-and-2 effect it is one blue, one gold twice and one gold, one blue twice, and in the 1-and-1 effect it is one blue, one gold and one gold, one blue three times. To further illustrate the arrangement of the warp pattern the complete draft is given at G of the first eight vertical spaces of design E.

In *Figure 16.10* each continuous line of colour in the plain stripe, indicated by the horizontal and vertical marks in design E, is taken as being formed by the working of two pile healds which combine the tufting of two pile threads of the same colour. The changing from one thread to the other in forming the pattern tends, however, to produce an irregularity in the continuity of each line of colour (which is not so apparent where the two threads are differently coloured), and to avoid this defect a separate pile heald, pile beam and let-off motion may be used for the threads which form the continuous lines. As these pile threads form pile all the time their take-up is greater than those that form the discontinuous stripe, while under the plain stripe in the bottom fabric three-ply slack ground ends must be used. B in *Figure 16.9* illustrates the type of fabric described above.

Motif A and the corresponding design B in *Figure 16.11* show, on a small scale, a development of the preceding example which can be produced by the same mounting, viz. a duplicate set of pile threads drawn on two healds, one

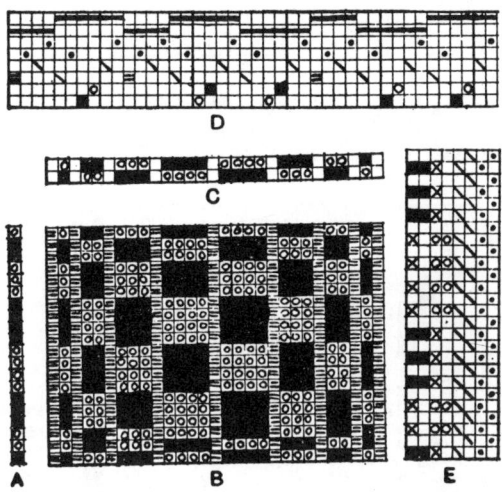

Figure 16.11

pile beam and pile let-off motion, and two compensating rods. The order as to colour in which the pile threads are drawn on the two healds is represented at C, the colour shown on the first heald preceding that indicated on the second heald. In this case, however, it is assumed that the continuous lines represented by the horizontal marks in design B are produced by threads which form a repp effect. These threads therefore are drawn on the same healds as, and in place of, corresponding slack ground ends, as shown in the draft given at D, and it is usual to weave the repp threads from a separate beam provided with a negative let-off. For convenience empty mails may be left on the pile healds to correspond with the repp threads, and there are no pile threads in the corresponding split. The arrangement of the pile threads in the check figure effect is as follows:

| Dark | 1 1 2 1 1 1 1 1 2 1 1 1 1 1 2 1 1 | 20 |
| Light | 2 1 1 1 2 1 1 1 1 1 1 2 1 1 1 2 | 20 |

D shows the complete draft to correspond with the first eight vertical spaces of design B, while E represents the lifting plan for the first 13 horizontal spaces, two vertical spaces in E being used to show the lifts of each pile heald.

B in *Figure 16.12* is the motif of the simple figured stripe design given at A in which there are three different orders of working, but the figure is so arranged that in each longitudinal line an equal number of figuring and ground tufts is formed, viz. 16 of each in the repeat of 32 tufts. Like the previous example the design requires two series of pile threads, one pile warp beam and one pile let-off motion, but each series of pile threads requires three healds. In addition to the four ground healds, therefore, six pile healds are necessary, and as the threads operated by the latter form pile at different times, six compensating rods to correspond are necessary.

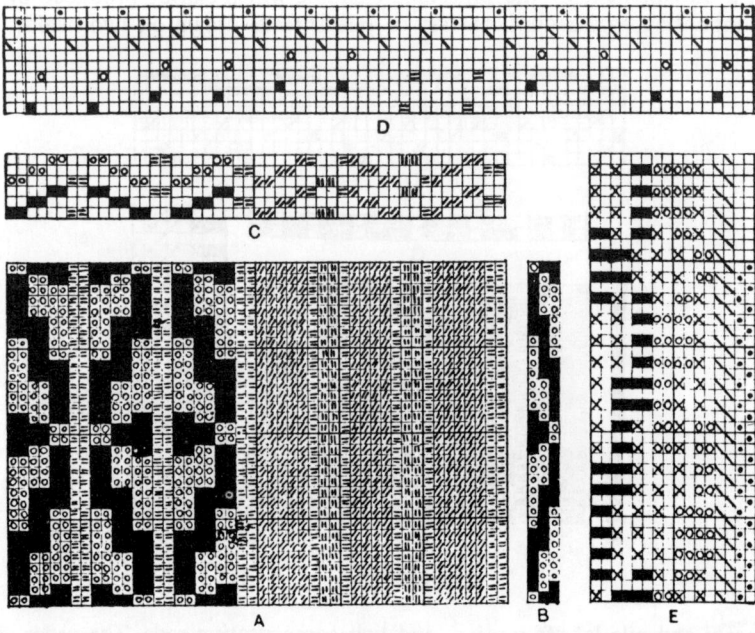

Figure 16.12

C in *Figure 16.12* shows the colour arrangement of the pile threads as they are distributed over the six pile healds, and to simplify the heald order the threads in the plain stripe which form continuous lines of colour are indicated on the healds in such a way that an equal number of threads is drawn on each heald. The change from one thread to the other in the plain stripe occurs at different times so that the possibility of cracks or irregularities showing in the continuity of the lines is reduced.

The complete draft of the first 12 groups of threads of design A is given at D, and the lifting plan of the lower portion of the design at E in which the method of marking is the same as in the previous example.

In the figure of design A in *Figure 16.12*, the pile threads are arranged one dark, one light throughout, but additional variety of effect would have resulted if, say, the middle strip had been changed to one light, one dark, as a counterchange pattern would have been formed. This method has been employed in the design given at A, *Figure 16.13*, the motif for which is shown at B with the result that the central line of figure appears different from the outer lines without an increase having been made in the number of different orders of working. In this example, in each vertical line of figure, more tufts of pile are formed by one colour than by the other in the proportion of 15 to 9 in the repeat of 24 tufts. Four series of pile threads drafted upon four healds and four compensation rods are needed, and as two of the series of threads require a very much greater length of pile yarn than the other two, two pile warp beams and two pile let-off motions are necessary.

The colour arrangement of the pile threads and the order in which they are distributed upon the four pile healds are shown at C in *Figure 16.13*, and as

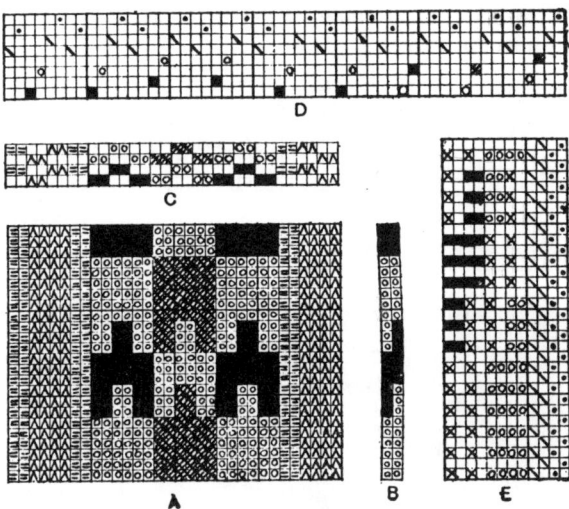

Figure 16.13

before each vertical space of A and C represents two pile threads. The complete draft of vertical spaces 9 to 17 of design A is given at D, while E shows a portion of the lifting plan.

Face-to-face looms are usually built to accommodate three pile beams and three pile let-off motions, and opportunities are thus provided for increasing the size and diversity of designs which require two series of pile threads in each line of the figure, while special colour effects can be introduced by employing where necessary a third series of pile threads. A multiplicity of pile beams, let-off motions, compensation rods and healds complicates the working of the loom, however, and care needs to be taken that the resulting pattern is worth the means employed in producing it.

The use of three series of pile threads, instead of two, in each line of the cloth greatly facilitates the production of variety of design and at A, in *Figure 16.4*, for which B is the motif, a style is illustrated on a small scale which can be woven with a comparatively simple mounting. Three colours are represented by the blanks, shaded squares and solid marks, and the spaces of the design are arranged both vertically and horizontally in the order of six, four, two, two, four repeated. An equal number of tufts of each colour, viz. 18, is formed in a vertical direction in the repeat of 54 tufts. There is only one order of working the three series of pile threads so that the required mounting consists of three pile healds, one pile warp beam and let-off motion, and three compensating rods. The change of pattern in a horizontal direction is due to changing the order in which the colours are arranged on the principle illustrated with two colours in *Figures 16.10* and *16.11*. Thus, assuming that the blanks represent light, the shaded squares mid and the solid marks dark, in the first section the colour arrangement is one light, one mid, one dark for six times; in the second section one mid, one dark, one light for four times, and in the third section one dark, one light, one mid twice. This is indicated at C in *Figure 16.14*, which shows the order of colouring and the arrangement of the threads on the pile healds of the first 16 vertical spaces of design A, while D shows the complete draft of the ninth to the sixteenth space.

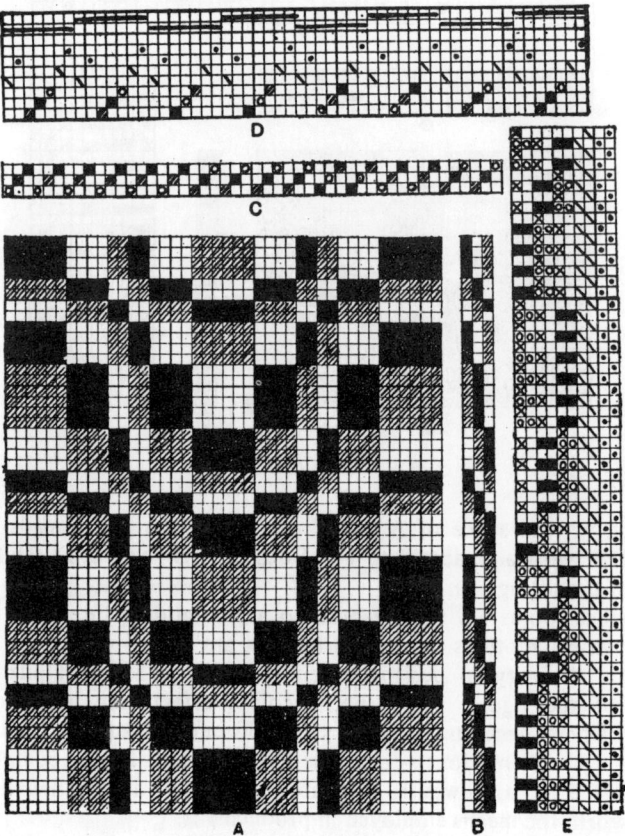

Figure 16.14

The lifting plan for the lower portion of design A in *Figure 16.14* is given at E and, as it is assumed that two dobby jacks are used for each pile heald, the lifts of each are indicated on two adjacent vertical spaces. Where two solid marks are followed by two blanks the corresponding pile heald moves the threads the full depth of both sheds for the purpose of tufting. The crosses show where the pile healds are moved, each by one jack, to the intermediate or centre position. Thus, when a cross on an odd pick is followed by two blanks the corresponding threads are interwoven in the bottom fabric when not forming pile, while where a cross on an even pick is preceded by two circles the pile threads are similarly interwoven in the top ground fabric. In this manner each series of pile threads, after forming pile, interweaves in the bottom fabric and then in the top fabric so that both cloths appear the same, and a saving of two-thirds of the slack ground ends can be effected in both ground warps.

Jacquard figured constructions

Jacquard shedding is employed in producing many different kinds of figured pile fabrics by the face-to-face system, and the machine and harness are varied in construction and arrangement according to the class of pile texture required while the pile threads are brought from bobbins or cheeses carried in a creel. A variety of pile weaves may be used in the figure, but when the ground is a light texture, such as voile, or crepe-de-chine, the 'fast' or 'W' form of pile weave is used in order to bind the pile firmly. This type of structure may be conveniently woven on the single-shuttle principle, using an ordinary double-lift jacquard for the pile threads, and ordinary healds for the ground threads. Cloths in which the foundation is dense and compact may have the pile threads interwoven into the ground when they are not forming figure, and they may be woven equally into both the upper and lower fabrics. In some cases, however, it is more convenient for the pile threads to be interwoven only in the ground of the lower fabric.

Using single-shuttle, multi-frame constructions suitable for upholstery or velvet pile carpets are produced on the one-shot principle, i.e. a tuft is anchored around every pick. A two-frame structure of the above type is depicted by the weft section at A in *Figure 16.15* but any number of frames could be employed if desired, up to a maximum of five. A plain weave ground structure is used in both the top and the bottom cloth arranged so that all the odd picks form the upper and all the even picks, the lower fabric. The ground ends are controlled by healds and only the pile ends are jacquard operated. To produce the design selected pile ends from each longitudinal row of pile are raised over the odd picks. The unselected or dead pile ends float on the underside from which they are removed by mechanical means during finishing. It will be noted that on the even picks all the pile ends remain down. In consequence the jacquard needs to operate at only half the loom speed the cylinder presenting the cards on the even picks and the knives lifting the selected ends on the following odd picks. A jacquard of this type has been shown in *Figure 15.9*.

At B in *Figure 16.15* a small portion of a condensed design is shown in which each vertical row represents one longitudinal row of pile, i.e. two pile ends, and each horizontal row represents two picks. A fully worked-out weave for the first two vertical rows is given at C in which the first row corresponds to the section at A. In the condensed design the solid marks show the lifts of the pile ends from frame 1 and the circles the ends from frame 2. At C solid marks and circles represent the lifts of the pile ends from frames 1 and 2 respectively, whilst the dots and diagonal marks respectively indicate the lifts of the ground ends in the top and bottom cloths.

In a higher frame cloth each longitudinal row contains three, four, or five pile ends of which only one is lifted on any given odd pick and the design for such a cloth must be painted in an appropriate number of colours. Colour planting is often carried out and sunk or pile-less places can be easily created by deliberately missing out pile lifts where required. If figuring by means of sunk places is undertaken then the portions at which it occurs must be indicated by a special colour or mark in the condensed design to denote that at such portions none of the pile ends are to be lifted.

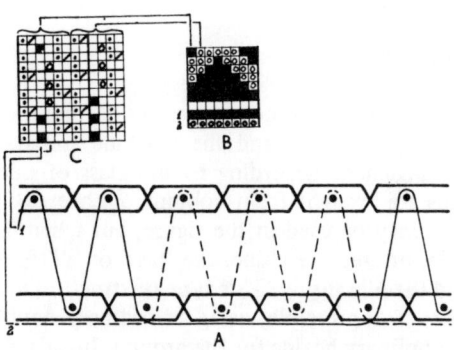

Figure 16.15

As the tuft anchorage in the structure described above is not very secure, especially in the bottom cloth, such fabrics are usually finished with a latex or other adhesive backing. Their main advantage lies in the fact that their rate of production is high, one horizontal row of tufts being produced on every pick as opposed to one for every two or three picks in other constructions, but being single shuttle they are confined to fabrics in which the pile height is limited to only about 5 mm. For this reason, and also due to a certain amount of lack of dimensional stability, the one-shot structures are infrequently used for carpets for which double-shuttle looms are preferred.

In the double-shuttle looms two to five frame designs are produced in two different major classes of structures. In the first one all the pile yarns are allocated to, and interwoven with, the bottom cloth, from which the selected pile ends are raised to the top cloth level only to form the tufts whereupon they immediately drop down. Thus, a difference in the weight of the two cloths exists which has to be compensated for by running in the top cloth additional stuffing or ground yarns but the construction permits the use of a simple jacquard system. In the second structural class the pile ends are equally distributed between the top and the bottom cloths and the ends allocated to the top cloth make the tufts by descending down to the bottom cloth level whilst those allocated to the bottom cloth ascend to make the tuft. Both fabrics are of equal weight but a more complex jacquard system is required. Each class of structure can be employed for both upholstery moquettes and carpets, and is produced in machines which operate in accordance with the same principle, irrespective of the end use of the cloth. It must be realised, however, that although the same in principle the machines differ in weight and size, much more robust constructions being required for carpetings than for upholstery cloths.

The first class of construction is shown by the weft sections A and C in *Figure 16.16* in which the former represents a 3-frame moquette and the latter a 3-frame Wilton carpet structure. It will be noted that although the ground weaves in the two sections, A and C, differ the pile yarns are operated in an identical fashion. On odd picks all the pile ends are down, on even picks the selected pile ends are raised by the knives to the high level to form tufts whilst the unselected or dead pile ends are lifted by the working comber-board to the centre position to be interwoven with the picks of the bottom cloth. Consequently, the same type of jacquard can be used in both cases the operation of which has been given in *Figure 15.9* in the previous chapter. The method of designing is the same as explained with reference to multi-frame pile fabrics produced with the aid of wires (see pp 305 to 314).

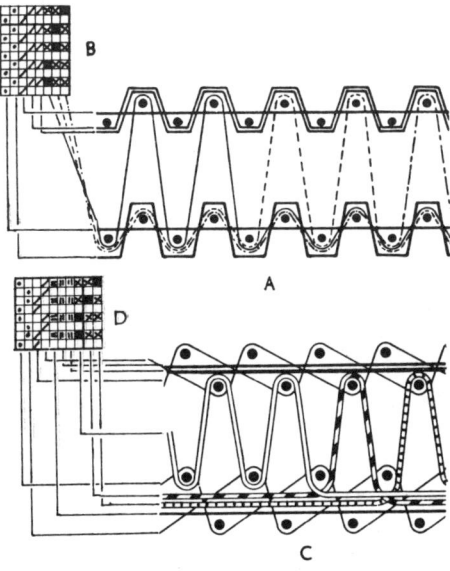

Figure 16.16

B in *Figure 16.16* represents the fully worked-out weave for the longitudinal row of pile given at A with which it is connected by fine lines. It will be noted that to compensate for the difference in weight between the top and the bottom cloths the top cloth contains two slack ground ends working as one. If greater degree of weight compensation is required, as it may be in 4 -frame or 5-frame structures, three or even four slack ground ends may be incorporated into each longitudinal row of the construction. In the carpet structure shown at C the weight compensation is achieved by introducing into the top cloth additional stuffer yarns as clearly indicated by the fully worked-out weave at D. In both B and D the solid squares represent the high lift of the selected pile ends, the crosses the comber-board lifts of the dead pile whilst the diagonal marks indicate the ground end lifts in the top cloth and the dots, the ground end lifts in the bottom cloth. At D the lifts of the stuffers are represented by the double vertical marks in the top, and by the double horizontal marks in the bottom cloth.

The second class of structures is represented in *Figure 16.17* by the weft sections A and C. Both sections depict 4-frame effects but whilst the former represents an arrangement suitable for upholstery moquettes the latter one shows a two-shot Wilton carpet structure. In this class of construction the pile yarns, as has been stated, are equally apportioned between the top and the bottom fabrics. Therefore, as there is no need for the equalisation of weight the ground structures in the top and in the bottom cloths are identical. This is clearly shown in the sections and in the fully worked-out weaves at B and D in which the same marks have been used as those in *Figure 16.16*, except that for the top pile ends, which move down to make sheds, the circles indicate the full distance drop of the selected yarns whilst the shaded squares represent the half distance drop of the dead pile yarns. The jacquard which controls the pile yarns must have a sufficiently flexible action to provide various levels of lift '

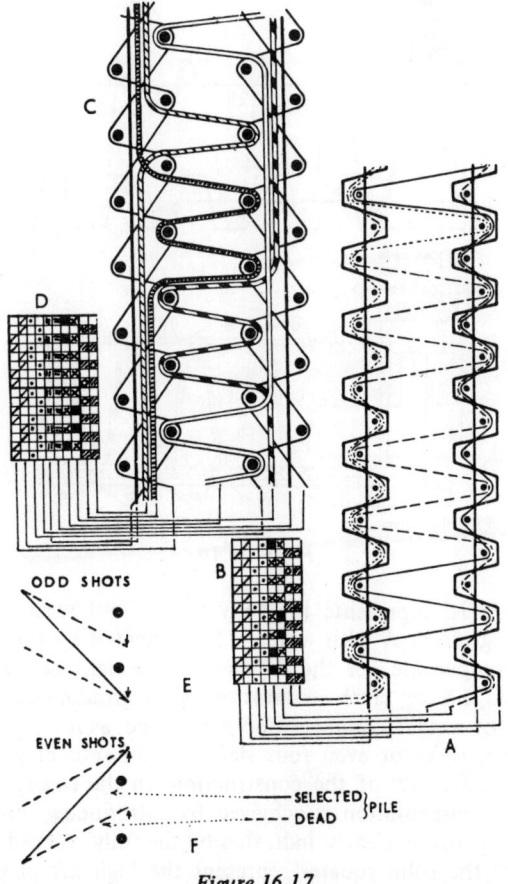

Figure 16.17

(or drop) simultaneously. Studying the sections, A and C, and the shed diagrams, E and F, in *Figure 16.17* it will be seen that the pile yarn allocated to the top cloth—in this instance frame 1 or 2—makes a figuring tuft by moving from the top to the bottom position through two shed heights, and by converse, the bottom pile yarn—frame 3 or 4—produces a figuring tuft on the following shot by moving from the bottom to the top cloth level. Each yarn returns to its normal allocated position after the figuring movement. On odd shots, the jacquard will have to provide the following movements: Top selected pile to drop right down; top dead pile to drop half-way down; all bottom pile to stay or return right down. On even shots the movements are: All top pile to remain or to return right up; bottom selected pile to lift right up; bottom dead pile to lift half-way up.

A schematic diagram of one type of jacquard which is used for the production of multi-frame face-to-face cloths is given in *Figure 16.18*. The jacquard is arranged on the inverted hook principle with the hooks which control the top cloth pile yarn facing the cylinder and the bottom cloth hooks inverted. For convenience only two hooks per short row are shown in each—the top and the bottom cloth machine although normally there are eight. Every hook is operated by its own needle and selection of the pile ends is given by a blank in the card. Thus, the cards are perforated except for the tuft forming pile ends. Any pile ends not selected by the blanks to form tufts are automatically treated as dead pile yarns. The operation of this machine is explained with reference to A and B in *Figure 16.18* which show the movement of the shed forming elements of the jacquard on the odd and the even shots respectively. On odd shots when the ends allocated to the top cloth make the figuring movement the griffe and the bottom boards descend. The selected hooks, TS, are pushed clear off the knives and fall upon the bottom board and are taken by it right down. At the same time the unselected hooks, TD, remain over the knives (because the needles which control them are opposed by holes in the card) and are taken only half-way down. On the same shot the pile ends allocated to the bottom cloth are all in the bottom shed line irrespective of whether on the previous shot they were fully up or half-way up. This arises out of the fact that on the odd shots, the bottom board and the knives, which between them control the bottom pile, are both at the lowermost positions in respect of the bottom pile yarns. On the even shots, when the ends assigned to the bottom cloth make the figuring movement, the griffe and the bottom boards ascend, as shown at B. The selected hooks, BS,

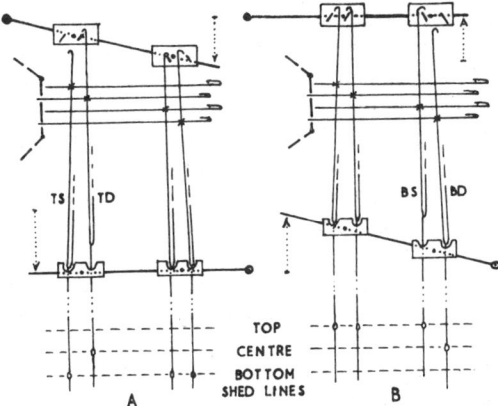

Figure 16.18

in the inverted hook section of the jacquard are pushed over the knives and are taken by them right up whilst the unselected hooks, BD, remain clear of the knives and are taken half-way up by the bottom board. On the same shot the top cloth pile ends are all taken right up irrespective of whether on the previous shot they were fully down or half-way down. This occurs because on the even shots the knives and the bottom board which control the top pile are both at their uppermost positions in respect of the top pile ends. Thus, it will be seen that the movement of the pile ends on each shot of the figuring sequence conforms exactly to the requirements stipulated in respect of the shed diagrams E and F in *Figure 16.17*. The original positioning of the griffes and the bottom boards and their movements are such that a uniform and correctly angled shed line is achieved for all the pile yarns between the front and the back harness rows in all the three shed positions. This is valuable in reducing the friction to which some ends would be subjected if the angle to which they were raised or lowered differed from that of other ends. It will be appreciated that the jacquard controls only the figuring or pile ends. The ground or chain ends and the stuffers, if any, are controlled by healds which are operated by suitably contoured positive tappets.

An equal apportionment of pile ends between the top and the bottom cloths presents no problems in 2-frame or 4-frame structures. However, in 3-frame or 5-frame effects equality of distribution in each longitudinal row of pile is impossible, therefore, to maintain the same weight and quality between both fabrics equalisation is achieved by reversing the number of frames allocated to the top and the bottom cloth over two adjacent longitudinal rows. Thus, in a 5-frame structure in odd rows two pile ends may be allocated to the top cloth and three to the bottom cloth. In even rows this allocation is reversed, i.e. there are three pile ends in the top and two in the bottom cloth.

A short row in the jacquard described in the foregoing consists of 15 needles and 16 hooks, eight in the normal section and eight in the inverted section. The one hook which is in excess of the number of needles is left out in each short row alternately from the normal and the inverted section on alternate short rows as shown at A and B in *Figure 16.19*. *Figure 16.19* shows the order in which the needles control the hooks and the tie of the harness on odd and even short rows. It will be noted that the needles, for ease of card cutting, control the pile ends from the different frames in a consecutive order, i.e. in each short row needles 1 to 5 control ends from frames 1 to 5 in the first longitudinal row or course of pile, needles 6 to 10 frames 1 to 5 in the second row, and needles 11 to 15 frames 1 to 5 in the third row of pile. Thus, each short row of needles is used to select the tuft forming ends in three adjacent longitudinal rows. If a fewer number of frames is employed in a design then the jacquard is simply cast out in long rows and the excess needles in each group of five are not utilised. For example, in a 2-frame design in each short row only the needles 1 and 2, 6 and 7, and 10 and 11 would be in use. The method of designing is the same as described in the previous chapter in connection with wire-produced multi-frame pile fabrics. To show the order of card cutting for the face-to-face jacquard a small portion of one horizontal row of a 5-frame design with a gamut is given at C in *Figure 16.19* and this is connected to a corresponding portion of a card at D.

In addition to the structures described above the same form of jacquard, with a slight change in the order of movement of the shedding mechanisms, can be used to produce carpets in which the selected pile yarns are woven through to the back. The design is fully visible on the underside which in certain markets is preferred as it resembles more closely the effect produced in hand-knotted carpets. A section of a 4-frame structure in which the design is visible on the back is given at E in *Figure 16.19*. It will be seen that the structure is similar to the one given at C in *Figure 16.17* except that prior to and upon each figuring movement the selected pile end is placed around the back shot of weft. The dead pile yarns are retained in the same position as in the standard construction.

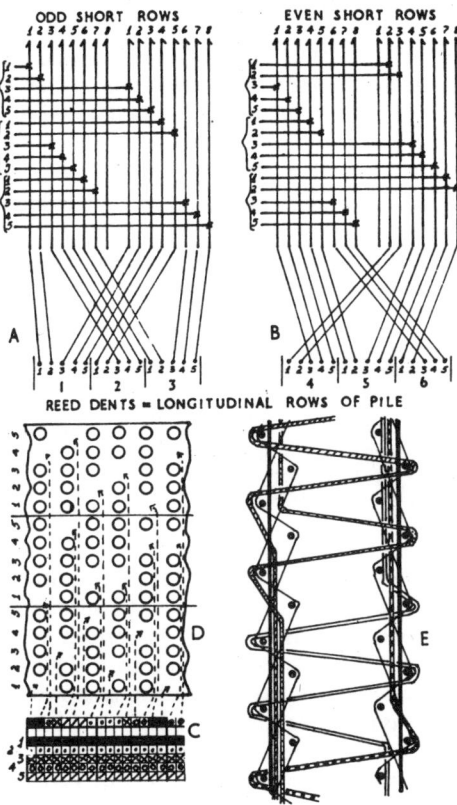

Figure 16.19

In all the multi-frame structures described in this chapter auxiliary ornamentation by using the sunk places can be easily carried out. It consists simply of not operating any pile ends in selected areas which must be distinctly marked in the design so that the card cutter leaves them unselected. To enable manufacturers to diversify their production even further, face-to-face machines may be built as convertible looms which can be used either for face-to-face weaving or, if necessary, for wire-woven pile effects with looping or cutting wires. The conversion involves removal of the knife mechanisms and the twin jacquard and substitution of the wire motion and a single jacquard and it is claimed that the complete change-over in either direction can be accomplished within a day and a half. The main advantage of this system lies in the fact that a manufacturer may follow any changes in the fashion without keeping a proportion of specialised machinery under-utilised when for a given time the demand is for one type of cloth rather than another.

In multi-frame face-to-face weaving, using the structures described in the foregoing, a considerable saving in the pile yarn can be effected compared with similar wire-produced fabrics. This is due to the fact that the pile yarns in face-to-face weaving are shared between two carpets whilst in the structures made with the aid of wires all the pile yarns are contained within one carpet. Thus, in a 5-frame structure produced by the latter method under each tuft there are four dead pile ends but in a similar structure woven face-to-face there are, on average, only two dead pile ends. The saving is of considerable importance because the pile yarns are the most expensive item in the make-up of a carpet and in a high quality structure the cost of materials may represent as much as 75 per cent of the total factory cost of an article. If required exact replicas of cut pile constructions achieved by the wire method can be produced on the face-to-face system as shown at A in *Figure 16.20* in which duplicate threads, simultaneously alternating between the two cloths, are used. However, such structures are rarely made as their manufacture results in the loss of an advantage inherent in the face-to-face method.

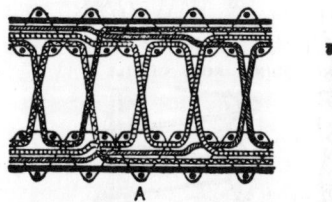 

Figure 16.20

It will be noted from the study of the standard multi-coloured effects produced by the face-to-face system of weaving that although the design between the top and the bottom cloth is identical in respect of colour it differs in respect of construction. The difference is due to a slight displacement of the tuft anchorage point between the two cloths and is shown at B in *Figure 16.20*. Normally this variation is of no significance but to avoid difficulties matching of strips of the top to the bottom piece is not recommended.

The range of materials and yarn settings in this system is as wide as that described in respect of pile fabrics produced with the aid of wires and the details of construction given in Chapter 15 can be equally well applied to fabrics for similar end uses woven face-to-face.

17

Spool and Gripper Axminster Carpets

It has been shown in previous chapters that excellent figured effects can be produced in carpets in which the tufts are formed from continuous lengths of pile yarn controlled by a jacquard. Unfortunately, such carpets, whether produced with the aid of wires or on the face-to-face principle, suffer from a limitation in the number of colours which can be employed in each longitudinal row of pile, and their production involves a wastage of the expensive figuring material in the form of dead pile yarns. It is this wastage, inherent in the construction, which for reasons of practical economics limits the maximum number of colours per row of pile to five. Admittedly, the total number of colours in a carpet can be increased appreciably by planting but excessive or unskilled colour planting may result in the formation of stripes which detract from the excellence of a design. There is, therefore, a limit to the extent of planting and when it is realised that carpet designers, especially in connection with floral designs, often speak in terms of, say, eight shades of brown and six shades of green for the background even before the main colours are considered, it is obvious that the limits of planting are soon reached.

No such limitations existed in the fore-runner of all pile carpets, the hand-knotted carpet, where every tuft could be of a different colour without creating any wastage in respect of the dead pile yarn content, with the additional benefit of a very firm tuft anchorage. The two chief systems of hand-knotting are illustrated in *Figure 17.1*, in which A shows the Ghiordes or Turkish knot which is mainly used in Turkish and Caucasian carpets. Each knot is formed on two adjacent warp threads and both ends of each tuft come between the two threads. B in *Figure 17.1* represents the Sehna or Persian knot chiefly used in Persian, Central Asiatic, and Chinese carpets. The pile yarn encircles one warp thread while the two ends of each tuft pass separately between the two threads, so that a thread and single tuft alternate. This method produces a more uniform surface of pile and enables a closer texture to be made than with the Turkish knot. Thus, while in Turkish carpets the knots range from 4 to about 16 per cm^2, Persian carpets are made with up to 80 tufts per cm^2.

Although the hand-knotted carpets are still produced their share of the market is very small and is likely to decline even further in view of the rapidly increasing labour costs. Due to the structural advantages of the hand-knotted carpets efforts were made to produce machines which could imitate the knotting action and thus improve the rate of production of such carpets. Eventually, machinery capable of reproducing faithfully either Turkish or Persian knots was devised and is still used in a number of European countries. Mechanisation of the knotting action has resulted in a severe limitation of the colour range and although it

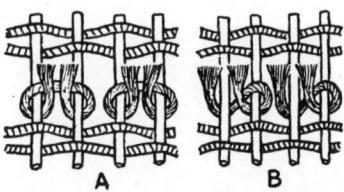

Figure 17.1

represents an improvement in the rate of production compared with hand-knotting the operation is too slow to compete on equal terms with spool or gripper systems of carpet weaving. Consequently, it is confined mainly to the production of high quality, luxury articles, usually in designs which reproduce the traditional Eastern styles of ornamentation. The inventions of the spool and, later, of the gripper Axminster systems have overtaken the machine-knotting looms and between them they now represent the major means of production of multi-coloured woven carpets with no dead pile yarns in the ground structure.

In spool and gripper carpets a horizontal row of pile tufts is formed, usually at intervals of three double picks, by passing the tufts around one of the double picks, as illustrated in *Figure 17.2*. In the warp each group of threads consists of two chain ends and a stuffer end, the length of the former being alike in the structure shown at A so that, in addition to the stuffer warp beam, only one chain beam is necessary. In the structures B, C and D, however, the chain ends vary in length so that the two chain beams are necessary. The double pick arrangement

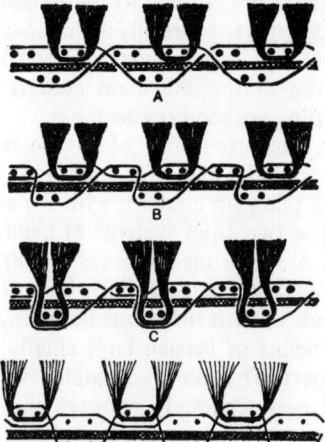

Figure 17.2

of the weft occurs because, instead of a shuttle, a rigid rapier or needle carries the weft across the full width of the warp. The weft, which is usually jute, is taken from a large cone placed near the floor, the method of supply having the advantage that a large amount of fabric can be woven without the weft being replenished.

Diagram A in *Figure 17.2* shows the three-shot structure, known as *Imperial,* which is generally woven by the spool method. The fine chain ends interweave over and under three double picks and a slightly sloping tuft, which gives good cover, is produced by the pressure of the double pick that is in the corresponding shed to that of the double pick that holds the tufts. B in *Figure 17.2* differs from A as regards the interweaving of the chain ends. This structure was first woven on a Crompton spool loom and the term *Crompton* was applied to it, but it is now produced on the gripper system and is known as the *Corinthian* structure.

Structures A, B and D in *Figure 17.2* give the most economical use of the pile yarn in the woven fabric, as each tuft forms part of the surface design and none of the pile material is concealed in the body of the fabric, except what is used for attachment to the binding threads. The structure D, however, results in a rather poor tuft anchorage and is only suitable for very densely tufted cloths.

In the structure represented at C in *Figure 17.2* the chain and stuffer ends interweave the same as in B, but the tufts are inserted round a different double pick on which the stuffer ends are raised. The pile tufts are held more firmly than in the other structures and they show on the back of the fabric somewhat the same as in a hand-knotted texture, which is an advantage. The gripper system can be adapted to weave the structure, but the pile material is not used so economically as in structures A, B, and D and the style is employed to a lesser extent. The construction C is known as the wool-back or Kardax weave.

Most of the machines are built to produce carpets in a standard pitch of 28 per 10 cm (7 per in.) i.e. 28 longitudinal rows of pile per 10 cm width. For very fine carpets looms are constructed in a higher pitch of 31 or 35 per 10 cm (8 or 9 per in.). On the other hand, for so called Berber rugs, in which very coarse pile yarns are used, special machines are employed with a pitch of only 12 or 16 per 10 cm. As the looms are built to a pre-determined warp setting the quality is changed by varying the density of weft spacing, i.e. in effect, the number of horizontal rows of pile tufts per unit length. Usually, the horizontal rows of pile per 10 cm vary between 18 to 48. Normally, the length of yarn required to form one tuft varies from 18 to 30 mm but for high pile rugs, known variously as shag pile, rya or Berber, machines are available to provide twice the above length of yarn per tuft. Modifications have also been made in the gripper system to permit production of carpets with two different heights of pile.

The ground warp yarns are usually cotton or staple viscose rayon in linear densities ranging from 180/3 tex to 250/3 tex, although other materials, such as hemp, are also used and recently some manufacturers employed polypropylene filament yarns for chain warps. The pile yarns are normally woollen spun and may be all wool, or wool in varying blends with man-made fibres or entirely synthetic and range in linear density from 500/2 text to 620/2 tex for standard qualities of carpets. For gripper Axminster carpets three-fold instead of two-fold pile yarns are sometimes used. The weft is almost invariably jute of about 520/2 tex but sometimes the bottom shot consists of low quality woollen yarn to improve the resiliency of the construction.

THE SPOOL AXMINSTER SYSTEM

A design for spool Axminster carpets is painted in the usual manner on squared paper ruled to correspond with the pitch of the machine and the number of horizontal rows of tufts per unit length. Thus, each square corresponds to one tuft. Normally the colours used in the design are the same as the actual colours of tufts in the carpet. Apart from aesthetic considerations and difficulties of dyeing and matching a huge range of colours there are no limitations as to the number of colours used in a design and some contain 40 or more different shades and hues. No colour gamut is necessary.

The system is a two-stage process comprising:

(1) Spool setting, which consists of winding pile yarn on to spools and during which the design formation and colour sequence is determined.

(2) Carpet weaving, where tuft insertion takes place in an order already determined by the previous operation.

Spool setting

A general view of a machine used for spool setting and of the associated flat table creel is given in *Figure 17.3* and an empty and a full spool are represented in *Figure 17.4*. The design is pinned on to a drum above the machine and the first horizontal space is set to a straight edge, the surface of which is divided into spaces of the same pitch as the vertical ruling of the design. The straight edge enables the order in which the colours are indicated on the horizontal space against which it rests, to be readily followed. The bobbins of coloured yarn are then creeled to pattern in the same order as the colours are shown on the first horizontal space, and the threads are drawn forward, passed in the desired order through an open reed which is the same sett as the weaving reed, and are wound under suitable tension on to the first spool, as illustrated in

Figure 17.3

Figure 17.3. A separate spool is wound for each horizontal space of the design, that is for each horizontal row of tufts, and as many pile threads are wound on each spool as there are tufts of pile to be formed in the width. Thus, a design

1 m long for a 1 m wide carpet with 25 horizontal rows of tufts per 10 cm and 28 tufts in width per 10 cm, requires 10 x 25 = 250 spools and 10 x 28 = 280 pile threads on each spool. The number of spools is, of course, influenced by the form of the design; thus, a figure that reverses at the centre requires only as many spools to be wound as there are horizontal spaces in the half-repeat, the second half of the design being woven by running the spools in the opposite direction. The spools are numbered from one upwards to coincide with the horizontal spaces of the design, and for each succeeding spool the bobbins of pile yarn are rearranged as to colour in the creel in the same order as the colours are indicated in the corresponding horizontal space, in which order they are wound on to the spool. At each creeling the length wound on is regulated according to what it is estimated will be required for the length of fabric to be woven, and if the length is considerable, two or more spools may be wound from each creeling of the bobbins. When the desired length has been run on to each spool the threads are cut and secured in position.

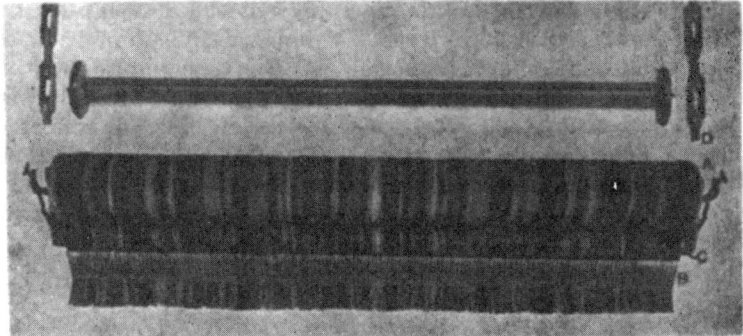

Figure 17.4

Each spool is capable of accommodating a sheet of between 10 to 15 m of pile yarn depending on the thickness of the yarns used and on the flange diameter. If it is assumed that, for a carpet requiring 2 cm of pile yarn per tuft, each spool contains 12.3 m of yarn of which 0.3 m represents waste (end wrap, trimmings, etc) then the spool will yield 1200 ÷ 2 = 600 tufts. As each spool is representative of one horizontal row of tufts in a repeat it is clear that 600 design repeats will be produced from one spool and, of course, if the length of the repeat is equal to the length of a carpet square then, effectively, one set of spools will yield 600 identical carpets. In fact, it is not normally economic to wind just one spool after each creeling and usually at least two or three spools are wound which duplicates or triplicates the number of identical repeats or carpets which need to be produced before the point of economic viability is reached. The main reason why it is necessary to wind more than one spool, and the more the better, for each creeling lies in the fact that it is the creeling operation which is time consuming; winding on is very rapid and modern spool setting frames are capable of winding the sheet of yarn on at 15 to 25 m per min. It is, therefore, undesirable to expend an appreciable length of time upon creeling for the purpose of just one minute's running on time.

After winding the threads from each spool are drawn through a series of tubes which are shown with the pile yarn protruding through them at B in *Figure 17.4*. The tubes are situated alongside each other and are secured to a tufting frame or carriage C. This is shown in *Figure 17.4* together with end bearings that support the journals of the spool in such a manner that the spool is free to rotate. The number of tubes attached to each frame coincides with the number of pile threads on each spool, and they occupy the same width, the pitch of the tubes thus corresponding with the sett of the fabric. When each spool A, with the pile threads extending through the tubes B— as shown in *Figure 17.4*—has been attached to a tufting frame C, the pile yarn is ready for the second stage of the process—that of tuft insertion.

Spool presentation

A spool Axminster loom, of the Platt split-shot type is illustrated in *Figure 17.5*, in which a prominent feature is a gantry and endless chains that support the tufting frames at each side. *Figure 17.4* shows at D the form of the chains which are made with alternate single and double links, and it also shows at the sides of the frame C the end brackets and the flat springs by which the frame is retained in the chains. The spools and frames are placed in correct order according to the design upon the chains, which are mounted on sprocket wheels supported by the gantry, and are driven either from the main shaft of the loom or by means of a separate motor. The movement of the chains brings each frame

Figure 17.5

in turn, with its spool and tubes, a short distance above the fell of the cloth, and the chains then remain at rest while the frame is lowered, the pile threads are entered into the fabric and cut off, and the frame is returned to the chains.

Figure 17.6

Only a portion of the chains is stopped intermittently, as the rear sprocket wheels are driven continuously in order to secure easy movement. In *Figure 17.5* a spool with its tubes is shown in its lower position, and the same is represented in *Figure 17.6* which shows the parts after the free ends of the tuft yarn have been entered between the warp threads and before the attached ends have been severed.

Loom operation

Figure 17.5 illustrates a typical, medium-width, split-shot loom, the distinguishing features of which are the formation of a double shed and the simultaneous insertion of three double picks of weft by means of two rigid rapiers or needles as shown at A in *Figure 17.7*. An enlarged view of the Imperial structure which the loom is built to produce is given at B in *Figure 17.7*, and shows that the order of shedding remains the same while each group of three double picks is put in. In C, *Figure 17.7* the two needles are shown at N in their respective sheds, the extreme top and bottom lines of which are formed by the chain ends. These are brought from the chain warp beam and change position after the insertion of each group of three double picks, while the middle line is formed by the stuffer ends which remain permanently in the centre of the shed. No heald is required for the stuffer ends.

The heavy jute weft which is usually precision wound on large cones is placed in three cans at the side of the loom. One end of weft is threaded through an eye of the bottom needle whilst the top needle is threaded with two ends of weft. As the rapiers (or needles) move forwards the weft is drawn from the can and is deposited in the form of a double strand—one double strand for each of three ends of weft—in the two sheds as shown at A in *Figure 17.7*. Immediately

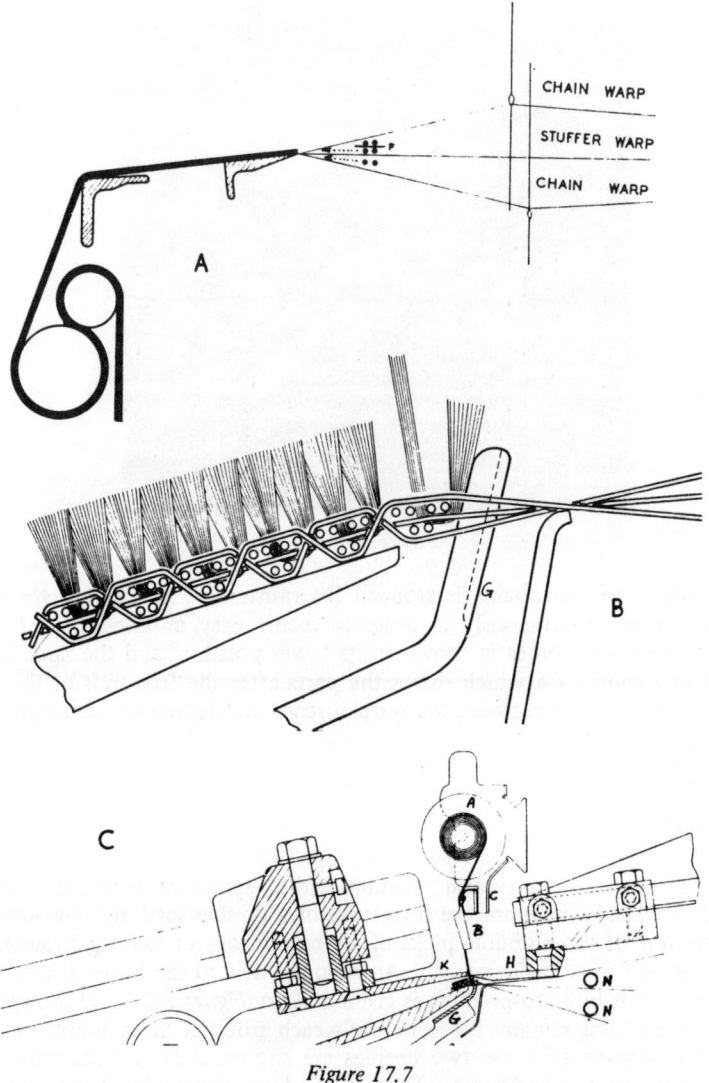

CHAIN WARP
STUFFER WARP
CHAIN WARP

A

G

B

C

Figure 17.7

after the rapiers reach the end of their stroke a separator finger, designated F in diagram A, enters the top shed and separates—splits—the two top strands into two levels. The loops or strands of weft are held by a selvedge shuttle motion at the far end of the loom, the rapiers withdraw and an auxiliary reed comes up through the warp and beats up the two lower strands of weft to the fell as indicated by the arrows and dotted lines at A in *Figure 17.7*. The uppermost strand of weft is left isolated in the middle of the top shed in a position in which a pile tuft can be manipulated around it.

Owing to the weft being inserted two picks in a shed by means of needles, a special selvedge motion is required in order to secure the picks in the selvedge and hold them at proper tension as the needles are withdrawn. The method employed in the double-needle loom is illustrated in *Figure 17.8,* where the top needle is shown entering the top selvedge shuttle race the bottom needle

Figure 17.8

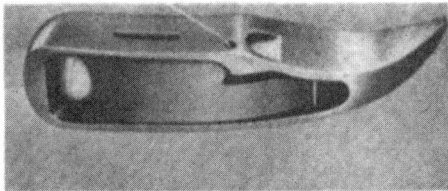

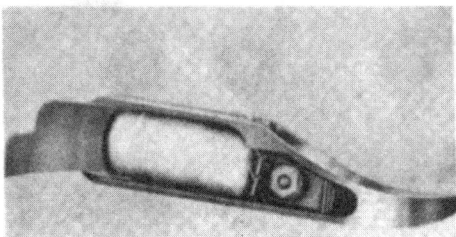

Figure 17.9

operating in an identical manner in respect of the lower shuttle. The shuttles containing strong linen or polyester catch cord move through the loops of the double picks and hold them as the needles commence to withdraw: The cords are suitably tensioned to prevent them being pulled into the body of the cloth as the needles withdraw further so that a straight and firm edge is formed.

Two views of a small selvedge shuttle containing a polyester cop are given in *Figure 17.9*. Larger shuttles which give longer uninterrupted runs have been developed for the modern broad looms whilst for narrow looms, due to high speed of operation, a stationary form of selvedge shuttle has been devised.

Tuft insertion

The operations involved in tuft insertion are illustrated in six consecutively numbered stages in *Figure 17.10*. The chain healds change position at the commencement of each cycle of operations, and the two needles insert the three double-picks of weft in the top and bottom sheds, as shown previously in *Figure 17.7* and also at 6 in *Figure 17.10*. The auxiliary reed beats up the two lower double picks and then retires. In the meantime, two arms of a transferring motion, one of which is seen in *Figure 17.6*, grip the frame C at the sides, remove it from the chains and move it downwards so that the spool A is lowered and the tubes B pass down between the groups of ends and take the tuft threads below the warp in front of the two double picks that have been beaten up, as shown at 1 in *Figure 17.10*. The insertion of the tubes B between the groups of ends is facilitated by the dents of the reed which divide them. The parts are then moved by the transferring mechanism so that the tubes are in a vertical position, as shown at 2 in *Figure 17.10*. The tubes continue to be raised, so that the upper portion of the tuft threads is brought above the level of the warp, while leaving the remainder of the threads passing vertically down between and below the ends of the warp a sufficient distance to form the right side of the tufts, in which position the main reed E beats up the third double pick against the tuft threads, as shown at 3 in *Figure 17.10*. The pressure of the reed then holds the tufts firmly, while the transferring mechanism raises the frame, spool,

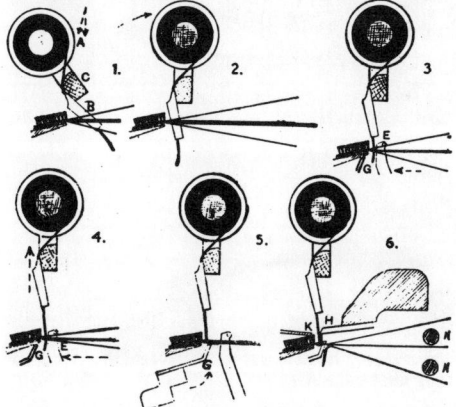

Figure 17.10

and tubes a sufficient distance to draw off the spool the required length of pile yarn to form the next row of tufts, as shown at 4 in *Figure 17.10*. During the draw-off of the pile yarn the shed commences to close and becomes fully closed as the sley moves back, while an under comb G below the fabric moves upwards, as shown at 5 and also, more clearly, in the enlarged view given at B in *Figure 17.7*. The tips of the under comb G pass through the goups of ends when the shed is closed (i.e., when the chain ends are midway of their movement as the shed changes), and they turn or double the lower portion of the tuft threads vertically round the third double pick as shown at 6 and also at B in *Figure 17.7*. As the new shed opens for the next group of picks the comb supports the tufts firmly while the attached side is cut by the knife motion shown at 6 and also at C in *Figure 17.7*. This consists of a fixed straight or ledger blade H that is parallel to the fell of the fabric, with which an inclined knife K operates on the guillotine principle at the required height to give the required depth of pile in the fabric, and at the same time sufficient length of thread is left projecting below the ends of the tubes B to form the next row of tufts. The transferring mechanism then replaces the frame with the spool and tubes in the chains which move forward and bring the succeeding spool and tubes into position in readiness for the next operation of tufting. As these movements are taking place the shed is opened and the two needles are inserted carrying the three double picks of the next group as shown at 6 in *Figure 17.10*, and the under comb continues to support the tufts until the auxiliary reed, in beating up the two lower double picks, is almost at the fell of the fabric; otherwise the tufts would fall towards the back of the loom. The under comb then moves below the fabric, and the parts assume the position shown at 1 in *Figure 17.10*.

General features

The split-shot spool Axminster looms are built in twelve different widths ranging from 46 cm to 366 cm and cannot be conveniently adapted to weave any width of cloth other than the one for which they were specifically constructed. The narrowest machines are capable of inserting 30 horizontal rows of tufts per min and the widest 14 horizontal rows of tufts per min. In favourable circumstances these speeds can be increased but when two or more looms are operated by one weaver it is advisable to reduce the rapidity of operation for the sake of better overall efficiencies. The wide looms can be made to use either sectional or single-span tube frames, and most looms are constructed to give the cloth with the standard pitch of 28 per 10 cm.

Looms can be fitted with a variety of electrical stop motions which ensure safe and faultless operation. When one weaver operates several looms stop motions are essential and frequently a full range is installed which can stop the loom for the following causes: Breakage or exhaustion of weft; breakage or exhaustion of the selvedge shuttle catch cord; breakage of chain warp end; breakage of selvedge cord; failure of warp let-off motions; failure of tube frame to leave the chains; missing tube frame; tube frame falling due to faulty replacement action; uncut pile threads. Due to the large number of stop motions they are usually connected to a central indicator panel which lights up to show the cause of any stoppage thus saving the weaver's time.

The spool system is capable of operating at very high efficiency but it is suitable only for long runs. As has been explained, the preparation processes are so time consuming that in a short run it may be impossible to recover the cost involved. The system is particularly advantageous for the production of continuous length cloth as opposed to carpet squares, especially if a short length, repetitive form of design is used. In such circumstances it is possible to achieve considerable savings in preparatory costs by running say, six or eight spools from a single creeling during spool setting.

THE GRIPPER AXMINSTER SYSTEM

A general view of the narrow gripper Axminster loom from the front and side is given in *Figure 17.11*, while the common gripper Corinthian structure is illustrated on an enlarged scale in *Figure 17.12*. (The thread interlacing corresponds with that shown in the section B in *Figure 17.2*.) The fine ends are of unequal length and require two warp beams, while a third beam is required for the stuffer ends, and the three ends which constitute a group of ground threads are operated by three healds: The reed, shown at H in *Figures 17.12* and *17.13* consists of arched dents attached to the sley which are open at the top.

On the edge of the breast plate a slotted, horizontal bar or comb L, *Figures 17.12* and *17.3*, is mounted, the slots of which correspond with the sett of the reed and admit the dents of the latter when they are in the forward position. The slots are formed to the line P in *Figure 17.12*, and are widened at the top inside the dotted line R to provide room for the warp threads which are thus prevented from being cut when the reed beats up. Incorporated with each section of the breast comb is a lip M, *Figures 17.12* and *17.13*, over which the weft is driven by the reed when beating up. The lips M serve to prevent the weft from rebounding after it has been beaten up, and they also assist in turning up the tufts after the grippers have opened.

Figure 17.11

In the gripper system the preparation of the pile yarn is the same in principle as in Wilton or Brussels carpet weaving, the different coloured pile threads being wound separately upon bobbins which are placed in creel frames behind the loom, as illustrated on the right of *Figure 17.11*. Fourteen horizontal frames or rows of bobbins are shown and each frame consists of a set of bobbins of the same colour except when modifications are made by 'planting'. The number of colours in each longitudinal line of the fabric is limited to the number of frames used, but the restriction is not so great, as in Wilton and Brussels structures, for it is possible for as many as 16 frames to be employed. Generally however, not more than about 10 frames are used, but as each frame can be planted the diversity of colour effect obtainable is very considerable. The pile yarns are selected for presentation to the grippers by a special form of jacquard.

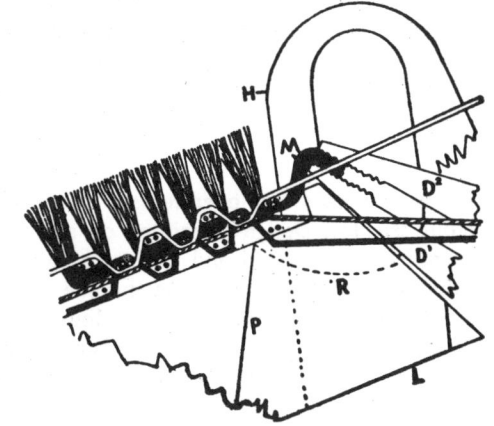

Figure 17.12

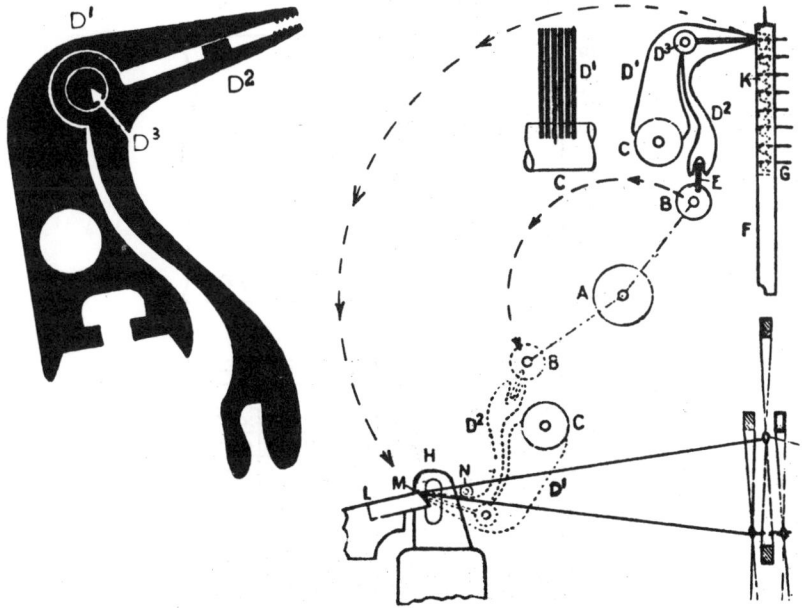

Figure 17.13

From the bobbins in the creel frames, shown in *Figure 17.11*, the pile threads pass between horizontal guide bars and through holes in a guide board to slots in yarn carriers, shown at F in *Figure 17.13*. As many yarn carriers F are provided as there are longitudinal tufts of pile formed in the width, i.e., one to each longitudinal row of pile and to each split of the reed. Each carrier consists of a narrow vertical strip of metal, grooved back and front, and contains as many slots as frames used, one above the other, through which the pile threads G pass. Suitable tensioning holds the threads in position and steadies them while they are being drawn forward and then cut to form the tuft lengths. The threads are passed through the slots in the order of the frames, those from the top frame passing through the highest slots, and those from the bottom frame through the lowest slots.

Selection of pile colours

The jacquard in this system does not operate as a shedding mechanism but simply as a pile colour selector. Design is painted on squared paper ruled in accordance with the number of vertical and horizontal rows of pile per unit space. Each square thus represents one tuft. As the pile yarns are arranged in frames it is necessary to show under the design a colour gamut to indicate to the card cutter what colours are placed on which frames—a procedure similar to the one used in Brussels and Wilton carpet designing. A gamut for eight frames, with a portion of the first horizontal row of design above it, is given in *Figure 17.14*. It will be noted that frames 1 to 3 are planted with various additional colours but by referring the design colours down to the gamut it can be easily established to which frame a given colour belongs.

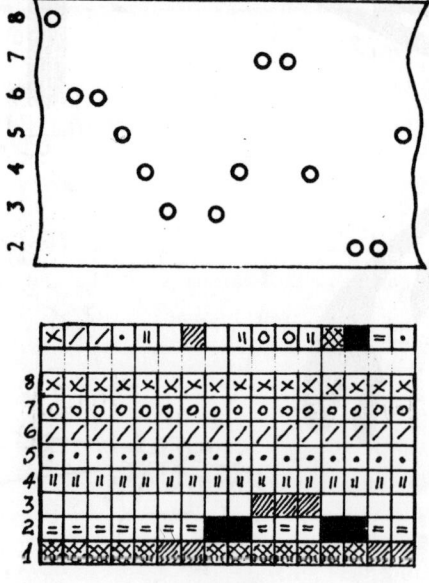

Figure 17.14

Jacquards can be built in various sizes to provide selection for 8, 10, 12, 14 or 16 frame designs. The function of the jacquard is to move the pile yarn carrier (F in *Figures 17.13* and *17.15*) to the level of the gripper which operates at a fixed height and will draw off such colour as is presented to it. A schematic diagram of an 8-frame jacquard in which a hole in the card provides selection is shown in *Figure 17.15*. It will be noted that it has only seven needles—this is so because the gripper operates at the level of frame 1 when the yarn carrier is at rest and, therefore, pile yarns from frame 1 are presented to the gripper automatically without any need for jacquard selection. This is clearly indicated by the portion of card in *Figure 17.14* where against the seventh and thirteenth tufts in the design, both of which are obtained from frame 1, no holes are cut. Obviously in the bigger jacquards there will be also one needle less than the number of frames which it can operate, thus a 10-frame machine has 9 needles, a 14 frame, 13 needles, and so on. Either coarse pitch machines with cards or fine pitch machines of the Verdol type are in use and they can be constructed to provide selection by means of either a blank or a hole. In the machine shown in *Figure 17.15* selection is obtained through the agency of a hole in the card.

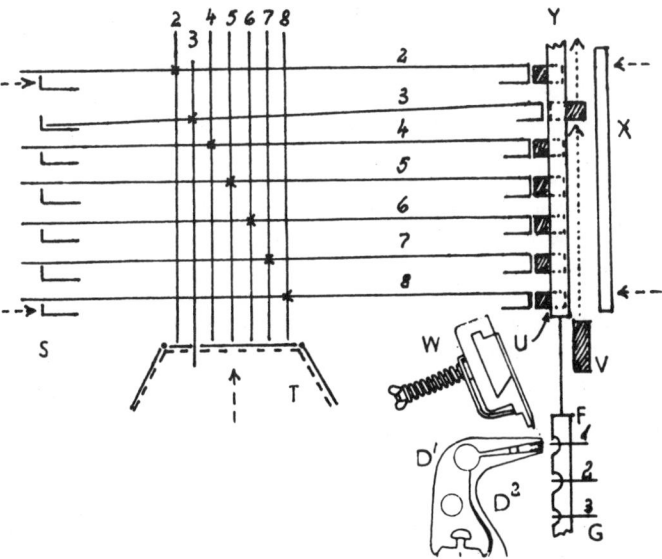

Figure 17.15

Upward pressure of the card cylinder, T, causes contact between the vertical feelers, numbered 2 to 8, and the card. Feelers opposed by blanks in the card are pushed up thus lifting the tails of the corresponding needles clear of the advancing pusher plates, S, and, therefore remain inoperative. Feelers opposed by holes penetrate into cylinder apertures and their corresponding needles remain in the path of the pusher plates and are, therefore, moved to the right—note feeler and needle numbered 3 in *Figure 17.15*. The needle pushes a corresponding lifting peg, U, carried in a slot of the selector, Y, out into the path of the ascending lifting bar, V, which has a fixed stroke. The selector is thus raised to varying heights depending upon which lifting peg is engaged. As the yarn carriers, F, are connected by wire to the selectors, *Figure 17.15*, they follow the movement of the selectors and present the selected yarns to the level at which the grippers, D, operate. In the example illustrated in *Figure 17.15* the lifting bar will engage the peg pushed out by needle 3 thus lifting the selector and the associated yarn carrier to the second height level which results in the presentation to the gripper of pile yarn from frame 3.

Tuft insertion

Carried in one casting there is a central shaft A, *Figure 17.13*, a moveable shaft B, and a fixed shaft C. Shaft A is operated from a camshaft that runs at one-third the speed of the loom so that tuft insertion takes place every three double picks. The feature of the tuft motion consists of the grippers D, *Figure 17.13*, which in side view appear very similar to the neck and beak of a bird as shown by the enlarged view on the left. As many grippers as there are longitudinal lines of pile in the fabric are arranged in a horizontal row (a front view of six is shown on the left of the top portion of *Figure 17.13*) and each consists of a fixed jaw D^1, and a moveable jaw D^2 which are privoted together at D^3. Each fixed jaw D^1 is mounted on the fixed shaft C, while each moveable jaw D^2 is forked and straddles a flat joint or blade E which is connected to moveable shaft D. By means of cams, shaft B is rocked so that at the correct time the jaws D^2 are moved privoted together at D^3. Each fixed jaw D^1 is mounted on the fixed shaft C, while each moveable jaw D^2 is forked and straddles a flat joint or blade E which is connected to moveable shaft D. By means of cams, shaft B is rocked so that at the correct time the jaws D^2 are moved privotally at D^3, and the jaws of the grippers are caused to open and close as required.

Central shaft A and shafts B and C revolve in about a semi-circle round the centre of shaft A and cause the grippers D to move between their top and bottom position. In their upward movement the grippers are open and on reaching the top position the beaks are inserted just into the grooves of the carriers F and close upon the projecting ends of the pile threads. Then a sufficient length of thread is formed in front of the carriers to give the required height of pile by moving the grippers the necessary distance away from the carriers. A vandyked bar comb, of the same pitch as the gripper shown at W in *Figure 17.15*, descends with its points between the threads and holds the latter steady whilst a travelling knife traverses the comb and cuts off the tuft lengths. The position of the grippers immediately after cutting is shown in *Figure 17.16*. In wide looms several sliding cutters are provided, one for every 45 cm width of carpet. This ensures that the grippers do not need to stand waiting for all the selected pile ends to be cut which would be the case if only one knife was used.

Shaft A then rotates and the grippers D move forwards and downwards until they pass below the upper line of the warp shed, as shown by the dotted lines in the lower portion of *Figure 17.13*, and at the same time lay the free ends of the pile tufts against the fell of the carpet. The rigid rapier or needle is inserted and passes over the grippers through the shed, as shown at N, and inserts a double binding pick which is then beaten up, the open dents of the reed passing between the grippers as they lie in the warp. The grippers then move backwards and upwards and draw the secured end of the tuft threads round the double

pick, and as the grippers rise above the surface of the fabric the beaks open
and release the threads. In *Figure 17.12* part of an open dent of the reed H
is shown in its forward position in the next slot of the comb L to that in which

Figure 17.16

a group of three warp threads is indicated. The grippers commence their backward and upward movement as the reed completes the beating up of the weft,
and *Figure 17.12* shows the serrated working points of the jaws D^1 and D^2 of
a gripper about to release a tuft of pile. As the reed moves back, a rake, which
normally is in front of the fell, moves over the row of tufts on to the warp, and
draws them into a vertical position. Two more double picks are inserted and
beaten up, and the grippers move in their semi-circular path upwards to seize
the next row of pile threads which have been in the meantime selected and are
projecting from the yarn carriers.

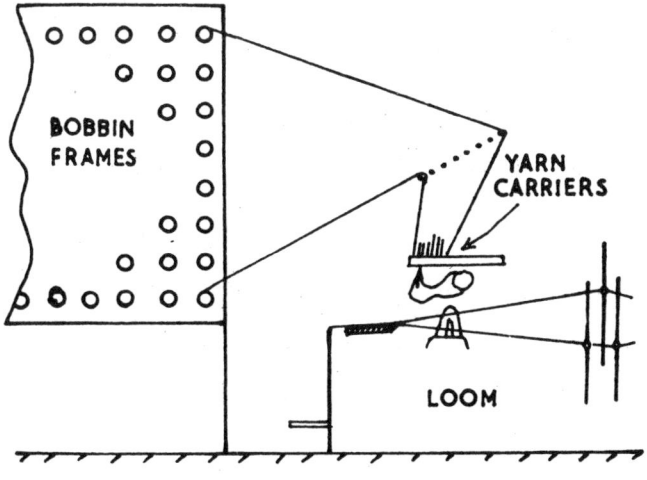

Figure 17.17

Figure 17.13 illustrates the basic principle of tuft insertion, however, the very considerable arc of travel (about 150°) of the grippers represents a very serious impediment to any improvement in the speed of operation. In modern gripper looms, although the basis of tuft insertion remains the same, the arc of travel of the gripper assembly is substantially reduced with a resultant improvement in the tuft insertion rate. In one version, yarn carriers are placed horizontally above the cloth fell and slide backwards and forwards to present selected yarns to the grippers which have only a very short distance to move to insert the tufts. This is shown schematically in *Figure 17.17* in which, it will be noted, that the pile yarn frames are placed in front of the loom, immediately behind the weaver's platform which facilitates the mending of breaks in pile yarn, etc.

General features

The take-up motion is arranged to coincide with the structure of the cloth, the take-up roller being caused to dwell while the bottom shot and the tuft shot are inserted and beaten up. The advantage of this is that the cloth is stationary whilst the pile tufts are being inserted, and the ground ends remain at the same tension. The weft, precision wound on to large cones and often magazined, is inserted in double strands by a single rigid rapier (or needle). A selvedge shuttle and catch cord, similar to the one described with reference to the spool system, holds the weft on the side opposite to the insertion side. High speed of operation is facilitated by the use of weft accumulator drums which markedly reduce the frequency of weft breaks. Accumulator drums also permit the use of lower qualities of weft yarn.

Due to the manner of tuft insertion the gripper carpet is remarkably rigid and dimensionally stable and requires little back finishing. The system is suitable for both, short and long runs and sometimes one design is produced in several different colour versions, without expensive changes and long machine down time, by partial re-creeling in the frames. As the pile yarn is wound on individual packages, either bobbins or cheeses, no excessive waste is created.

Looms are built in varying widths up to 366 cm and can operate efficiently at very high speeds. A modern 366 cm wide loom is capable of producing cloth at the rate of 18 to 20 horizontal rows of tufts per min.

THE SPOOL-GRIPPER SYSTEM

Spool-gripper looms combine some merits of the spool and the gripper system. The weaving of the cloth and the insertion of the tufts are the same as in the gripper loom, while the differently coloured pile threads are wound on spools which are supported in frames, that are carried in chains in the same way as in the spool method. The spool frames, however, are not taken out of the chains during the process of weaving, and they are specially constructed to accommodate the seizing and drawing-off of the pile yarn by the grippers. As each frame in turn is brought to the draw-off position it is held clamped to locating blocks while the required length of yarn is drawn forward and the tufts are cut off by the knife. The above situation is illustrated in *Figure 17.18;*

the gripper, D indicated in dotted lines, is shown commencing to draw a yarn from the clamped spool, S, whilst the comb and knife unit, K, is ready to come forward and sever the drawn-off length of tuft. The tuft insertion position of the gripper is indicated at D^1 which shows clearly the very short arc of movement of the gripper assembly in this system. This helps to achieve a high speed of operation which is comparable with that of the most modern of the gripper looms.

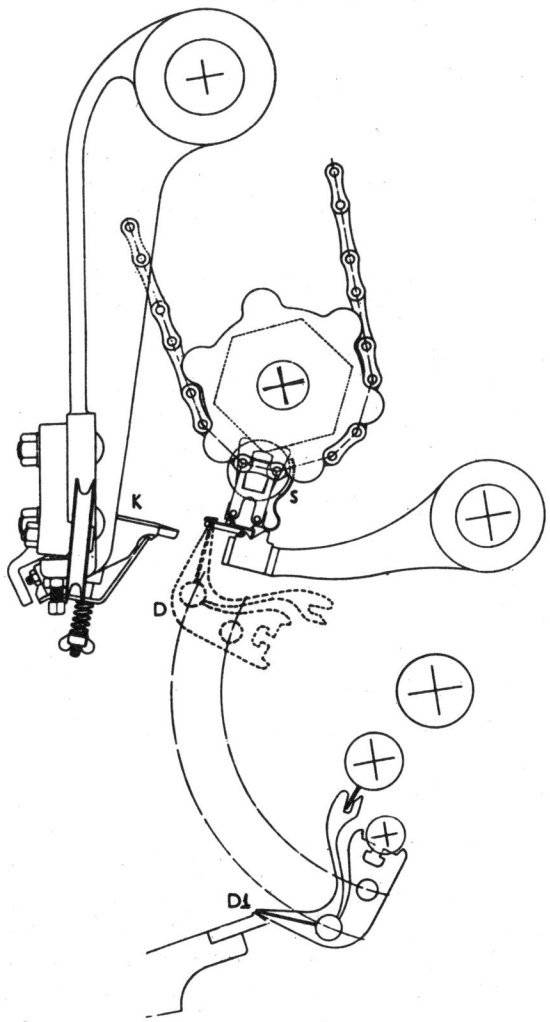

Figure 17.18

The main advantage of employing spools as the design forming unit lies in obtaining an unlimited colour scope. Some simplification of designing procedures also occurs, as no colour gamut is required, and there is no fear of stripiness which in limited colour jacquard controlled systems results from misjudged or unskilled colour planting. However, the use of spools inevitably results in some disadvantages inherently associated with the spool systems. The most important of these are the need for the time and space consuming process of spool setting and the unavoidable creation of waste in pile yarns.

The spool setting, as explained in connection with the ordinary spool system, makes it necessary to recognise the fact that only long design runs will be economically acceptable. However, as the spools in the combination systems are not the instruments of tuft insertion, the overall effort of design preparation is easier because the process of tube threading is cut out. Instead of the tubes the spool frames are made with self-threading yarn slots shown at A and B in *Figure 17.19* which represent the end and the front elevations of the frames respectively.

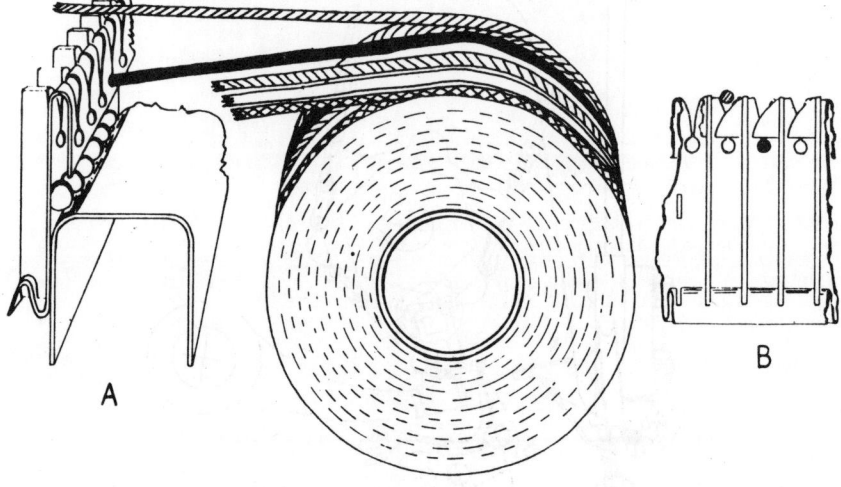

Figure 17.19

In respect of pile yarn waste, which represents an appreciable loss in ordinary spool Axminster weaving, considerable savings are achieved in the combined system. Apart from spool setting and final cropping, waste is created in the ordinary spool system at several points during weaving. The first source occurs as a slight overcut of the upturned leg of the tuft during the cutting action after tuft insertion as shown at 6 in *Figure 17.10*. The overcut is very small but it takes place at every tuft insertion and the resultant fluff which gathers at the knife edge has to be cleaned periodically as shown in *Figure 17.20*. This source of waste is eliminated completely in spool-gripper weaving as the tuft is inserted in its entirety by the gripper action. The second reason for unavoidable waste in the spool system is due to the combined effect of differences in yarn diameter in some pile yarns across the spool length and the impossibility of applying high tension to the yarns. The diameter differences are due mainly to varying dye percentages between the light-coloured and the dark-coloured yarns. As such yarns are wound on side by side the larger diameter yarns form larger diameter rings and as the yarn unwinds the spool is turned, in effect, by the thinner yarns thus creating an excess of length in the thicker ones. This excess

eventually would cause entanglements and, therefore, has to be trimmed periodically. The trimming consists of pulling the extra length available through the
tubes and cutting it off. Nothing can be done to eliminate the diameter differences but the excess occurs because the spool is comparatively weakly

Figure 17.20

tensioned so that it can be turned easily. The brake acting upon the flange of the
spool is clearly shown in the centre of *Figure 17.6* and the reason for the weak
braking action is that when the frame moves up to measure out the tuft length
(stage 4, *Figure 17.10*), the inserted portion of the tuft is held in the ground
structure only by the pressure of the reed and if the tension on the spool
were too high then instead of unwinding yarn from the spool the upward movement would tend to pull the tuft from the ground. In the spool-gripper method
the grippers are used to pull the yarn off the spool, therefore the spool tension
can be higher so that a combined pull of all the ends on the spool is necessary
before it turns. Thus, the length released is the same for all ends and no build
up of an excessive length results on the spool. Admittedly, before the spool
turns some of the thinner yarns may be stretched which means that, upon
relaxation following cutting, the tufts formed from them are marginally shorter.
However, this difference is so small that it is normally easily equalised during
the final cropping operation. The final source of waste in spool weaving is in
the form of yarn remnant wraps upon the barrel of the spool. These cannot be
eliminated in the spool-gripper system but as the spools are not detachable
they can be made to hold a greater length of pile yarn which means
that the remaining wraps represent a smaller proportion of the total than in
ordinary spool weaving. Thus, the employment of the grippers for tuft insertion
permits them to achieve worthwhile reduction in the amount of waste associated
with the use of spools.

The gripper also enables the formation of a rigid and dimensionally stable
construction, usually in the Corinthian weave, which helps to reduce the cost
of back finishing.

In the spool-gripper machine considerable attention has been paid to the
chain drive for the spools and the gantry arrangements. This was necessary
because of the great weight of a full spool complement which may reach 10 tonnes.

The chain is driven by its own independent motor with electromagnetic clutches controlled by automatic compensator devices. Provision is made for driving the spool frames forwards or in reverse direction and also in skip fashion to present alternate spools. This permits the loading of the gantry with two different designs, one on odd spools, the other on even ones. After the first complement is exhausted the second can be engaged immediately without any loom down-time otherwise necessary for the loading of the gantry. Often two gantries are used—one in operation and the other one for loading away from the loom. In this manner the loading of the gantry can proceed without the loom standing idle.

Most of the spool-gripper looms are built in the standard pitch of 28 per 10 cm, some are, however, constructed in the pitch of 24 per 10 cm for coarser work and some in 35 per 10 cm for contract quality carpets.

Appendix I

Traditional Loom Mountings and Special Jacquards

Most of the shedding arrangements described in this appendix are no longer employed or are used only to a very limited extent. As has been explained in Chapter 1 they have been devised to improve either the comparatively small figuring capacity of coarse pitch jacquards or the laborious operations of the design painting and card cutting. In most cases the special systems were capable of working only at a slow and therefore, uncompetitive speed and with the advent of a new generation of fast jacquards with high figuring capacity and complete versatility they had to be discarded.

Apart from historical interest the special mountings represent a stage in the development of shedding motions which in its ingeniousness is also technologically interesting.

HEALD AND HARNESS MOUNTINGS

Healds are used in association with jacquard harness to relieve the jacquard of the need to control such ends in compound structures which are required to weave in a closely defined repetitive order with a repeat length not exceeding eight to twelve picks. Their use results in both an increase in the figuring capacity of the jacquard and in the simplification of design painting and card cutting. In certain constructions it is still convenient to employ healds in conjunction with the jacquards as described with regard to leno and warp pile fabrics.

The number of healds used normally varies between one and four depending on the weave repeat of the ends which they control and the density of setting. They can be mounted either in front or at the back of the harness and their operation can be controlled by negative tappets, positive tappets, a dobby or the jacquard itself by utilising the spare hooks.

The functions of a heald assembly can be classified under three broad headings:

(1) To operate ground threads so that a simple and unchanging foundation structure be provided for the display of the figuring thread elements.

(2) To operate auxiliary threads such as stitching, stuffing or wadding ends in a constant order; the former for the purpose of binding the figuring

371

weft to the ground structure or to stitch together cloth layers in multi-ply constructions, and the latter to retain the wadding material in the centre of the cloth.

(3) To introduce binding weaves into cloths in which the jacquard only determines the areas in which warp float or weft float effects are produced.

Heald control of ground ends

In modern practice healds are still used to operate the ground ends in the production of jacquard figured warp pile fabrics, as described in Chapters 15 and 16, and also in jacquard lenos, as shown in Chapter 12. In other constructions they are not normally employed due mainly to their restricting influence upon the structural versatility of the jacquard. At one time, however, their use was widespread and typical examples are given in *Figure A1.1*.

A in *Figure A1.1* shows a mounting with one heald suitable for warp rib brocades or figured repps (see Chapter 6) in which the odd ends weave continuously one tabby of the plain structure as indicated by the dots in the fully worked-out design B. Figure is developed by floating the even ends as desired which is represented by the diagonal marks at B. It will be noted that the jacquard controlled even ends are all raised on the odd picks to make the opposite tabby to that produced by the heald, therefore, in the simplified design at C it is necessary to indicate the lifts of the even ends on even picks only. The card cutter will then be given the design as at C with the following instructions: Cut all marks, lace-in fully perforated cards for odd picks. The cards cut according to the design are laced-in for even picks whilst the fully perforated cards, which will have been previously punched on a repeater machine, are introduced for the alternate picks as indicated by the arrows at C.

D in *Figure A1.1* represents a mounting suitable for extra warp figured fabrics (see Chapter 2) which was extensively used, particularly for the making of Alhambra quilts. The harness mails were frequently double so that two figuring ends were operated as one to achieve better cover. The two healds operate the odd ends in plain weave order producing a complete ground structure, represented by the dots in the fully worked-out design at E, whilst the jacquard-controlled even ends form the figure by floating on the plain ground where required. The simplified design, in which only the floats of the even ends need be indicated, is shown at F. The card-cutting instructions are simply: Cut marks.

As the healds in both mountings, A and D, complete their sequence of operation on two picks they can be conveniently controlled from a plain tappet assembly on the bottom shaft of the loom.

The mounting given at G in *Figure A1.1* is suitable for book muslin structures (see Chapter 5) and although two healds are shown both are controlled from the same tappet and operate as one. It will be seen by studying the interlacing diagram H and the corresponding fully worked-out design I that the structure is a stitched extra weft figured effect in which the extra picks are retained by selective lifts of the jacquard-controlled even ends. Where the extra weft is not required it is permitted to float freely on the surface of the plain ground cloth the free floats being severed in finishing. The weft is inserted in a constant sequence of 2 ground, followed by 2 extra picks. The plain ground is formed by

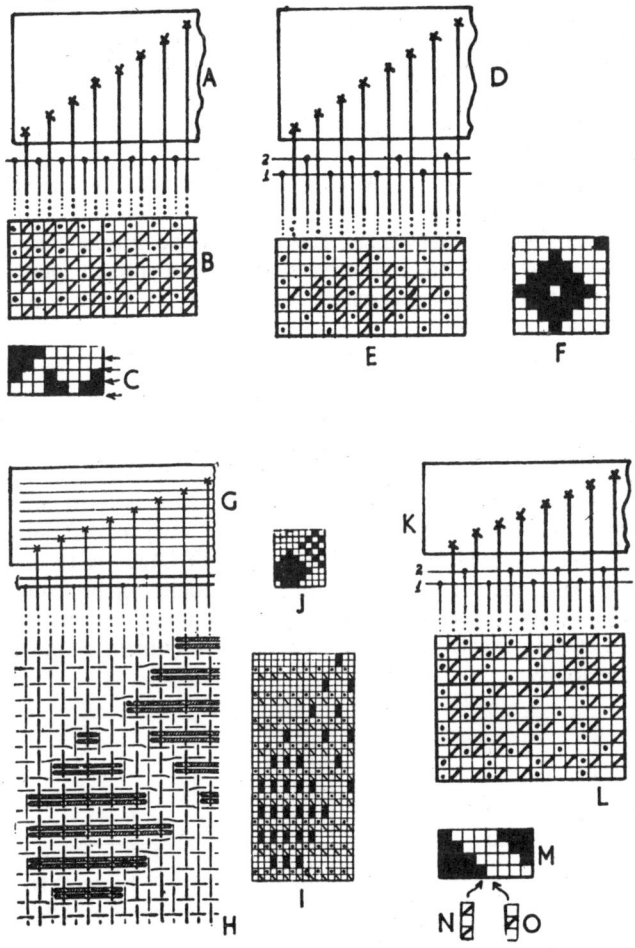

Figure A1.1

the alternate lifts of all the jacquard-controlled (even) ends on the first pick and all the heald-controlled (odd) ends on the second pick. The healds are tappet controlled and operate continuously 1 up, 3 down being up on the second pick of the 4-pick sequence as shown by the dots at I. On the first pick a fully perforated card causes a lift of all the jacquard-controlled ends, on the second pick the heald is raised and a blank card is introduced to keep the jacquard-operated ends down. Picks 3 and 4 are the extra picks on which the heald is down and the jacquard-controlled ends are raised selectively according to a simplified design given at J. Thus, the instructions can be formulated as follows: Cut two cards from each horizontal row of design (picks 3 and 4) cut marks; lace-in a fully perforated card for pick 1 and a blank-card for pick 2.

K in *Figure A1.1* shows a typical mounting used for figured terry pile fabrics (see Chapter 14). The pile alternates between the face and the back whilst the healds weave the terry ground structure, the first one operating 2 up, 1 down, and the second one 2 down, 1 up. The simplified design M is condensed by 3 weft-wise a filled square representing a pile loop on the face as given at N, and a blank square, a pile loop on the reverse side as given at O. To produce the full structure from the simplified design M the following cutting instructions must be formulated: Cut three cards from each horizontal row; first card—cut marks, second card—cut blanks, third card—cut marks. The same design could be used to produce 4-pick or 5-pick structures with suitably amended cutting instructions.

The healds in the mountings G and K were usually controlled from a negative tappet assembly mounted on a counter shaft with a suitable pinion to give a ratio of bottom shaft to counter shaft rotation of 2:1 in the former, and 3:2 in the latter case.

Heald control of stitching ends

In modern jacquard weaving healds are not normally used to operate the stitching ends but other auxiliary yarn elements are still sometimes heald controlled. One example of this occurs in the weaving of figured warp pile fabrics where the stuffer yarns are operated by a heald as shown in Chapters 15 and 16. Heald control of the stitching yarns in association with coarse pitch jacquards resulted in both, the increase of figuring capacity and the simplification of design painting and card cutting exactly as in the case of heald control of the ground ends described previously.

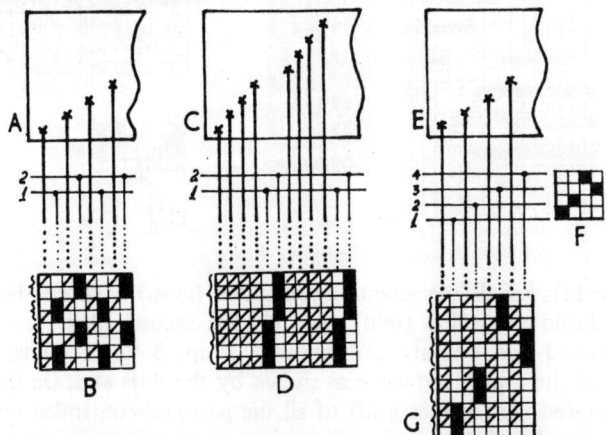

Figure A1.2

The stitching ends bind the weft yarns in a set order and the three examples given at A, C and E in *Figure A1.2* show arrangements suitable for weft tapestry structures (see Chapter 11). At A the ends are arranged 1 ground, 1 stitching, and two healds are used which operate alternately 2 up, 2 down binding the figuring wefts in a regular order as shown in the fully worked-out weave B by the solid marks.

At C there are also two healds but, as the structure is a 4-weft tapestry, they are operated in 4 up, 4 down order. The arrangement of ends being 4 ground, 2 stitching, the binding ends tend to produce a vertically ribbed surface in the cloth. In the fully worked-out weave at D the lifts of the stitching ends are indicated by the solid marks.

In the example E four healds are employed, the stitching being carried out in a satinette order indicated at F. As the structure is a 3-weft tapestry the actual heald lifts are 3 up, 9 down, as shown clearly in the full weave at G. In view of the long weft-wise repeat of the stitching ends in some of the above structures the healds which control them are not operated by negative tappets. Sometimes positive tappets are used but frequently it is most convenient to employ a dobby for this purpose as it permits of an easy change-over to suit different structures (2-weft, 3-weft or 4-weft tapestries) without the difficulty of changing complete tappet assemblies, drives, etc.

Other examples of the use of healds in the control of stitching ends are shown in subsequent sections of this appendix where healds are employed together with other shedding elements in more complex mountings.

Insertion of binding weaves by healds

In hand-loom weaving of figured damasks a system known as the pressure harness was at one time used. In this mounting the jacquard and the healds controlled the same ends but whilst the jacquard mails controlled several ends at a time the healds controlled the ends singly. The jacquard cards merely determined where a warp float or a weft float area would be formed in a block type of selection. The healds were used to bind such areas structurally by introducing individual end lifts or drops usually in a satin and sateen order. As the dual control of ends resulted in the healds countermanding some jacquard selections double sheds were formed with the resultant excessive strain on the ends which were pulled one way by the jacquard and the other way by the healds. Some attempts were made in the late nineteenth century to adapt this system to power looms but as it was based on an unsound technological principle the efforts were unsuccessful and the system is now only of historical interest. When multiplication of pattern in damask weaving is required now, jacquards of the self-twilling type are used (see Chapter 6).

THE BANNISTER HARNESS

The bannister harness (also known as the split or scale harness) was developed for weaving of fabrics which were very finely set in the warp in order to double, treble or quadruple the width of repeat. Non-reversible silk damasks are one example of structure for which this arrangement was frequently used.

In the most common form of split harness, illustrated in *Figure A1.3*, an ordinary single-lift jacquard is used, but some distance above the comber-board C each single cord D from the neck cords is connected to two or more double harness cords E, each of which is passed through a separate hole in the comber-board. A knot F is tied in each double harness cord, so as to form, above the mail, a loop G which is sufficiently long to allow the proper depth of shed to be made. Also, the comber-board is placed high enough above the knots F to permit the cords to be lifted the proper height without obstruction. The diagram on the left of *Figure A1.3* represents an 8-row machine, in which the scale is doubled—i.e., two looped harness cords are connected to each single cord D, giving 16 rows of harness cords in the comber-board C. On the right of *Figure A1.3*, three looped harness cords are shown connected to each single cord, which in an 8-row machine, gives 24 rows in the comber-board.

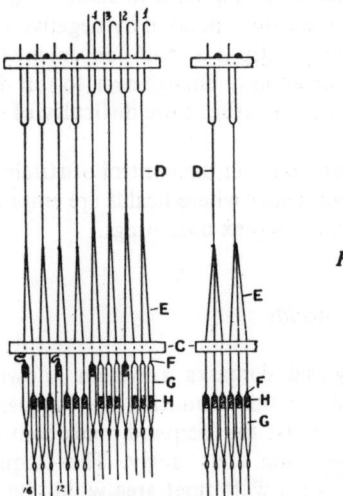

Figure A1.3

A bannister shaft or rod H, which is rather longer than the width of the harness, is passed loosely through the loops of each long row of harness cords, so that each rod is capable of lifting one end in every sixteen or every twenty-four etc., according to the number of rods employed, quite independently of the figuring cards. The arrangement does not prevent the jacquard from lifting the ends in forming the desired figure, but they are necessarily raised by the hooks in groups of two or more to correspond with the scale of the harness. By lifting the rods, the ends that are left down by the jacquard may be raised singly and produce any ground weave (plain, twill, sateen, etc.), which repeats upon a number of ends that is a factor of the number of rods employed. Thus, in the diagram on the left of *Figure A1.3*, the hooks 1 to 4, which are shown raised by the jacquard, lift up the harness mails 1 to 8, but the rods are raised in 1-and-3 order, and lift up one-fourth of the mails—viz. the twelfth and six-teenth—which are left down by the jacquard. Only warp figures can be formed on the surface as the cloth is woven, and a weft figure is therefore produced by weaving the texture face side down.

The rods H may be operated by means of a dobby, but it is generally con-venient to use a number of hooks in the jacquards, cords from which are passed through guide holes in the comber-board to each end of the rods. If the card cylinder is at the back or front of the loom a row of special hooks should be used at both sides of the figuring hooks, in order that the weight will be evenly distributed on the machine. In some cases the needles and hooks, by which the lifting of the rods is governed are situated a sufficient distance from the figuring needles and hooks to enable them to be operated by a separate small set of cards. This method has the advantage that a design may be woven in different ground weaves simply by changing the small cards.

The split mounting is sometimes arranged with two neck cords (which pass separately through a board) to each hook, and with the loops, through which the rods are passed, formed in the neck cords. It is claimed for the arrangement that the rods are situated where there is most space and are out of the way of the weaver. The double neck-cord system is also used in conjunction with a double-lift single-cylinder jacquard machine.

System of Designing

In painting out designs no ground weave requires to be filled in, as this is produced by the lifting of the rods, but the long floats of the figure require to be stopped in the usual manner. Thus, A in *Figure A1.4* illustrates the method of preparing a design for the card cutting, the instructions for which are: Cut marks. Assuming that the rods are raised in 1-and-3 twill order—as indicated at B—the full design will be as shown atC in double-scale mounting, and as represented at D in a treble-scale mounting. At the edge of the figure each step of one in A corresponds to a step of two ends in C, and three ends in D; while similarly, each single binding point in the figure represents two ends in C and three ends in D. In the ground however, the ends are operated singly, as shown by the dots.

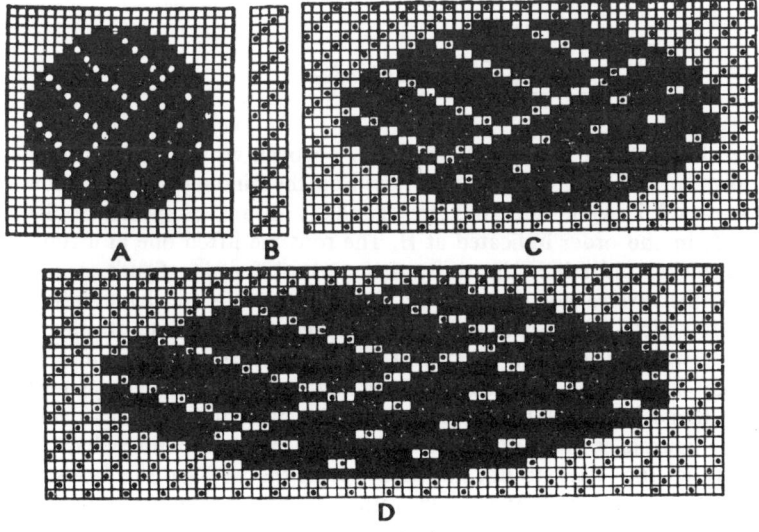

Figure A1.4

It is necessary to take into account that the lifts produced by the rods are liable to occur where ends have been left down by the jacquard for the purpose of binding the figure. The dots inside the figure in B and C indicate such lifts, and it will be seen that the binding of a warp float is neutralised at each place. In fine cloths which have a weft figure upon a warp sateen ground (produced by weaving the cloths wrong side up), the defective warp float is on the wrong side, and the fault is considered of such little importance that it is generally ignored.

Lifting rod and heald mounting for repp-stitched weft tapestries

A special form of bannister mounting used in conjunction with a heald-and-harness system and suitable for repp-stiched weft tapestries (see Chapter 11) is shown in *Figure A1.5*. In this case the rods are used simply to enable the back-stitching harness to be operated independently of the figuring harness. The hooks and needles are arranged the same as in an ordinary machine and it will be seen that there are twenty ends in the cloth to each row of eight hooks and needles.

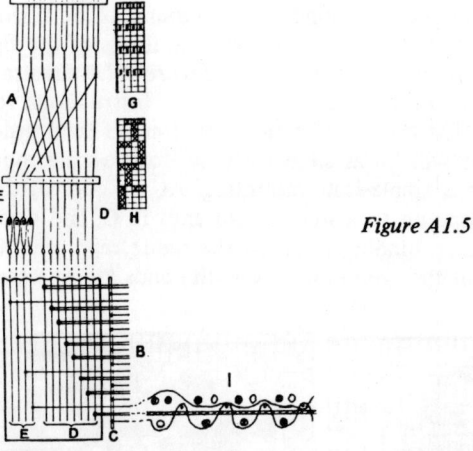

Figure A1.5

The cloth, which is a 3-weft tapestry, is woven wrong side up; the heald is left down on the figuring picks and raised on the binding picks, as shown at G in *Figure A1.5*, while the rods are raised by means of special hooks in the jacquard, in the order indicated at H. The rods are lifted one at a time on the figuring picks, and stitch the wefts that are not forming figure in a satinette order on the reverse side of the cloth. Three of the rods are always left down on the figuring picks, and it is to automatically prevent the ends controlled by them from interweaving with the weft that forms the pattern on the under surface (the right side), that the back-stitching harness cords are connected to the figuring harness cords. Where a weft forms the pattern on the right side of the cloth the figuring ends are raised, and with them the corresponding back-stitching ends, so that the latter do not interweave with the weft on the face side of the cloth. The back-stitching ends really form a loosely woven cloth with the weft on the reverse side, which, however, is stitched to the face texture by the interweaving of the back-stitching ends with the binding picks. This is effected by leaving one rod down on each binding pick, as shown by the blanks on the picks, 4, 8, 12 and 16 in the plan H, and also in the weft section I.

SECTIONAL JACQUARD AND HARNESS ARRANGEMENTS

Sectional systems of mounting are used in the manufacture of cloths which are composed of two or more different kinds of warp threads—arranged alternately, or in 2-and-1 order, etc., with one another—each of which has a separate function in forming the design or the structure of the fabric. Except when employed in conjunction with a special harness mount (e.g., working comber-boards), the object of a sectional arrangement is solely to simplify the processes, and reduce the cost of design painting and of card cutting. There is no saving, as compared with an ordinary form of jacquard and harness, in either the number of hooks, or the number of cards required for a design.

The different kinds of warp threads must follow each other in the harness in the order in which they are required in the cloth; the sectional arrangement enables each kind of warp to be governed by a separate section of the needles, so that the lifts of each warp can be cut independently upon a corresponding section of the cards. Three methods of accomplishing the result are illustrated in *Figures A1.6* and *A1.7*.

Sectional harness ties

In the method shown in *Figure A1.6*, the hooks and needles are connected in the ordinary manner, but a special system of tying up the harness is employed. A separate transverse section of the hooks is allotted to each kind of warp, the number of hooks in the respective sections being in the same proportion as the threads of each kind. From each section of hooks the harness cords are passed through a separate *longitudinal* section of the comber-board to correspond, and each kind of warp is drawn through the harness mails of the section allotted to it. The hooks are divided into two equal parts, A and B, and the harness cords that are tied to the hooks A are passed through the front longitudinal section A of the comber-board, while those tied to the hooks B are passed through the back section B of the comber-board. In the warp draft, which is represented in the lower portion of *Figure A1.6*, the even ends are shown drawn through the harness mails of the front section A, and the odd ends through the mails of the back section B. One half of the needles, taken consecutively, thus governs the even ends, the lifts of which are cut on the corresponding half of each card, while the other half of the needles governs the odd ends, the lifts of which are similarly cut on the other half of each card. In weaving designs, which are so large that two machines placed side by side are required, one machine will govern one series of ends, and the other machine the other series; the two machines being operated as one. This is very convenient for the card cutting, as the provision of two separate sets of cards enables one warp to be cut quite independently of the other.

Other proportions of the warp threads are arranged in the same manner as the foregoing. Thus, if two series of threads are arranged in the proportion of 2 to 1, the hooks of a 600-machine will be tied up in two sections of 1-400 and 401-600 to correspond, and the harness cords will be passed through longitudinal sections of the comber-board which are respectively 12 holes and 6 holes deep, or 8 and 4. For a three-thread arrangement in 1-and-1 order, the

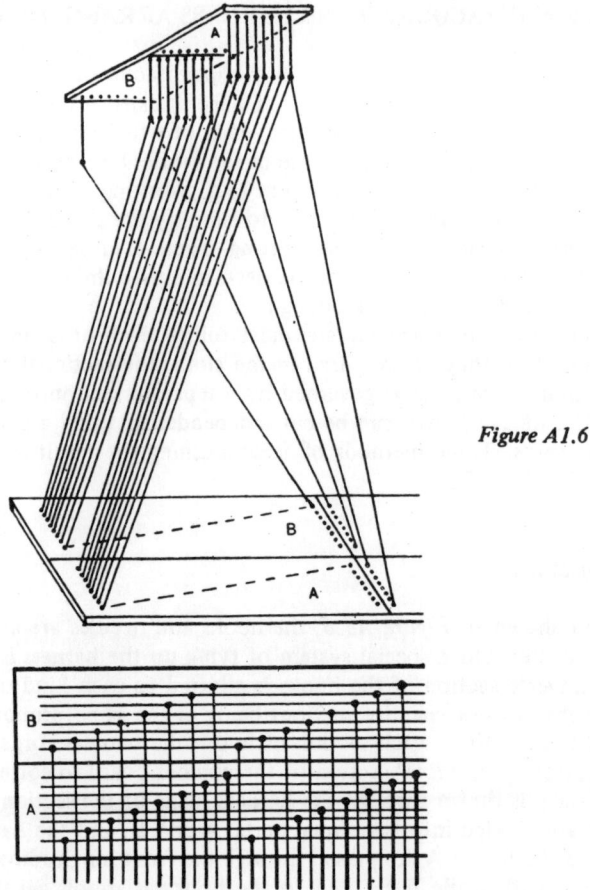

Figure A1.6

hooks and needles of a 600-machine will be in three equal sections—viz., 1-200, 201-400, and 401-600; but if there are two threads of one to one of each of the others the sections will be arranged 1-300, 301-450, and 451-600 and so on.

Special connection of hooks and needles.

Two methods are illustrated in *Figure A1.7*, either of which may be employed in place of a sectional harness tie for achieving the same results as regards the simplification of the designing. In the method illustrated by the diagram on the left the hooks C and needles D are so arranged that the four bottom needles A are connected to the odd hooks and the four top needles B to the even hooks. The harness tie and the draft of the warp threads, which are represented at E and F respectively, are exactly the same as in an ordinary machine, and it will be seen that the odd threads are controlled by the four bottom needles in each row and the even threads by the four top needles. In this system each card is divided into two longitudinal sections, as shown at G, and the lifts of the odd threads are cut on the section A which presses against the four bottom needles, and of the even threads on the section B which presses against the four top needles.

Special draft of the warp threads

This method consists simply of drawing in the warp threads in such a manner that one series passes through the front half A of each short row of harness mails, as represented at H in *Figure A1.7*, and the other series through the back half B. With the needles and hooks and the harness tie arranged in the ordinary manner, the lower half of each row of needles controls one kind of warp, and the upper half the other kind, so that the system of card cutting is exactly the same as in the previous method. An advantage of the last method is that the usual form of jacquard and harness can be adapted to the special system of designing by drawing-in the warp to suit the arrangement of the ends.

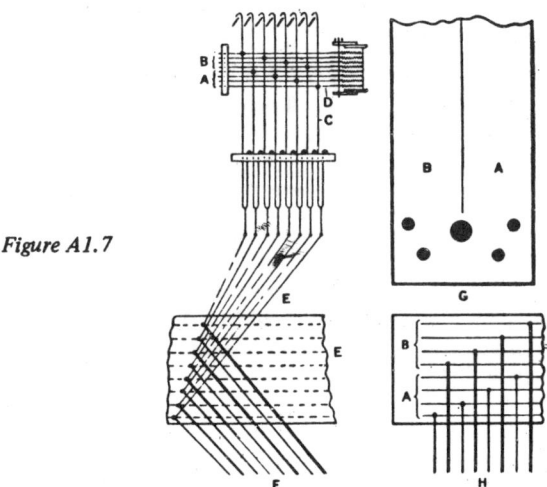

Figure A1.7

In *Figure A1.7* the arrangement of two kinds of warp in 1-and-1 order, only, is illustrated, but either of the methods may be applied when more than two series of ends áre used.

Designing and card cutting for sectional arrangements

The examples given in *Figure A1.8* show how the painting out of a design is simplified by means of a sectional arrangement, and also illustrate a method of ascertaining the card-cutting particulars by which the desired structural effects will be produced in the cloth. Four different double plain weaves are given in full at A, B, C, and D which it may be assumed, are required to be combined in a design. Two series of ends and picks in the order of a thread of each alternately, are employed, and a jacquard and harness arrangement in two equal sections, as illustrated in *Figure A1.6* or *Figure A1.7*, is therefore suitable for the arrangement in the warp. The lifts of the odd ends of the respective double plain weaves, which are shown separately at E, F, G and H in *Figure A1.8*, will be cut on one

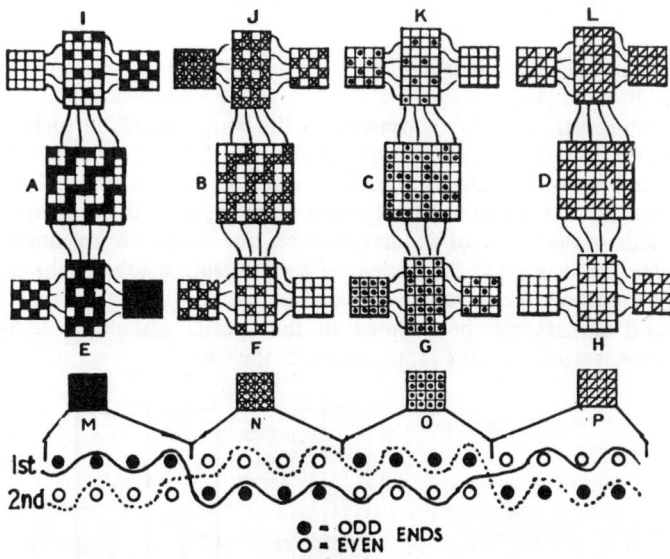

Figure A1.8

section of the cards, and those of the even ends, which are shown separately at I, J, K and L on the other section of the cards. On the left of each plan E to L the lifts on the odd picks are shown apart, and on the right the lifts on the even picks. The small plans on the left and right of the examples lettered E to L thus indicate the interweaving of each kind of warp with each kind of weft in the respective weaves A, B, C, and D.

To represent the effects shown at A, B, C and D a design would be painted solid in four different colours (or in three colours, the fourth effect being represented by the paper), as indicated by the different marks shown at M, N, O, and P in *Figure A1.8*. As there are two series of threads in both warp and weft, each vertical space in the design then corresponds to two ends, and each horizontal space to two picks. Two cards are therefore cut from each horizontal space, and further, the design is cut twice—first, for the section governing the odd ends, and then for the section that governs the even ends. The plans on the left and right of the examples E to L indicate the exact order in which the cards require to be cut from the plans M, N, O and P, and in order to enable comparisons to be readily made, bracketed references are made to the respective plans in Table 3 in which the card-cutting instructions are given.

If the cards are in longitudinal sections, as shown at G in *Figure A1.7*, for convenience in the card cutting, the design paper should be ruled in fours vertically for an 8-row machine, and in sixes if the machine is 12-rowed.

As each vertical space of a design corresponds to two or more ends (according to the number of sections) the number of spaces over which a design requires to be extended is only equal to one-half, or one-third, etc., of the number of ends in the repeat. Also, as shown in *Figure A1.8*, the arrangement frequently enables

Table 3

	Section governing odd ends	Section governing even ends
First Card	Cut M plain (left of E)	Blank M (left of I)
	Cut N plain (left of F)	Cut N solid (left of J)
	Cut O solid (left of G)	Cut O plain (left of K)
	Blank P (left of H)	Cut P plain (left of L)
Second Card	Cut M solid (right of E)	Cut M plain (right of I)
	Blank N (right of F)	Cut N plain (right of J)
	Cut O plain (right of G)	Blank O (right of K)
	Cut P plain (right of H)	Cut P solid (right of L)

the painting out to be done in such a manner that more than one card can be cut from each horizontal space, so that the design is simplified in length as well as in width; furthermore in most cases the weave structure need not be indicated.

INVERTED HOOK JACQUARDS

This type of machine is used with great advantage in weaving large designs in which two series of ends, arranged in 1-and-1 order, work exactly opposite to each other. As shown in *Figure A1.9* the jacquard is made with two sets of hooks, A and B, to correspond with the two series of ends. The hooks A have their

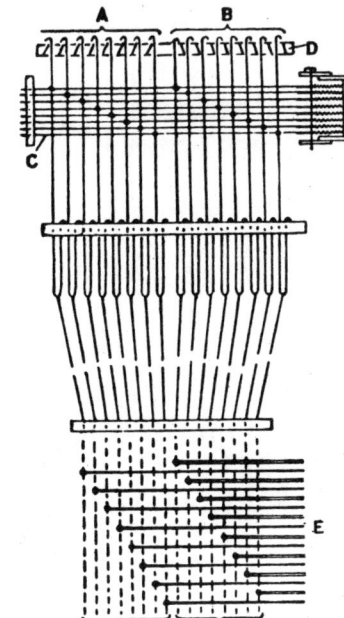

Figure A1.9

crooks turned towards the card-cylinder in the ordinary manner, whereas those of B are turned towards the spring-box. One griffe D is employed carrying 16 knives in two sets of 8 knives each, which are inclined towards the hooks that they govern. When in the normal position the hooks A are over their knives, whereas the hooks B are clear of the other set of knives. The harness cords are tied up in the ordinary manner, but in the warp draft, which is represented at E, one series of ends is drawn upon the harness cords connected to the hooks A, and the other series upon the cords connected to the hooks B. Only one set of needles is used, but each needle is connected to a hook of each set, and thus controls an end of each series. A blank in a card presses a hook A away from the path of its lifting blade, and places the corresponding hook B in position for being raised, while a hole in a card leaves a hook A in position for being lifted, and a hook B out of action. Therefore, where ends of one series are raised, corresponding ends of the other series are left down, and *vice versa*.

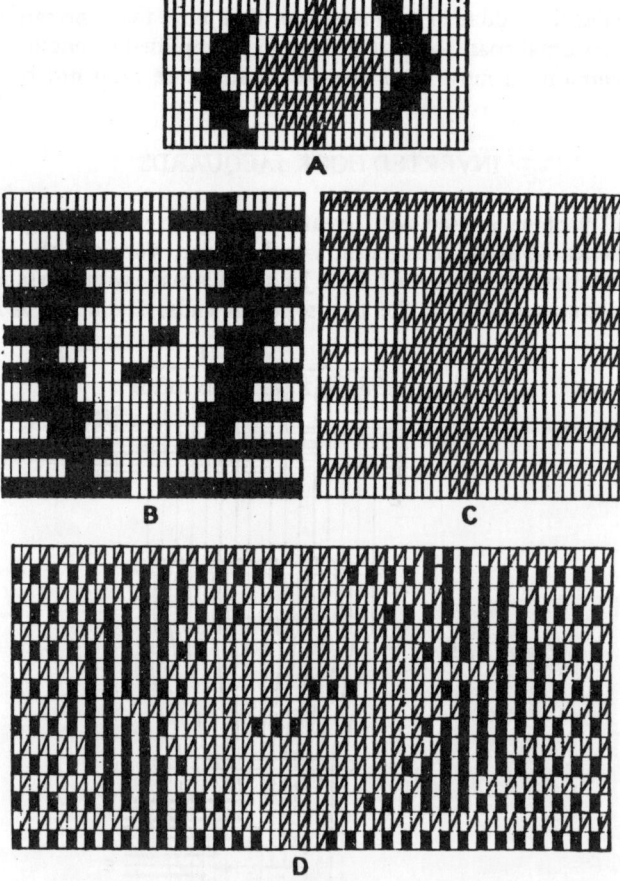

Figure A1.10

A class of fabric for which the arrangement is particularly useful is a reversible warp rib structure, in which a warp figure is produced in two colours upon plain or rib ground on both sides of the cloth. In order to show the special use of an inverted hook jacquard plans are given at A, B, C, and D in *Figure A1.10*, which illustrate the development of a portion of a design from the solid system of marking to the complete reversible rib structure. The figure formed by each colour of warp is painted solid, as shown by the different marks in A, each vertical row of which represents an end of both series, and each horizontal row two picks. Assuming that the dark figure is required to be produced by the ends which are controlled by the ordinary hooks, two cards are cut from each horizontal row, as follows:

> First card—cut all except the marks of the light figure.
> Second card—cut only the marks of the dark figure.

B in *Figure A1.10* shows the lifts that are cut on the cards and are formed by the ordinary hooks, while C, which is exactly opposite to B, shows how the other threads are raised by the inverted hooks. As the ends are drawn through the harness in 1-and-1 order, an end of B is followed by an end of C, and the complete weave is, therefore, as indicated at D.

The inverted hook arrangement not only enables a very simple method of designing to be employed, but a design is produced that repeats upon twice as many ends as there are needles in the machine.

The system is also ideally suited for the production of interchanging figured terry pile fabrics (see Chapter 14) where one needle produces a loop on the face in one colour and a loop on the back in another colour, the construction of one loop being exactly opposite from the other. In terry weaving the inverted hook machine is assisted by a heald mounting controlling the ground ends.

The principle of inverted hook operation is used at present in the production of some jacquard leno fabrics (see Chapter 12) and also in the weaving of certain figured warp pile structures (see Chapter 16) whilst the inverted keyhole system represents an adaptation of the same principle for cordage jacquards described in Chapter 15.

WORKING COMBER-BOARDS

In this system each harness cord is knotted in such a position that the knot rests on the comber-board when the harness mail is at the bottom line of the shed. The knots do not prevent the cords from being raised individually by the jacquard in the ordinary manner, whereas by lifting the comber-board all the cords, whose knots rest upon it, are raised together. The use of a single working comber-board to achieve solid lifts of all figuring threads on certain picks is described in connection with figured warp pile fabrics in Chapter 15. For patent satin and figured pique fabrics a system was developed in which two working comber-boards acted in conjunction with healds and a special method for card presentation to increase the figuring scope of the jacquard, save cards, and to simplify the designing and card cutting.

The arrangement suitable for patent satin structures (see Chapter 5) is shown in *Figure A1.11*. In these structures the weft is inserted in pairs, two fine ground picks being followed by two coarse figuring picks. The stitching ends are controlled by two healds, A and B; the ground ends are controlled by the harness which is knotted above the comber-boards, C and D. Two adjacent ground ends are attached to the same hook, 1 to 8, but pass through separate comber-boards.

To produce the structure the various shedding mechanisms are operated in a sequence indicated at I. The two healds, A and B, operate alternately 2 up, 2 down, changing places between the two fine, and again between the two coarse picks which in the fully worked-out weave F is represented by the crosses. On the fine picks the two comber-boards, C and D, lift alternately producing plain weave between the ground weft and ground warp, as indicated by the dots at F, and as seen in the weft section at G. On the two coarse picks (picks 3 and 4 of the sequence) the comber-boards are inoperative and the lifts of the ground ends are obtained from the jacquard hooks, E, upon presentation of the same card for both the coarse picks, as shown by the solid marks at F. Due to the fact that all

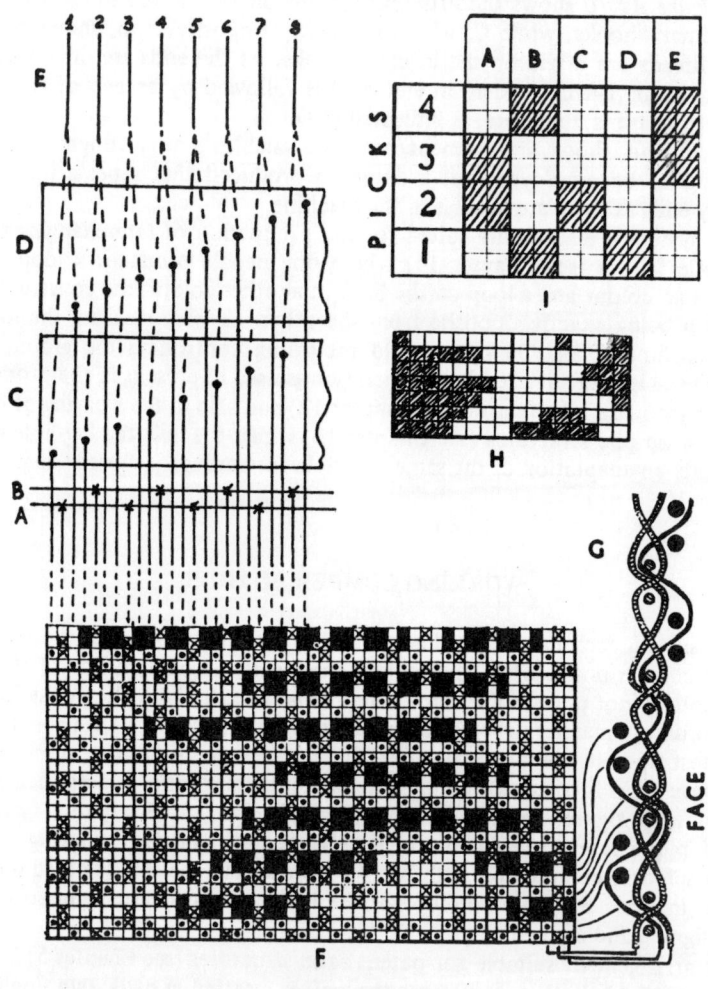

Figure A1.11

the lifts, apart from the figuring lifts of the ground ends, are independent of the jacquard the designing is considerably simplified. This is shown at H where the areas in which the figuring weft forms a float on the surface are painted solid, as indicated by the shaded squares. The design is cut according to the following simple instructions: Cut one card from each horizontal row, cut blanks. As each card acts for two figuring picks in succession and each vertical row of the design represents two ground ends, the floats of the figuring weft are automatically enlarged by two in each direction which is clearly indicated in the full weave given at F. The total enlargement of the simplified design which occurs in this system is by 3 warp-wise and by 4 weft-wise with the result that an area of 16 vertical by 7 horizontal rows represented at H in the actual cloth contains 48 ends and 28 picks.

The working comber-board arrangement for figured piques, the structure of which is described in Chapter 4, although superficially similar to the one used for patent satins, is employed for an entirely different purpose. As shown in *Figure A1.12*, the ground ends in this mounting are operated by the four healds 1, 2, 3 and 4 in which they are skip drafted. Due to this method of drafting only

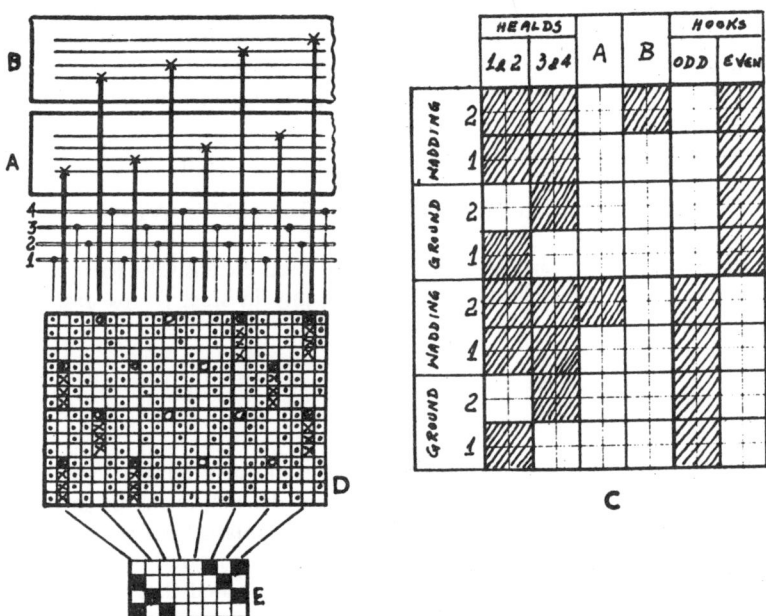

Figure A1.12

two tappets are required, healds 1 and 2, and 3 and 4 being operated as one in pairs. The stitching ends which in this structure are, in fact, responsible for the formation of figure are jacquard controlled. One card acts for a number of picks equal to the length of stitch, i.e. for two to six picks depending on the construction. When the stitching ends are not required to form a figuring stitch on the face the jacquard leaves them down to float idly on the back. As the back floats may be very long, it is desirable to stitch them in a regular order to the back of the structure. This, however, cannot be done through jacquard selection because each card is pressed in for several picks in succession and a jacquard stitch invariably shows on the surface as a figuring stitch and that is the reason for using the working comber-boards. The latter are raised only on wadding picks in plain order and ensure that no long floats are formed on the underside.

The lifts of the comber-boards A and B are synchronised with the figuring lifts of the odd and even hooks so that the understitch never occurs in the wrong place. The sequence in which the various shedding mechanisms operate in a 4-pick structure is depicted at C in *Figure A1.12*. In the fully worked-out weave at D the heald lifts are shown by the dots, the comber-board lifts by the circles, and the jacquard lifts by the crosses. The full weave area at D corresponds exactly with the simplified design E and shows clearly the degree of magnification which results. The instructions for the design E are simply: Cut one card from each horizontal row, cut marks. The healds and the comber-boards in this, as well as in the previously described system, are usually operated by positive tappets.

A different system, using four working comber-boards in conjunction with a twin inverted hook jacquard, was at one time used for the production of double plain Kidderminster carpets. The comber-boards were employed to produce the two plain cloth layers whilst the jacquard only determined which layer at any given time occupied the top position and which the bottom position. Although the system achieved considerable degree of simplification in designing and card cutting, it was so inflexible that it was discarded in favour of the sectional harness systems when more varied forms of ingrain carpets became fashionable in the nineteenth century.

STRING DOUP MOUNTINGS FOR LENO WEAVING

The string doup mounting was at one time used extensively for many leno constructions; at present lenos are woven mainly in steel doups described in Chapter 12.

The doup

In *Figure A1.13* at X, Y and Z the different types of string doups are represented, X and Y showing two methods of connecting the half-heald to an ordinary heald, while Z shows how it is connected to a jacquard harness. The crossing ends pass through the loops of the half-heald where indicated by the dots, and in the direction shown by the arrows; and it will be noted that while in X and Y the half-heald is permanently connected to the ordinary heald, in Z the loop is held in position by the warp thread. The first method is more convenient in drawing in the warp and in repairing broken ends. The advantage of the latter method is that the doup, which wears out much more rapidly than the harness, can be readily replaced; but on the other hand, if the crossing ends are absent, the loops will slip out of the mail eyes.

Bottom and top douping

The half-heald may be placed with the lath below the warp, as shown in *Figure A1.13*, or above, the position in the latter case being represented by the diagrams in *Figure A1.14*. Bottom douping is more commonly employed, as it is simpler to follow, is more conveniently applied in the loom, and the lift is not so heavy.

This method, however, makes it necessary for certain styles to be woven wrong side up. In top douping the parts are better under the observation of the operative, and the repairs to the half-heald are more readily effected; but if a loop breaks, as it hangs down, it is liable to become entangled with the warp and cause breakage, while the tension on the crossing ends, when a spring-reversing motion is used, makes it difficult in some cases to get a level bottom shed line.

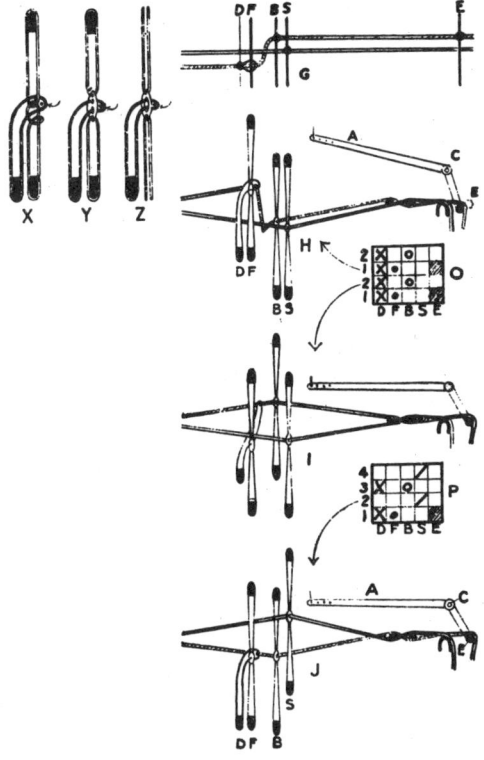

Figure A1.13

In practice, the use of the top doup is generally limited to styles which require to be woven right side up, and can only be thus produced with the top doup. For some patterns which require two doups it is found advantageous to use both methods together.

Leno drafting

For the simplest style of doup weaving the following healds are necessary: A half-heald or doup, lettered D in *Figures A1.13* and *A1.14*; a front-crossing heald F, to which the doup is attached; a back-crossing heald B, through which the crossing ends are drawn; and a standard heald, S, which carries the standard ends. (In practice the terms very largely used for the healds are respectively— doup, front standard heald, back standard heald, and standard heald, or sometimes ordinary heald. It is considered that the former designations are less liable to lead to confusion, the term standard being used only for the healds which carry the standard ends, crossing for the front and back healds which operate the crossing ends, while ordinary will be applied to healds which are not used for the douped effect, but to produce some other weave). G in *Figure A1.13* shows a bottom doup draft in which one end crosses one end. In drawing in the warp the ends are drawn through the back crossing heald B and the standard heald S in the ordinary manner. Then the front crossing heald F and doup D are placed in front, the standard ends are drawn between the leashes of F, while each crossing end is passed under the standard end and drawn through a loop of the doup D. In top douping the crossing ends are passed over the standard ends, as shown at K in *Figure A1.14*. The crossing ends may cross the standard ends either from the left or from the right. No ends are drawn through the mails of the front crossing shaft, the purpose of this heald being simply to support the half-heald.

Relative position of the healds

In mounting the healds in the loom, in order to reduce the acuteness of the angle formed by the crossing warp when the crossed shed is made (shown at H and L in *Figures A1.13* and *A1.14*) it is customary to allow a greater amount of space between the front and back crossing healds than between the other healds. In the drafts G and K in *Figures A1.13* and *A1.14* the back crossing heald is shown next behind the front crossing heald. The position is a matter of opinion and by some it is preferred to have the back crossing heald behind all or a portion of the other healds as the further back it is placed the less acute is the angle formed by the crossing ends when the crossed shed is made, and there is, therefore, less strain on the crossing warp. However, so long as sufficient space can be obtained between the back and front crossing healds, the former may with advantage be placed next behind the latter, as then there is no liability of friction and entanglement of the crossing ends with the leashes of the other healds.

In addition to the healds an easer bar (or bars) is provided at the back, labelled E in *Figures A1.13* and *A1.14*, which is operated in the same manner and for the same purpose as described in Chapter 12 in connection with the steel doups.

Sheds formed in doup weaving

These are illustrated for bottom douping, at H, I and J in *Figure A1.13* and for top douping at L, M and N in *Figure A1.14*. The formation of the *crossed* shed, which is the chief feature in doup weaving, is shown at H and L in the two figures. When this shed is formed the crossing ends are moved out of their normal position to the opposite side of the standard ends. In bottom douping the doup D and the front crossing heald F are raised, and the back crossing heald B and the standard S are left down, as shown at H; while in top douping the position of the healds is exactly the reverse, as shown at L. The crossing ends, being held by the back crossing heald in one line of the shed, and by the doup and front crossing heald in the other line, pass almost at right angles from one to the other; hence a greater length of crossing warp is required from the fell of the cloth to, say, the lease rods than when these ends are in the normal position. The easer A, is, therefore, operated at the same time as the front crossing heald, and the easing bar E is moved in from the position represented by the dotted circle; the additional length of crossing warp required thus being given in.

The formation of the *open* shed, in which the crossing ends are operated in their normal position, is illustrated at I and M in *Figures A1.13* and *A1.14*. In bottom douping the doup D and the back crossing heald B are raised, and the front crossing heald F and the standard S are left down; the loops of the doup being drawn under the standard ends and lifted with the crossing ends when the

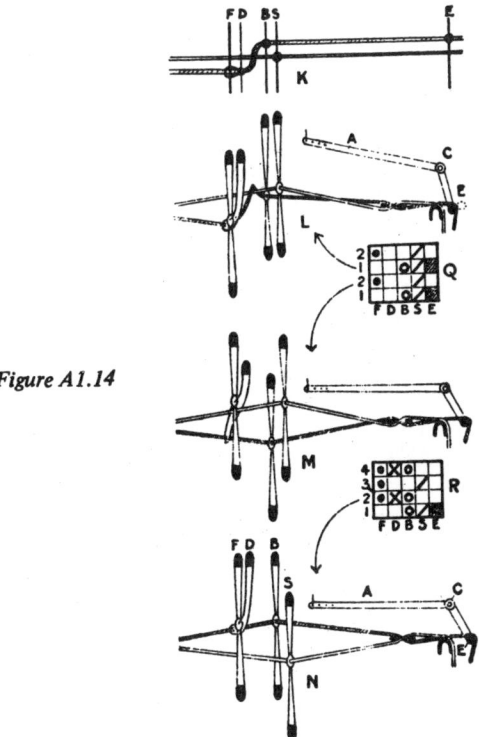

Figure A1.14

latter rise on the normal side of the standard ends, as shown at I. In top douping exactly the opposite conditions prevail, as shown at M. When this shed is formed the lever A is depressed and the easing bar E is moved outward, the stretch of the crossing warp thus being increased to the normal.

J and N in *Figures A1.13* and *A1.14* show the formation of a plain shed in which the standard heald only is raised in bottom douping, while only that heald is depressed in top douping. The easing bar E is again in its outward position.

The correct setting of the doup is of the greatest importance as regards the prevention of broken ends and undue wear of the loops. Its height should be carefully regulated, and it should move exactly in accordance with the movements of the front and back crossing healds. Thus, in bottom douping, if it is not raised sufficiently the loops will drag on the crossing ends, or on the leashes of the front crossing heald. On the other hand, if it is raised too high the loops will slide through the eyes of the front crossing heald when the cross shed is formed, while they will hang slack and be liable to become entangled on the open shed. Also, in order to facilitate the crossing movement, the standard heald should be set slightly higher in bottom douping and rather lower in top douping than the healds which operate the crossing ends.

Construction of lifting plans

O and Q in *Figures A1.13* and *A1.14* show the lifting plans for bottom and top douping respectively when the crossed and open sheds are formed alternately, the picks numbered 1 in the respective plans corresponding with the drawings shown at H and L, and those numbered 2 with I and M. The plans P and R similarly show the order of lifting when a plain shed is formed between the crossed and open sheds, the picks numbered 1 respectively corresponding with H and L, 3 with I and M, and 2 and 4 with J and N. The crosses represent the lifts of the doup, the dots of the front crossing heald, the circles of the back crossing heald, and the diagonal strokes of the standard, while the shaded squares show when the easer is operated. It will be noted that four spaces, D, F, B and E, are provided for showing the operation of the crossing ends. The easer is always moved when the front crossing heald is brought into action, while the doup is operated with both the front and the back crossing healds. In practice, therefore, the lifts of the easer and the doup are frequently omitted from the lifting plan, as the other marks of the plan readily indicate when these should be operated. Also, in order that there will be absolute certainty of the movements being in unison, in bottom douping especially, the easer is sometimes connected to the shedding lever that controls the front crossing heald; while the lath of the half-heald is connected to the back crossing heald lever, and also by cords to the lath of the front crossing heald on the opposite side of the shed.

In jacquard lenos the string doups are connected to douping harness which performs the same function as the front crossers in dobby weaving. The main part of the harness acts in the manner of back crossing and standard healds whilst another part is reserved for the easing operations. The range of jacquard structures produced with string doup is similar to that described in respect of steel doups in Chapter 12.

Appendix II

Uncommon Woven Structures

LAPPET WEAVING

Lappet fabrics can be basically classified as extra warp structures in which the extra material forms an opaque figure on an open, semi-transparent ground. The fabrics are particularly popular in the Middle East where they are often used as shawls and other traditional items of attire. Due to the difficulties of manufacture however, they are produced by only a small number of specialist firms.

The ground weave is usually plain and is constructed in two or four healds with the aid of a negative tappet assembly. The ornamentation of ground fabric consists of crammed or cord ends, coloured stripes and other such devices which do not call for the use of additional shedding mechanisms in the form of dobbies or jacquards. On this plain ground the extra warp threads, known as whip threads figure in a manner entirely foreign to warp threads by traversing horizontally across the ground ends, each such traverse forming one float of a figure

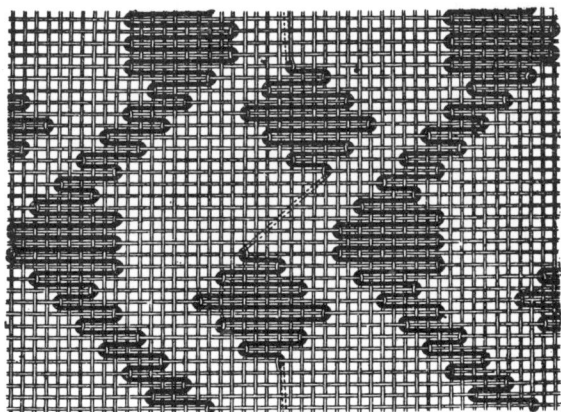

Figure A2.1

which is built entirely from a succession of these transverse laps as shown in *Figure A2.1*. As the crosswise movement of the whip thread takes place under the ground warp line, and the action occurs between picks, two points become obvious: (1) The fabric is woven face side down; (2) no interlacing of the float is possible in the middle of its traverse. The float can be bound only at each extremity which clearly imposes a certain limit on the extent of each traverse.

From this description it is clear that two distinct and independent movements of the whip thread are necessary—the horizontal or figuring movement which produces the float, and, at the end of each traverse, a vertical or stitching movement which binds this float to the ground cloth and thus determines its extent and position.

The figuring movement

The figure-forming elements of the lappet system are represented in the schematic diagrams in *Figure A2.2*. The whip threads, W, are drawn upon needle bars, D, which are thin slats of wood with eyed needles, N, spaced at intervals. These are the chief control and shedding elements for the extra yarn. The needle bars are fixed on shifter bars, R, in such a way that any horizontal movement

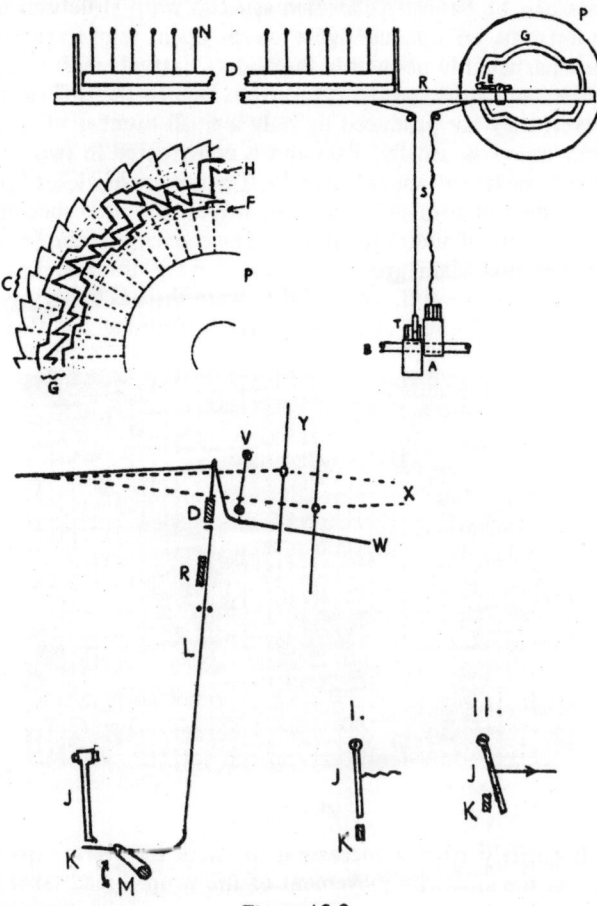

Figure A2.2

of the shifter bar has to be followed by the needle bar. The horizontal oscillating movement of the shifter bar is obtained through strap connections, S, and treadles, T, from a set of tappets, A, and these are timed to give a traverse from left to right between, say, picks 1 and 2, and traverse from right to left between picks 2 and 3, in this way giving the required sideways deflection of the shifter bar, the needle bar, and therefore the whip yarn. The tappet stroke and the treadle leverage are designed to give to the shifter bar a maximum movement of about 10 cm, and if the shifter bar were left at the mercy of the tappet alone, the whip thread would merely produce 10 cm traverses in both directions, resulting in a broad vertical bar composed of excessively long floats. This is where the patterning mechanism takes control.

The pattern or selection mechanism is so designed that it is capable of arresting the tendency of the shifter bar to move from one extreme of its stroke to the other; in other words, the selection mechanism is able to form the pattern by over-ruling the dictate of the tappet assembly. When the tappets demand the shifter bar to move 10 cm to the left, the pattern mechanism will let it move only, say, 8 mm to the left; when, between the next pair of picks, the tappets dictate a shift of 10 cm to the right, the selector may allow a movement of only, say, 4 mm to the right.

The pattern-forming unit does not, therefore, conform to the usual concepts of such a mechanism. It is, in fact, a novel type of selection device, and consists of a large wooden wheel P, in *Figure A2.2*, in which a deep groove, G, is cut. The groove is the actual pattern track for a peck or feeler, E, which is rigidly connected to the shifter bar and restricts its movement when it stops first against one wall of the groove, then against the other, in this way ignoring the tendency of the tappet to move farther. Obviously with such an arrangment there cannot be a rigid connection between the shifter bar, the treadle, and the tappet. The arrangement is such that the treadle follows the tappet merely by resting upon it, and is itself connected to the shifter bar by a leather strap. When the peck is stopped by the pattern groove wall, the weighted treadle lever is prevented from following the diminishing diameter of the tappet and hangs suspended on its leather strap, as shown in respect of the left-hand strap, S, in the upper diagram in *Figure A2.2*.

The maximum patterning capacity of lappet looms is four independent needle bars; therefore, up to four shifter bars may be employed, each with its peck working in a separate pattern groove of the wheel. Each pattern groove may give a distinct order of lapping and therefore four different figures may be formed simultaneously.

The pattern wheel itself is rotated one tooth (C in *Figure A2.2*) in two picks by a spring-loaded hook operated from a separate tappet on the bottom shaft. This means that the peck will perform both the traverse to the left and the traverse to the right within the radial space represented by one tooth. This does not preclude the possibility of one traverse being longer or shorter than the other, as can be observed by following the path of the peck, H, in the enlarged view of the pattern wheel in *Figure A2.2*.

The stitching movement

At the end of each horizontal traverse the whip thread normally proceeds above the ground warp line in order to be stitched to what will finally become the underside of the fabric. This vertical lifts of the whip thread presupposes a vertical lift of the needle bar, and this is the reason why the needle bar is not permanently clamped upon the shifter bar. The shifter bar-needle bar connection is a rigid one only as far as lateral movement is concerned: the needle bar is quite free to move vertically. Since the shifter bar is incapable of imparting any vertical movement, a separate mechanism is required. This is provided by rods, L, in *Figure A2.2* which support each needle bar from underneath at each end. These rods pass between the shifter bars without obstructing them, and the needle bars can still move sideways freely under the impulse of the shifter bar by sliding upon the flattened tops of the rods. The rods are also quite free to move up and lift the needle bars so that the whip threads are forced into the upper shed line, the shuttle passing under them, in this way stitching the extra thread to the upper side of the cloth.

Normally the whip threads are raised after every pick. If, however, it is desired to create a longitudinal or diagonal float on the face of the fabric, vertical movement of the whip threads must not take place; therefore there must be means of control and selection that will enable the stitching lift to be governed at will. This selective control is provided by the pendants, J in *Figure A2.2,* which in their normal position lift the rods and the needle bars, but which can be moved sideways, in this way preventing a lift from taking place. The pendants, four at each side to serve the four needle-bar rods are connected by cords to hooks which work at the back of the pattern wheel. These hooks are in their rest position, with pendants vertical, when normal stitching takes place. In places where this is not required a semi-circular baffle plate will be inserted at the back of the wheel to correspond with a certain number of picks during which the stitching is to remain inoperative. When the baffle plate moves opposite the hook, the hook will be deflected and through its cord connection will move the pendant sideways, thus preventing a lift of the rod. The operative and inoperative positions of the pendants J are indicated respectively in supplementary diagrams I and II in *Figure A2.2.*

To achieve the stitching movement a short lever, M, mounted on the rocking shaft lifts the toe, K, of rods, L, against the pendant, J, as the sley moves back. The toe, K, pivots against the pendant which causes the upward movement of the rods, and the shifter bars which rest upon them. If the pendant is withdrawn no pivot point is provided for the toe, K, and no lift can take place as the heel of the rod L, rests against the rocking shaft. During the forward movement of the sley the rocking shaft turns the lever, M, downwards causing the rods, L, to fall so that the tips of the needles are below the cloth line in readiness for the next lateral figuring movement.

Auxiliary mechanisms

In this system of weaving there are certain important auxiliary mechanisms. One of these is the pin bar. The reed in this system is very far back because it has to accommodate in front of it the series of needle bars which move the whip threads horizontally or vertically, and these actions must take place in front of the reed to be effective. As a result, during shed forming, which is a combined operation shared between the healds controlling the ground ends and the needle bar controlling the extra ends, the reed cannot serve as a back support for the shuttle because of the presence of needles in front of it. Therefore, since the precision of shuttle flight is not sufficiently good, some other type of shuttle guide must be provided. This is given by the pin bar, which is similar to a needle bar but does not carry any yarn. Its sole function is to serve as a false reed during picking. It will therefore rise in common with needle bars just prior to picking and immediately after picking it will recede, again together with the needle bars, so that the reed can come forward without any obstruction tò beat up the last inserted pick of weft. Since the action of the pin bar is so similar to the action of the needle bar, it will be operated from the same type of rod and from an identical source. The only difference is that the pendant against which the pin bar is acting will be permanently blocked, since the pin bar will move up without fail before every pick.

Another auxiliary mechanism of some interest is the tensioning mechanism for the whip threads. Because of different uptakes, the yarn supplying each of the four needle bars must be placed on a separate small beam (known as a whip roll), and it must be separately tensioned. The tension is delicately balanced, because the yarn should be taut enough to form a clear shed and to prevent curling of the horizontal float, and yet no undue pressure must be exerted at

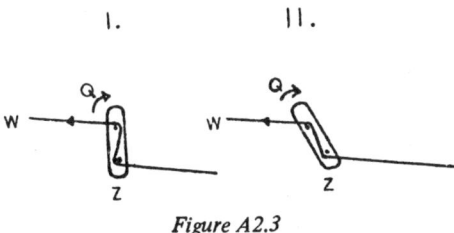

Figure A2.3

the end of each traverse, as this would pull and distort the ground ends. Also, the whip thread is subjected to sudden demands when the needle bar moves laterally and when it rises to form the shed, and the excess of yarn given at that moment must be readily removed when the needle bar moves down again. It would be difficult to obtain the necessary delicacy of balance together with the oscillation directly from the roll, and therefore a separate spring-loaded mechanism is provided between the small beam and the needle bar as shown in *Figure A2.3*. At I the whip thread, W, is represented in its normal position running between two cords in the frame Z which is springloaded in the direction shown by the arrow Q. When a sudden demand for extra length of yarn is transmitted the delicately balanced frame overcomes the springloading and swings to a horizontal position as shown at II. This position, apart from satisfying the sudden demand, permits the yarn to slip easily between the cords and deliver a length by rotating the roll.

Relation of movement of shifter bars to rotation of pattern wheel

The turning of the pattern wheel, which occurs only once in two picks, may be arranged to coincide with the movement of the shifter frames, either to the right or to the left. In most designs, however, it must agree definitely with one of the movements, according to the way in which the pattern grooves have been cut. There is no hard-and-fast rule; thus, a system may be followed, for both right-hand and left-hand looms, of having the rotation of the wheel coinciding with the movement of the pecks—(1) from left to right or (2) from the outside to the inside of the grooves.

Figure A2.4 illustrates how necessary it is, in certain patterns, for the rotation of the wheel, the movements of a shifter frame, and the manner in which a pattern groove is cut, to coincide with each other. In A, which illustrates the traversing of a whip thread, as viewed from the upper or wrong side of the cloth, the vertical spaces represent the splits of the reed, and the horizontal lines the picks of weft, the pattern extending over 20 splits and 30 picks. The whip thread is shown traversing 6 and 4 splits alternately, except where the pattern turns, in which positions consecutive moves of 6 splits are made. B shows a section of the pattern wheel in which the dotted concentric lines correspond with the splits, and the dotted radial lines with the moves to the left of the whip threads as shown by the connecting lines; while the thick solid lines indicate the edges of the groove, which repeats on 15 radial lines, or one-third of the wheel. The width of two splits only is, for convenience, allowed for the diameter of the peck, and the groove is, therefore, shown two concentric spaces wider than the distance that the whip thread is required to be traversed. The arrangement is for a left-hand loom, and as the movement from left to right (the odd horizontal spaces of A) is taken to coincide with the turning of the wheel, the centre of the peck will traverse the concentric spaces as shown by the solid lines within the groove. It will, of course, be understood

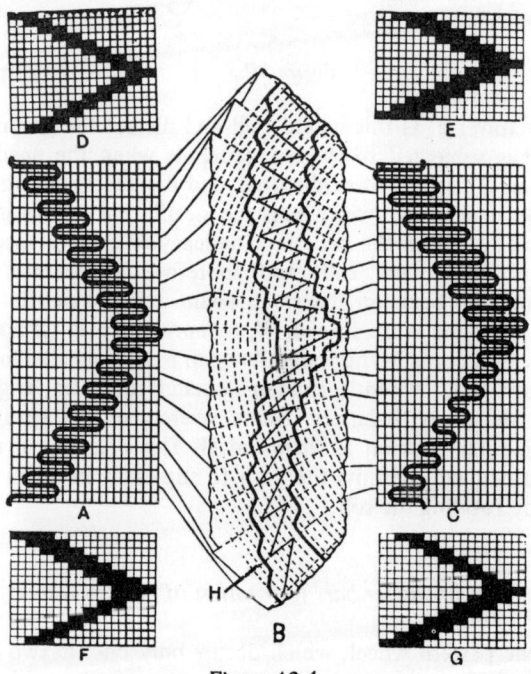

Figure A2.4

that the movement of the centre of the peck is always in a horizontal plane in line with the centre of the wheel. When the peck is moved from right to left (the even horizontal spaces of A), its centre follows a radial line and the lateral distance traversed by a whip thread is equal to the width of the groove minus the diameter of the peck, or $6 - 2 = 4$ splits in the lower portion, and $8 - 2 = 6$ splits in the upper portion of B. This does not apply to the movement from left to right, as while this is about to take place the wheel turns, and the peck, therefore, moves opposite a new position in the groove; the distance traversed being greater in this case where the inner edge is approaching the centre of the wheel and less where it is receding from the centre. It will thus be noted that in order to obtain the alternate movements of 6 and 4 splits in each half of the pattern, the groove is narrower in the lower than in the upper portion of B; and a representation of the form of the groove (leaving out of account the diameter of the peck) in solid marks on design paper, will not be as indicated at D or E in *Figure A2.4*, but as shown at F.

C in *Figure A2.4* shows how the pattern would be affected if the rotation of the wheel took place at the opposite movement of the shifter frame to what the groove has been cut for. In that case the turning of the wheel would coincide with the movement of the peck from right to left, and, compared with A, the traverse of the whip threads would be curtailed by the lower portion of the groove, and increased by the upper portion. With the latter timing of the rotation of the wheel, in order to produce the effect given at A, the groove would require to be cut according to the plan indicated at G.

Representation of lappet designs

In many cases the wheel cutter is simply provided with a sketch of the figure that it is desired to produce, and he prepares the plan from which the wheel is cut. In constructing a plan for the wheel-cutting squared paper may be used with advantage and different methods of representing a figure are shown in *Figure A2.6* in which each plan corresponds with the pattern given in *Figure A2.5*. The design repeats on 50 ends and 42 picks, or 25 splits of the reed,

Figure A2.5

and 21 teeth of the wheel, and the differently shaped figures, arranged in alternate order, are formed by one needle bar. A in *Figure A2.6* shows exactly how the whip threads are traversed in the cloth as viewed from the wrong side, each vertical space representing an end, and each horizontal line a pick. The dotted lines show the portion of thread which is cut away after the cloth is woven, leaving the figures quite detached from each other. If the count of the design paper is suitable for the proportion of ends and picks in the cloth, this method gives an accurate representation of the effect; but it is not convenient for the wheel cutter, since two vertical spaces correspond with one split of the reed, or one circular space of the wheel. By using paper in which each large square is divided into spaces in the same proportion as the splits per unit space are to the picks per unit space—as, for example, for a square cloth into 4 spaces vertically, and 8 spaces horizontally—a convenient representation of the design may be

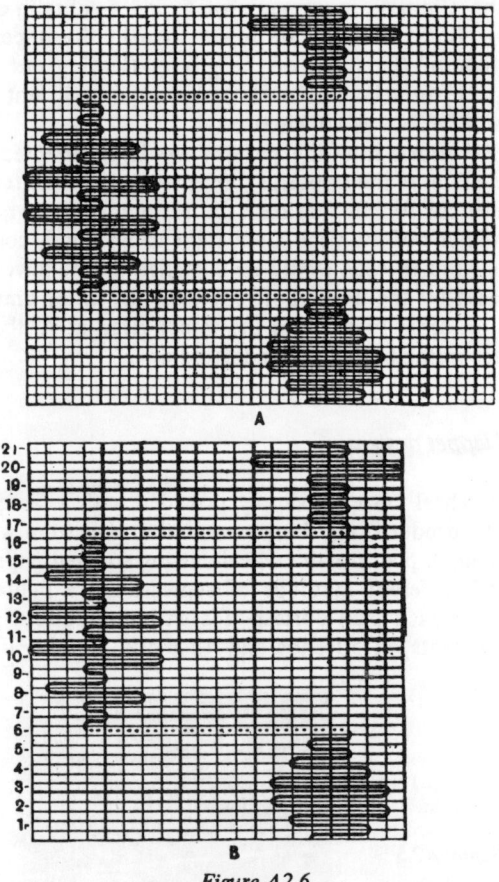

Figure A2.6

made. Thus, B in *Figure A2.6* shows the design A worked out on 4 × 8 paper, each vertical space of which represents a concentric space in the wheel and each horizontal line a pick. The full width of the repeat is not shown in this plan, as the wheel cutter is concerned only with the space over which a thread is required to be traversed. The accurate repetition of the figure in width is dependent upon the spacing of the needles in the bar. In the method shown at B the outline of a figure may be first drawn to scale on the paper in the ordinary manner, and then the required moves be indicated, as represented in the example.

Construction of a lappet pattern wheel

Lappet wheel is made from hard, fine grained wood. Its thickness is from 15 to 25 mm, and its diameter varies from about 20 to 60 cm, according to the number of picks in the repeat, and number of frames employed in forming a pattern. A hole is bored, about the centre of a piece of wood of suitable size, to fit the socket of a lathe, in which it is turned to the proper diameter. On the side of the wheel where the groove or grooves are to be cut, a steel comb is pressed while the disc is revolving, a number of concentric lines thus being made, the space between which corresponds with the pitch of the comb. Combs are made to suit different reeds, and for below about 14 splits per cm it is usual to use a comb of the same pitch as the reed for which the design is intended. With more splits per cm the fineness of the marking presents a difficulty, and a comb may then be used which is one-half the count of the reed, the half distance between the marks being judged by the eye in indicating the shape of a groove. The concentric lines may be marked to within 12 mm from the edge of the wheel, and another line is then marked about 5 mm from the edge to indicate the depth of the teeth. The circumference is next divided into as many equal parts as the number of teeth required, each tooth representing two picks, and radial lines are drawn from the divisions to the centre of the wheel. The teeth are then cut the required depth with the edges in line with the radial lines. Each engagement of the turning catch brings a radial line in a horizontal plane with the centre of the wheel and the centre of a peck moves upon this line.

A system of indicating the edges of a pattern groove is illustrated in *Figure A2.7*. The example corresponds with the effect represented in *Figure A2.5* and *A2.6*, and the marking of the groove will be readily followed by comparing it with B in *Figure A2.6*. The arrangement is for a left-hand loom, and the full repeat of the design is represented on one-half of the wheel. The moves to the left in B, *Figure A2.6*, are numbered to correspond with the similarly numbered radial lines of the wheel, and the first vertical space in the plan B coincides with the first concentric space in *Figure A2.7*. Commencing with the position marked 10, where the whip thread is at its farthest point to the left, the outer edge of the groove is marked on the first concentric line. At 11, the edge is marked on the fourth line, or three spaces inward, at 12 on the first, at 13 on the fourth, at 14 on the second, at 15 and 16 on the fourth, and at 17—where the groove changes position for the commencement of the other figure—on the fourth, and also on the sixteenth line, or 15 spaces inward. The position of the outer edge is thus indicated where the concentric lines cross the radial lines, until the complete circle of the wheel has been made.

In marking the position of the inner edge, it is first necessary to find the width of the groove at one position, by adding the number of spaces which the diameter of the peck is equal to, to the number of spaces traversed by the peck at this point. If the diameter of the peck is 5 mm with 10 spaces per cm, 5 spaces are added to the traverse; with 8 spaces per cm 4 spaces, and so on. In *Figure A2.7*, 4 spaces are allowed for the diameter of the peck, and commencing with the position marked 10, it will be noted that the traverse is 7 spaces; therefore the inner edge of the groove at this point is 4 + 7 = 11 spaces distant from the outer edge. When one position has thus been found on a radial line, the concentric lines are successively marked in the manner described with reference to the outer edge. When the lines of the groove have been completed, the wood between them is carefully bored out to the required depth, say 10 mm.

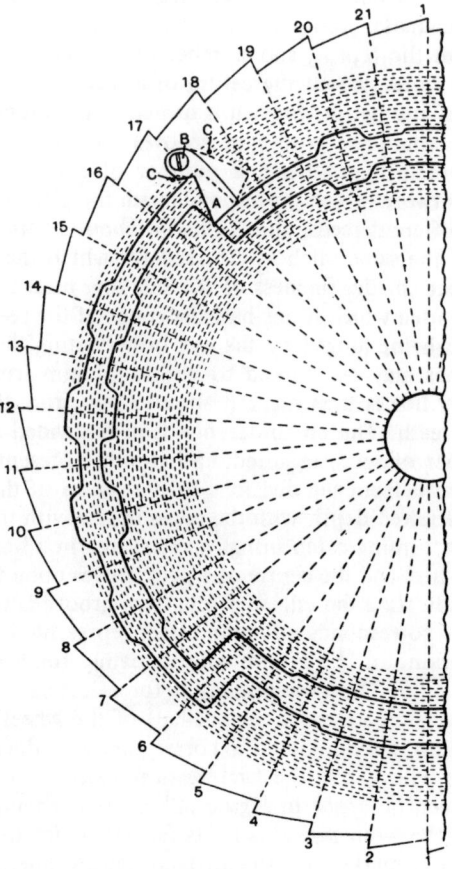

Figure A2.7

In an ordinary groove two concentric lines are sloped towards each other intermediate between two radial lines, but when the groove changes abruptly towards the centre of the wheel, as shown on the radial line numbered 17 in *Figure A2.7*, the peck is liable to catch against the approaching edge of the groove as the wheel revolves. This will be understood if the moves of the peck, in relation to the turning of the wheel, are followed. Thus, taking the radial line 16, the peck moves from the inner to the outer edge of the groove, then the wheel turns, and while this is taking place the peck is really in easy contact with the outer edge. After the rotation of the wheel the peck moves on the line 17 against the inner edge, then it moves back on the line 17, and this is followed by another rotation of the wheel. If allowed to pass the corner of the outer edge in moving back, the peck would lock the wheel, and in order to avoid this a small catch A, centred freely at B, is provided. When the peck is moving on the line 17 from the outer to the inner edge of the groove, it pushes up the

catch A to the position shown by the dotted lines; but when the return move-
ment takes place the catch has dropped and the peck moves against its edge.
The catch is shaped in conformity with the edge of the groove, and two pins
C are driven into the wheel to limit the extent of its movement.

There is a similar abrupt change in the position of the groove on the radial
line 6, but as the move is away from the centre of the wheel no catch A is
necessary. Thus, on the radial line 5 the peck moves from the inside to the outside of the groove; the wheel turns, and the peck moves against the inside on
the line 6, then on the same line against the outside, and while it is in this
position the wheel turns again. The catch A requires to be placed on the side of
the groove that the peck is in contact with when the wheel commences to
turn.

A feature to note in *Figures A2.4* to A2.7 is that each design repeats on an
odd number of teeth of the wheel. This is frequently necessary when a symmetrical effect is required. Thus in *Figure A2.4* in order that both turning
points of the waved line will be exactly the same, it is necessary for the half
repeat to be made on an odd number of picks, while in *Figure A2.6*, in obtaining
the moves from one spot to the other without the needle bar dropping, it is
necessary for an odd number of picks to be employed for each figure. An even
number of teeth could be employed for a style such as the latter by making one
figure 2, or 6, etc., picks longer than the other.

Multi-frame lappet designs

Figure A2.8 exemplifies a style of ornament produced by two frames working
in combination. It will be found that aspects of the designing procedure described in connection with this structure are also applicable to three-frame,
and four-frame styles by logically extending the discussed premises.

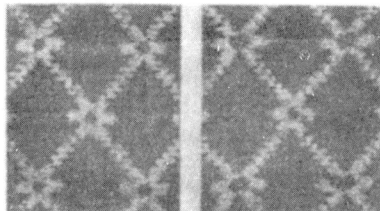

Figure A2.8

The corresponding plan is given at A in *Figure A2.9*, as viewed from the
wrong side of the cloth, the vertical spaces representing the splits of the reed,
and the horizontal lines the picks. The repeat is on 32 splits, and 70 picks, or
35 teeth an odd number of the latter being arranged for on account of the figure
being symmetrical. The full squares show the moves of the first needle bar,
and the dots of the second, while the circles indicate the moves of both bars.
The marks on the odd horizontal spaces represent the moves from right to left,
which decide the widths of the grooves. There are three features to note in this
example. (1) Where two whip threads unite to form a solid portion of figure it is
necessary for the traverse to overlap. If the threads approach each other without
overlapping, the side pull in opposite directions is liable to distort the ground

ends unduly, and make an open space between the two portions of the figure. In obtaining the overlap the needles do not cross each other, as both bars move in the same direction. (2) It is necessary for the distance from centre to centre of the needles in each bar to be exactly the same as the space occupied by the number of splits in the repeat. (3) The different bars require to be set so that the needles are in correct relation with each other. *Figure A2.8* illustrates good and bad setting, the pattern on the right showing the whip threads overlapping more in one central figure than in the other, while in that on the left the overlap is equal, and a perfectly symmetrical figure results.

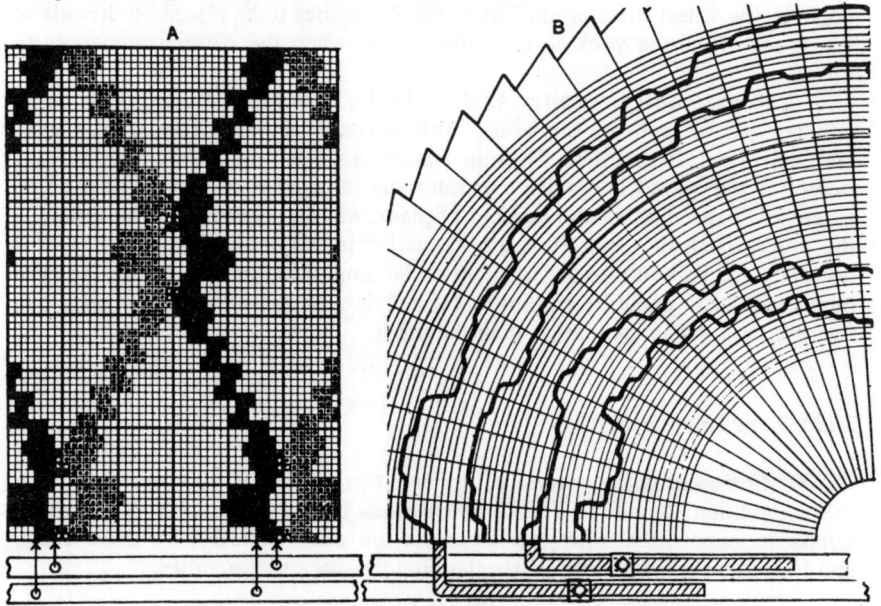

Figure A2.9

Although the traverses of the whip threads may require to overlap in the cloth, as in the example given in *Figure A2.8*, in the wheel it is necessary for some thickness of wood to separate the grooves at every point; therefore the relative position of the pecks is not the same as that of the needles in the bars. This is illustrated in the lower portion of *Figure A2.9* where the pecks are represented as being against the outer edges of the grooves, while the corresponding positions of the needles are indicated by the arrows at the completion of the first traverse to the left. The needles are only three spaces distant from each other, compared with 13 spaces from centre to centre of the pecks. When a new design is introduced, repeated adjustments are made by releasing the screws which secure the pecks and moving the shifter frames until the needles in the respective bars are in the correct relative position for producing the desired effect in conjunction with each other.

Spacing the needles in the bars

The correct spacing of the needles is of the greatest importance; and a method of marking the bars to show where the needles require to be driven in for the design D is illustrated at E,F,G, and H in *Figure A2.10*. Only rather more than one repeat of the pattern is shown; but in practice, in order to reduce the liability of error, it is customary first to measure off, by means of a reed scale and dividers, the width of several repeats. The spaces are indicated on a bar over the desired width, and then each space is divided up into the required number of parts. If more than one needle bar is employed, in order to ensure that all are equally accurate the spacing of all the needles is marked as represented at E, on a separate piece of wood, termed a 'pattern stick.' which is rather longer than the width of the warp in the reed. The number of the bar is indicated against

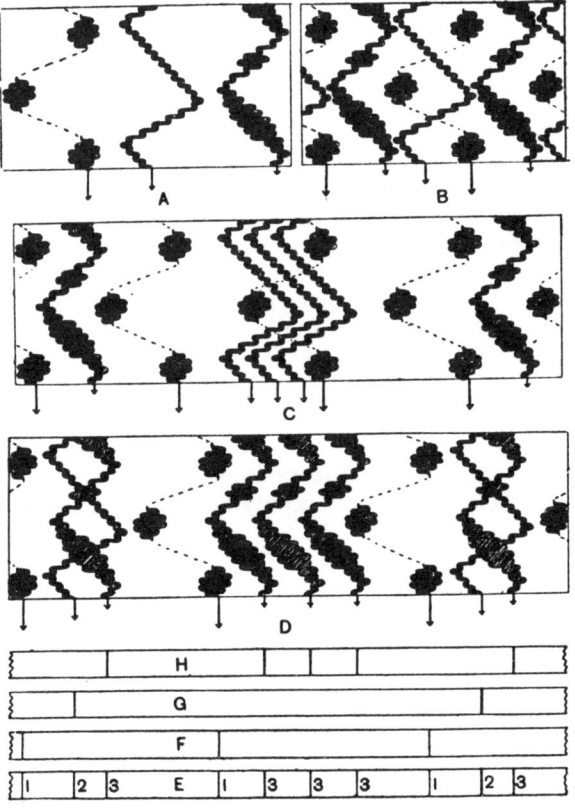

Figure A2.10

each mark upon E, and the piece of wood and the bars are placed together; then with the aid of a set-square the marks are indicated on the respective bars in turn, as shown at F, G and H. The punches used in driving in the needles are shaped so as to prevent the points from being damaged, and the lower end forms a projection which enables a needle to be driven in a vertical direction for the exact distance required, different punches being employed for the different lengths. After the needles have been driven in they are bent back, those in the rear bar being inclined until the points are directly above the edge of the bar, while those in the other bars are successively bent back a slightly greater distance.

Afterwards the spacing of the needles is again adjusted, but this time each bar is laid flat in a suitable position in relation to the pattern stick, and the needles are, if necessary, bent to right or to left until their points are in exactly the proper position.

The size of repeat

The maximum number of picks in a design is usually between 320 and 350 but if required it can be doubled by turning the pattern wheel once in every four picks. However this method is not often used as it has the effect of making the figure coarser in outline. The width of repeat of a single motif is limited to 9 cm, i.e. the maximum throw that the tappets can give to a shifter bar. Due to the possibility, however, of varying the spacing of the needles on each bar there is, in effect, no limit to the width of repeat.

For example, assuming that three grooves are cut in a lappet wheel to produce the three systems of traversing shown at A in *Figure A2.10,* the designs given at B, C, and D, and many others, may be readily produced from the same wheel. The spacing of the needles is indicated below B, C and D, by the arrows, which are shown of different lengths to correspond with the positions of the respective bars. B shows an effect which can be formed by spacing the needles the same in each bar, one needle being required in each for every repeat. The shifter frames and pecks will, of course, require to be so adjusted that the all-over design will result by the three bars working in combination. C shows a change of effect due solely to varying the positions of the needles, the bars being in exactly the same relation to each other as in B. The change of effect from C to D, however, is due not only to a variation in the spacing of the needles, but in addition the relative position of the bars will require to be changed; while the example is also illustrative of a scheme of applying differently coloured threads.

Presser wheel system

The presser wheel system is different from the common wheel system in that the wheel is rotated one tooth at every pick, and a peck is made to press continuously against the outer edge of the groove, which is the only side of the groove that requires to be shaped according to the pattern. In keeping the pecks constantly in contact with the outer edges, the straps, S, in *Figure A2.2* and the treadles, T, are thrown out of action. On the underside of each shifter frame which is in use, and near the centre of the loom, a hook is inserted to which one end of a light spiral spring is attached. The other ends of the springs are connected to a bracket which is fastened below the slay and passes under the frames. The springs are in line with the shifter frames, and the tension tends to draw the latter in the direction away from the centre of the pattern wheel; hence, as the wheel turns, there is always a certain amount of friction between the pecks, and the outer edges of the grooves. In some cases, in order to reduce the friction, larger pecks—up to 10 mm diameter—are used; or, when very long moves are required, the bent end of a peck may consist of a specially shaped spindle upon which a small anti-friction bowl revolves where contact takes place with the outer edge. As a rule, however, the ordinary size and form of peck is found to work quite satisfactorily, and is, therefore, most generally employed, as the use of a larger peck makes it necessary for the radial spaces and the pitch of the teeth to be greater, which increases the size of the wheel and restricts the length of the repeat. On account of a tooth being required for every pick, a presser wheel requires to be larger, and is more costly than a common wheel for the same number of picks in the repeat; nor can such long patterns be obtained. There is, however, greater scope for producing diversity of effect than with a common wheel. Patterns of a less massive or solid character may be formed, as in this case the return movement of a needle bar, on alternate picks, is not essential. Consecutive moves in the same direction can be made, and waved line effects be formed, each of which is of the same width as the thickness of a thread, as shown in *Figure A2.11;* or the whip threads may be used to form a fine outline to a simple figure, as is represented in *Figure A2.12.* The return movement of the needle bars may, however be readily arranged for, and variety

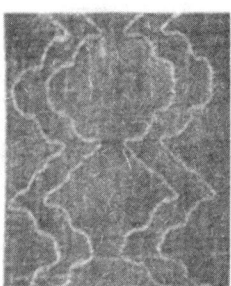 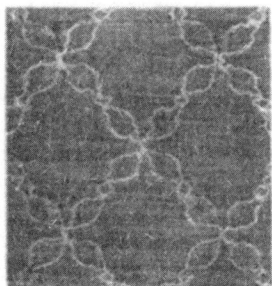

Figure A2.11   *Figure A2.12*

of pattern be obtained by combining solid figures with line effects, as shown in *Figure A2.13.* The traversing of the whip threads in *Figure A2.13* is shown in *Figure A2.14,* in which the vertical spaces represent the splits of the reed, and the horizontal lines the picks, the repeat extending over 24 splits and 62 picks.

Figure A2.13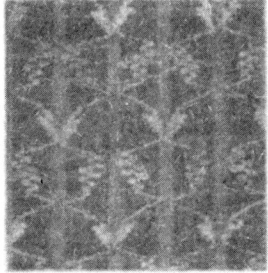

Two needle bars are required in forming the pattern, and the traverses overlap by one split, giving the design an all-over character.

Construction of a presser wheel

In constructing a presser wheel the concentric lines are marked according to the sett of the reed in the ordinary manner, but a radial line is drawn and a tooth cut for every pick in the repeat. Thus, in *Figure A2.15*, in which the thick lines represent the shape of the grooves for producing the effect in the lower portion of *Figure A2.14*, the wheel, which is arranged for a left-hand loom, is divided into 62 radial spaces. The radial lines correspond with the horizontal lines (or picks) of the plan, and are numbered to coincide, while a concentric space corresponds with a vertical space. As a peck is constantly in contact with the outer edge of a groove, the shape of the inner edge is of little account so long as sufficient space is allowed between the edges for the free passage of the peck. Every movement of a thread requires to be marked on the outer edge.

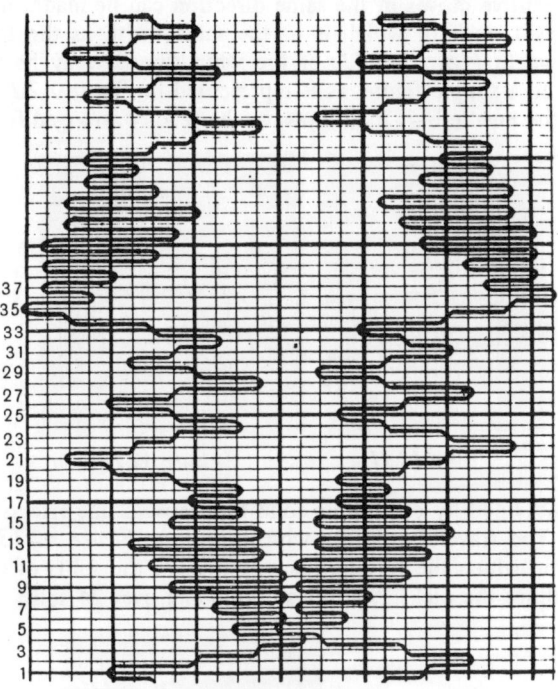

Figure A2.14

Thus, on the horizontal line numbered 17 in *Figure A2.14*. the first thread is 8 spaces inward; therefore, on the corresponding radial line in *Figure A2.15*, the outer edge of the first groove is marked on the eighth concentric space. On the following horizontal lines the first thread is 10, 7, 4, 2, 5, 7, 10 etc., spaces inward in succession, and comparison will show that the outer edge of the groove is successively marked on the corresponding concentric spaces where the radial lines are intersected. The shape of the second groove is similarly indicated, care being taken in commencing that sufficient space will separate the grooves at every part.

As the wheel is turned one tooth at a time, the outer edge of each groove presses a peck to the right, or permits it to be drawn to the left, according to its shape. Between the radial lines it will be noted that the shape of the outer edge varies according to whether the movement of a peck is from or towards the centre of the wheel. Where the traverses are from the centre (to the left in this case), the groves are so shaped that the movement is almost instantaneous, the springs being allowed immediately to contract. On the other hand, where the traverses are towards the centre of the wheel, during which the springs are distended, the outer edges are gradually sloped, which prevents the wheel from

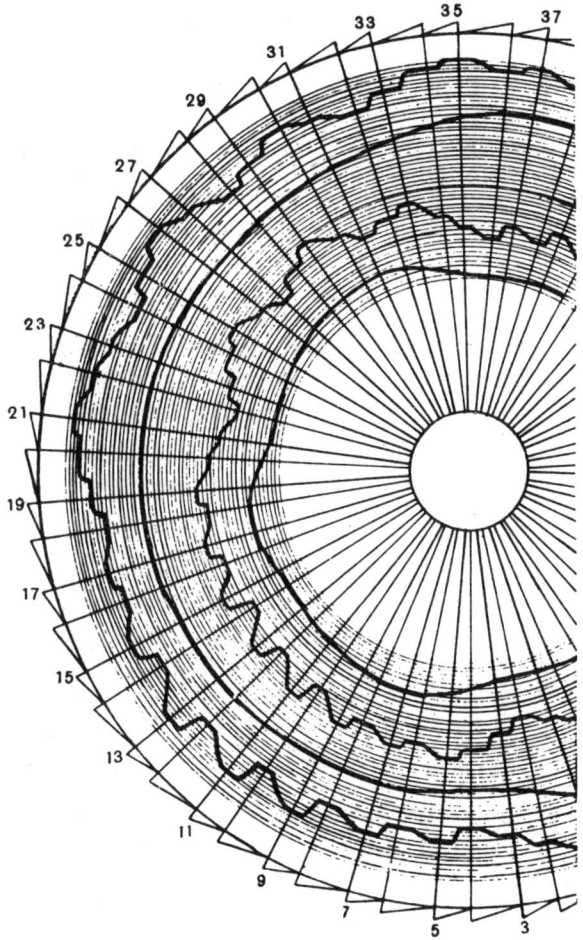

Figure A2.15

being locked, and at the same time reduces the friction with the pecks. A dis-advantage which arises from the grooves being thus shaped is that a wheel cannot be used for the opposite hand of loom to that for which it has been cut, as, if rotated in the reverse direction, it will be locked by the pecks.

In producing designs which include portions of a figure in which a warp thread is repeatedly traversed to left and right alternately (as on the picks 4 to 18 and 35 to 49, in *Figure A2.14*), the spaces between the radial lines of the wheel are usually made alternately of different sizes in the proportion of about 3 to 2. This is illustrated in *Figure A2.15*, in which the odd-numbered teeth are shown smaller in pitch than the even teeth; and it will be noted that in the solid portions of the figure the moves towards the centre of the wheel are arranged to coincide with the larger radial spaces. The friction with the picks is thus reduced when there is most strain, as a greater space gives more latitude for gradually sloping the outer edges of the grooves, and the engaging of a larger tooth provides more time for a movement. With the arrangement of different pitches of the teeth the two bowls, carried by the low shaft gear wheel, are of different sizes to correspond, the larger bowl lifting the turning catch high enough to engage the larger teeth, and the smaller bowl, the smaller teeth; but the leverage is so arranged that the small bowl is ineffective in operating a large tooth. If, therefore, the wheel gets an odd number of teeth out of proper rotation, it will remain stationary for a pick, and this gives the advantage that in the solid figure the to-and-fro movements of the needle bars are retained in correct time with the picking. It is, of course, only in the parts of a design where the traverses are alternately to left and right that each movement of a bar can be definitely arranged to correspond in direction with the pick that follows. In the other part of a pattern the direction of a traverse may or may not coincide with the direction of the following pick; and care is necessary here in arranging the moves, or undue friction may be caused. Thus, long moves may be more readily made if they correspond with the engaging of the larger teeth, and if each is in the same direction as the pick that follows.

Length of whip warp

Due to lateral displacement a whip thread requires to be very much longer than the ground warp. Assuming that the average length of the traverses of a whip end is 5 dents, and there are 10 dents per cm, 0·5 cm of whip warp will be required at every pick; and if 15 picks per cm are inserted, 0·5 x 15 = 7·5 cm of whip warp will be required for each cm of ground warp. The following formula will provide an idea of the proportional lengths of the whip and ground warps:

$$\frac{\text{Total dents traversed in repeat} \times \text{picks per cm}}{\text{Picks in repeat} \times \text{dent per cm}} =$$

=number of times the whip warp is longer than the ground warp

In finding the total number of dents traversed by a thread it is necessary to note the moves in succession and add them together—as, for example, in *Figure A2.14* the first eight moves of the thread on the left are 4, 3, 2, 3, 2, 3, 3, 5 splits or dents, which, added together, total 25. By continuing in this manner it will be found that the number of splits traversed in the full repeat by the first thread = 206. The picks in the repeat = 62, and assuming that there are 10 splits and 15 picks per cm the calculation will be:

$$\frac{206 \text{ total splits} \times 15 \text{ picks per cm}}{62 \text{ picks in repeat} \times 10 \text{ splits per cm}} = 5 \text{ of whip warp to 1 of ground warp.}$$

The calculation is more applicable to common wheel lappet designs than presser wheel styles, and only gives an approximate length of whip warp that is required, as the length can be varied by the tension that is put upon the whip threads.

SWIVEL WEAVING

The term swivel is sometimes applied to the type of loom in which several narrow fabrics, such as hat-bands, ribbons, tapes, etc., are independently formed alongside each other. In this machine a separate shuttle is employed for each fabric, but there is no fly shuttle, and the goods are now generally described as smallwares. In broadloom swivel weaving, however, a number of small shuttles work in conjunction with an ordinary fly shuttle, the latter inserting a ground weft which forms with the warp a foundation cloth upon which the swivel shuttles produce figures in *extra weft*. The chief purpose of the swivel arrangement is to produce the ornament with the least possible waste of the extra yarn. Each figure, and in some cases each part of a figure, in a horizontal line of the cloth, is formed by a separate shuttle; the extra weft thus being introduced only where required, with little material extending between the figures on the reverse side of the cloth. In addition to the great saving of the figuring yarn, the swivel method has the advantage over the ordinary system of extra weft figuring that each shuttle may control a distinct colour, while the figures have a richer and fuller appearance on account of the weft being thrown more prominently on to the surface. The addition of the swivel mechanism, however, makes the loom much more complex, consequently there is reduced speed and output. The cloths are woven wrong side up, and there is, therefore, the disadvantage that defects caused by broken threads more readily escape observation; but, on the other hand, weaving the cloth right side up would necessitate the bulk of the warp being raised on the swivel picks. Compared with lappet figuring, in which the floats of a thread cannot be stitched between the extremities, swivel figuring produces much neater effects, as any form of weave development can be applied to a figure. Effects are readily produced that appear and handle very similarly to styles in which the pattern is formed after weaving by embroidery. A distinguishing feature of the embroidered designs, however, illustrated in *Figure A2.16*, is that the figuring threads may be inclined at any angle in the cloth. In swivel effects the figuring threads are always traversed parallel with the weft threads of the foundation cloth, and at right angles to the warp threads.

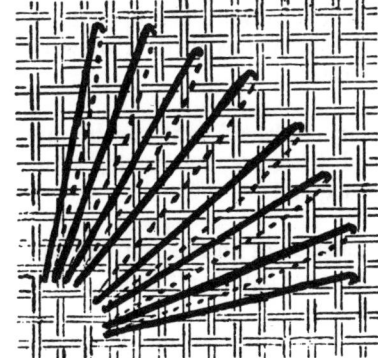

Figure A2.16

Despite some structural advantages, the swivel fabrics are produced at a very slow rate and have been superseded by similar constructions which can be more easily made on modern embroidery frames.

Basic operations in swivel weaving

The swivel wefts are wound on small bobbins which are placed in shuttles 4 to
7 cm in length. These shuttles are carried in a swivel frame attached to the sley.
When ground weft picks are introduced from ordinary shuttles, the frame is
kept above the ground warp with the swivel shuttles well clear of the top shed.
After each ground pick a shed is formed for the swivel yarn. The frame is lowered
and the raised ground ends fit into the recesses between the shuttle holders. The
shuttle holders are, therefore, lowered into empty portions where all the ground
ends have been left in the bottom shed line, as shown at 1 in *Figure A2.17*.

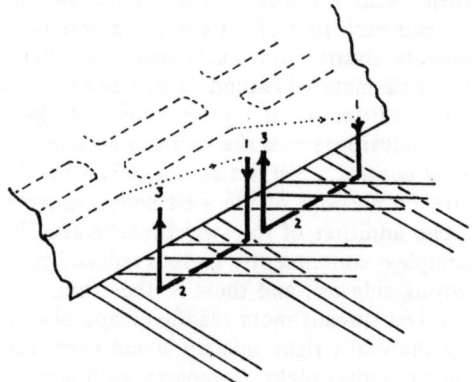

Figure A2.17

Whilst in this position the swivel shuttles are passed from one holder to another
underneath the raised positions of the shed, leaving a trail of weft in their wake
and this produces one figuring float (2, in *Figure A2.17*). Any interlacing can be
easily formed in the middle of the float because some ends in the top shed
line can be dropped without any interference with the passage of the small
shuttles. Having in this way introduced the extra weft picks, the frame withdraws
upwards (3, in *Figure A2.17*) a plain shed follows, and an ordinary pick of
weft is inserted. The downward and upward movement of the frame constitutes
the stitching sequence in this method of weaving. From this description two
aspects should be clear:

(1) No ordinary picks of weft can be inserted whilst the swivel frame is
    down, and therefore the picking is of a pick-at-will type controlled
    from the jacquard.

(2) Take-up must be of an intermittent type so that the cloth is not moved
    forward after the extra weft picks.

Two further points arising from the preceding description are that there
must be a mechanism to control the downward movement of the frame, and
another to control the passage of shuttles from holder to holder. Both mechanisms are jacquard controlled because there may be portions of the fabric
where the figure is not required, such as in isolated spot designs. The frame

itself is permanently spring-loaded to remain up. When it is required down, a jacquard connection releases a tappet which operates against a treadle attached to the frame and this treadle forces the frame down, overcoming the effect of the spring. As soon as the shuttles complete their traverse the tappet is withdrawn and the frame returns to its customary position.

Swivel shuttle propulsion

The shuttle traverse may be controlled by a variety of mechanisms, the rack-and-pinion arrangement shown in *Figure A2.18* being the simplest. A long rack, R, at the back extends through the full length of the frame, F, and its movement is controlled by levers operated from a jacquard. This rack is capable of rotating small pinions, P, in each shuttle holder. The pinions in turn operate against corresponding racks, U, at the back of the swivel shuttles, S. A pinion in one holder in full contact with the shuttle rack will cause the shuttle to move out of

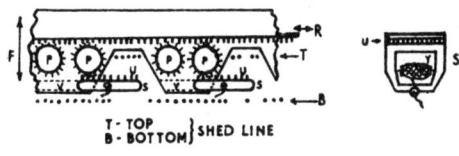

T - TOP
B - BOTTOM } SHED LINE

Figure A2.18

its holder and to traverse into the next along the track, V. Just before the first pinion loses control of the shuttle, the pinion in the next holder catches it and pulls it through, so that the shuttle is always under positive control. In the next series of swivel picks, the rack is operated in the opposite direction and the shuttles return to their original holders, in this way laying the opposite traverse of weft. Other mechanisms, such as circular track and pusher rod, can also be employed to control shuttle movement, and though they offer the advantages of closer figure spacing they are not as easily adaptable for two-frame or three-frame work as the rack-and-pinion device. Two-frame or three-frame work is, in fact, a rarity because of the slow speed of the cloth production; even with single-frame work the effective speed of weaving is reduced to one-half of normal loom speed. With two and three frames, two out of three, or three out of four picks do not add to the length of cloth produced, and therefore the actual speed of weaving in a loom running normally at say, 140 picks per min is reduced to 47 and 35 ground picks per min, respectively. This is far too low to offer any serious price competition to ordinary jacquard figured extra-warp effects even though there is some wastage of material in the latter method, or to power embroidery effects.

Elements of swivel design

The pitch of the shuttles should bear a definite relationship to the width of repeat that the jacquard will give; and there are three factors to take into account—viz., the pitch of the shuttles, the number of jacquard hooks tied up, and the number of harness cords per cm. For instance, a machine tied up to 600 hooks with 30 harness cords per cm will give a repeat of 20 cm in the reed. Therefore, if there are two swivel shuttles to each repeat, the pitch will be 10 cm; if four shuttles 5 cm; and if five shuttles, 4 cm. Conversely, a given pitch of the shuttles will determine what sett of jacquard is suitable for a certain number of hooks tied up—e.g., if the pitch is 8 cm, 40 ends per cm are suitable for a 320 tie giving one swivel shuttle to the repeat, and 24 ends per cm for a 384 tie giving two swivel shuttles to the repeat. Again a given number of harness cords per cm will determine the number of hooks to tie up to a certain pitch of the shuttles. For example, with 40 harness cords per cm and shuttles with a 5 cm pitch, the number of hooks tied up may be 200, 400, 600, etc., according to the number of swivel shuttles required to each repeat.

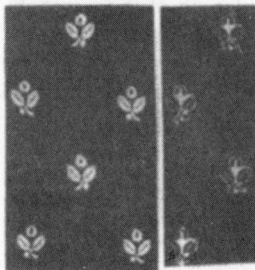

Figure A2.19

A typical swivel spot figure, on a plain foundation, is illustrated in *Figure A2.19*, the face side of the cloth being shown on the left, and the reverse side on the right. The squared paper design (on a reduced scale) is given at A in *Figure A2.20*; at B the face floats of the first figure are indicated with the swivel picks arranged alternately with the ground picks; while the corresponding interlacing diagram, shown at C, illustrates how a swivel thread is traversed in forming a figure. A complete spot is formed by one thread which is traversed alternately to right and to left on succeeding swivel picks; and as many swivel shuttles are employed as there are spots in a horizontal line of the cloth. Upon the completion of a line of spots, the swivel mechanism is thrown out of action until the commencement of the second line, when the carrying frame is situated so that the shuttles occupy the intermediate position, and the swivel threads are traversed again to right and to left in forming the figures which alternate with those in the first row. The mechanism is once more inoperative, until the shuttles are moved back to the original position in order to repeat the first line of figures; and, as shown by the dotted lines in A, *Figure A2.20*, a thread floats loosely on the reverse side of the cloth from one spot to another. The floating threads are afterwards cut away, and this is the only waste of the swivel weft that is made. It will be noted in A, *Figure A2.20*, that on the first and last picks of each figure the swivel weft is firmly interwoven. This is in order that the free ends of the threads will not be liable to fray out of the foundation. As the cloth is woven wrong side up, the marks of the plan A indicate warp, and are, therefore cut. A ground card is cut for each horizontal space in the full plan, hence there will be 64 ground and 50 figuring cards in the repeat of A, which will be arranged 1 ground card, 1 figuring card, for 25 times and 7 ground cards.

From the example given in *Figures A2.19* and *A2.20* it will be seen that each swivel shuttle can be employed to ornament the cloth over a certain area in a longitudinal line. In forming spot figures in which the width of the repeat is equal to twice the pitch of the shuttles, it is necessary for all the shuttles in a frame to be traversed from one holder to another, but the weft is withdrawn only from those which are passed through a warp shed. In such a case an alternate arrangement of spots can be woven without the carrying frame being moved laterally, the odd shuttles forming one row of figures, and the even shuttles the figures that are intermediate.

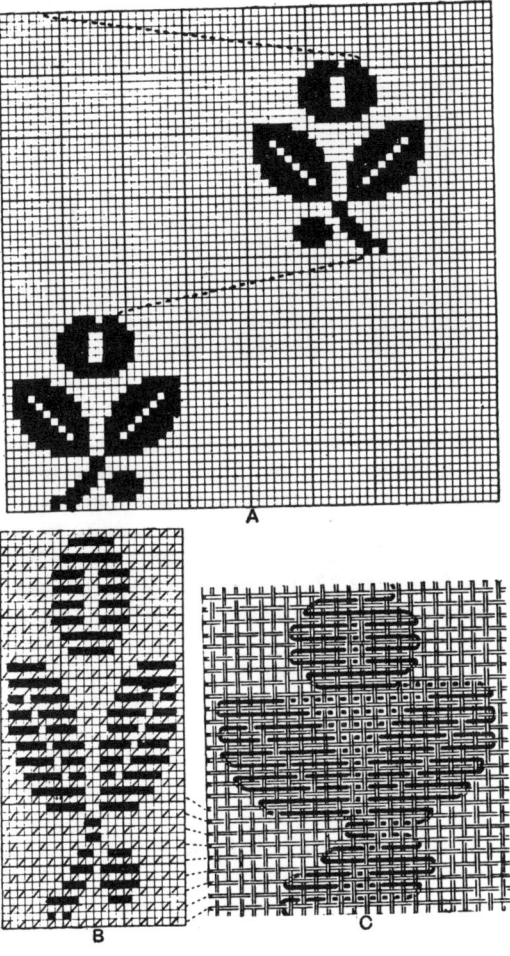

Figure A2.20

In addition to normal figuring it is possible to create full embroidery-type figures by racking the shuttles back to their original positions when they withdraw from the shed, in this way producing a backing float which adds to the solidity of construction. Other methods of embellishement consist of loading shuttles with weft of different colours and combining ground-cloth figure with swivel figure.

ONDULE FABRICS

All or a portion of the ends are made to form waved lines in the cloth, as shown in *Figure A2.21*, by means of a deep rising and falling reed in which the wires are not placed vertically, but are arranged at varying angles. For example, 30 splits of the reed may occupy a space 2 cm wide at the bottom and 4 cm wide at

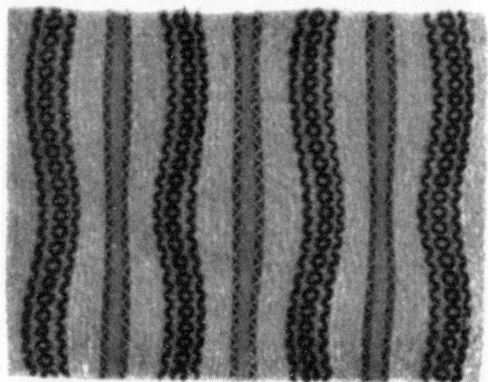

Figure A2.21

the top, followed by 30 splits in the space of 2 cm at the top and 4 cm at the bottom, 60 splits thus occupying 6 cm. The arrangement is repeated across the width, and, on account of its appearance, the term fan or paquet is applied to the reed. The wires are at an equal distance apart midway between the top and bottom, and when the reed beats up in this place the ends are in the normal position, as at A in *Figure A2.22*. By means of a special mechanism, however, the reed is slowly raised, as at B, and lowered, as at C and the ends (except those

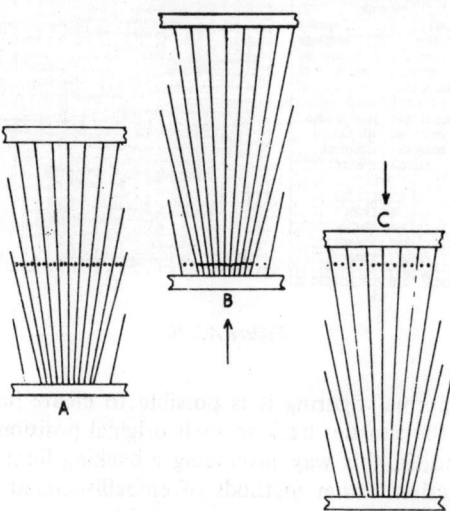

Figure A2.22

in the central splits) are gradually moved, some to the right and others to the left of their normal position, and then back again. An ogee shaped wave effect is formed, which usually extends over about 5 to 8 cm in length and width. All the warp is brought from one warp beam, so that additional strain is put on the ends which wave the most, while the straight ends in the centre contract more than they would under normal conditions.

A modification of the above style is made that is not an ogee shaped effect, but all the ends wave uniformly in a vertical direction. A weft ondule effect, also, is sometimes made by arranging the warp in alternate sections (each, say, about 2 to 3 cm wide), under the control of two easing bars, by means of which the odd sections of ends are gradually tightened while the even sections are slowly slackened, and then *vice versa*. Where the warp is held tight the picks lie closer together than in the slack warp sections, hence the changes in the tension on the ends cause the picks to form a horizontal waved effect.

WILTON PILE HOOK LOOM

In weaving wide, seamless Wilton pile carpets the transverse wire method of forming and cutting the pile limits the speed of the loom which, in proportion to the width of the woven fabric, also occupies a very large amount of floor space. The wire method of weaving is, therefore, neither so productive nor so economical as other systems of pile carpet manufacture. Wilton carpets, however, are very popular and the demand for wide seamless fabrics has led to the introduction of Wilton pile looms in which, in place of transverse wires, a series of hooks, placed longitudinally, is employed. There is one hook to each group of warp threads and to each split of the reed. The principle had been applied in different ways, and in the following the special features are described and illustrated of one hook and reed motion which enables wide Wilton carpets to be woven at a comparatively high speed.

In the upper portion of *Figure A2.23* a side elevation and a plan are given of a loop forming and cutting hook A, of which B shows the hooked end and C a

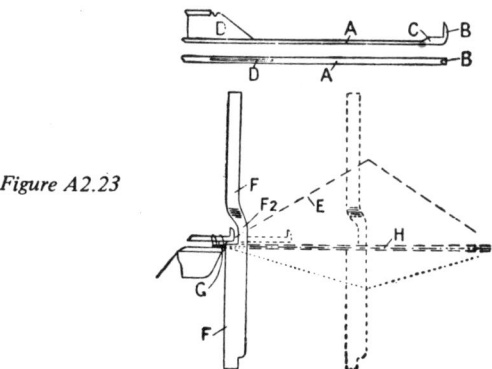

Figure A2.23

shank part which is slightly deeper than the body part that extends to the cutting knife D fixed on the hook A. The hooks are mounted above the carpet fell on the breast rail, and by means of positive cams fitted on the driving shaft of the loom, the whole body of the hooks is given a reciprocating movement towards and from the reed. Also, by means of two racks between which they are mounted the hooks are tilted sideways. Instead of the usual kind of reed wire a special form of blade, F, illustrated in *Figures A2.23* and *A2.24*, is used in conjunction with the hooks.

The order of shedding is the same as in a three-shot, wire-woven Wilton pile, and the chain healds work opposite to each other in 3-and-3 order so that the picks are in groups of three. The chain ends are raised only to the central line, shown at H in *Figure A2.23*, and the shuttle passes the weft between this line and the bottom shed line. On the second pick of each group of three the stuffer heald raises the stuffer ends, and the comber-board the pile threads to the central line, and the comber-board is tilted so that the rear part is moved through a greater distance than the front in order to place all the warp threads in the same horizontal plane. The jacquard griffe is raised and lifts one pile thread E in each group to a higher level, as illustrated in *Figure A2.23*. As the reed moves back following the beating up of the second pick, the pile threads E are held in their normal position on the right of hooks A, which also are in their normal

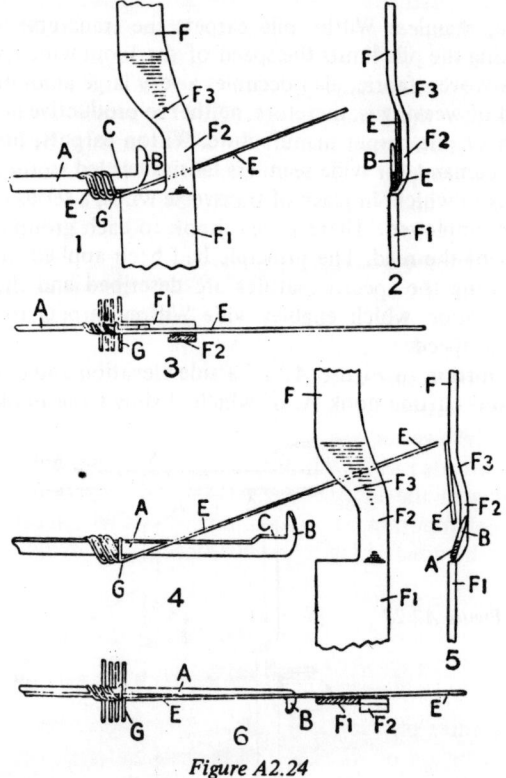

Figure A2.24

position, as shown in the enlarged diagrams given at Nos. 1, 2 and 3 in *Figure A2.24*.

No. 1 in *Figure A2.24* is a side elevation of a portion of a reed blade F and the end portion of a hook A with a pile thread E, and shows the respective positions when a pick of weft G has been beaten up. No. 2 is a front view and No. 3 a plan of the parts represented in No. 1. Each reed blade F has a recess, the base of which is formed of part of a broader portion F1 that extends downwardly in its forward position the broad portion is in advance of the outer end of the hook A below the lower edge of which the weft G is beaten up. This position is represented also in *Figure A2.23*, and, as shown at No. 2. in *Figure A2.24*, each reed blade is bent sideways so that the portion at F2 forms a recess which receives the pile thread E that extends from the right side of hook A. The pile threads are always on the right side of the hooks A during the beating up of the weft.

As the jacquard griffe rises and lifts the selected pile threads the hooks and the reed are moved back to the position illustrated at No. 4 in *Figure A2.24*, and the dotted lines in *Figure A2.23* and the hooked end B of each hook moves to a position where it is lower than the pile thread E. Each pile thread, as it rises, is moved to the left by the pressure of the bent portion F3 of the blade F, and at the same time the hook A is tilted to the right so that the hooked end B passes beneath the pile thread E which is lowered so that it descends on the other side of hook A. This is illustrated at Nos. 5 and 6 in *Figure A2.24*, which show a front view and plan respectively of the parts represented at No. 4. Immediately the pile thread has descended into contact with the hook A the latter is oscillated and brought back into its vertical position, and the thread, as it continues to descend, is guided by the bent portion F2 of the neighbouring blade F under the hook A and passes to its initial or normal position in the bottom shed line. The third pick of a group of three is then inserted and beaten up to the position shown in *Figure A2.23* and No. 1 in *Figure A2.24*.

In looping a pile thread over a hook A it is folded over the shank part C which is deeper than the body part of the hook, and it is the part C which determines the length of the pile produced. As the picks are beaten up succeeding loops of pile are forced forward on to the narrower part of the hook A along which they advance readily because of having been formed over the deeper part at C. The loops then come into contact with the cutting edge of the knife D, *Figure A2.23*, which cuts the threads on the forward stroke of the hook and converts the loops into velvet pile.

Due to improvements in wire Wilton looms and the development of high speed broadloom weaving in the face-to-face system the hook loom is not likely to achieve a position of importance in the manufacture of Wilton carpets, particularly as it suffers from the disadvantages of constructional inflexibility and rigidity of pitch and pile height.

CHENILLE AXMINSTER PILE

The distinctive features of chenille Axminster pile fabrics are: (1) A cut pile is produced without the aid of wires, (2) all the pile material is on the surface of a foundation cloth, (3) any number of colours can be employed. Two separate operations are required in producing the texture. In the first operation, which is termed 'weft weaving', the pile yarn in the form of weft, is interwoven with groups of warp threads that are placed some distance apart. This is followed by a process in which the fabric is converted into a number of long threads that form the chenille pile, which in the second operation of weaving (termed setting) is inserted as weft in such a manner as to form the pile surface of a foundation texture.

The production rate is extremely slow and the process is highly specialised but it is capable of achieving the greatest density of pile of all the machine woven carpets. Due to very high labour costs involved it is used only to produce small quantities of the luxury class of carpets. Recently a process has been developed in which the pre-woven chenille weft can be set upon a hessian backing cloth by adhesion. Very high rate of production can be achieved but the resultant cloth lacks the rigidity and the stability of a texture in which the chenille weft is inserted at the same time at which the ground fabric is woven. In the adhesion setting no design, apart from broken colour or marl effects, is possible, because the chenille threads are not introduced singly but in multiples in the longitudinal direction.

In addition to carpets chenille threads are used to produce curtainings and table covers termed chenille velvet. In these structures no question of designing arises as they are usually made in self colour styles. The chenille thread is often constructed by a twisting process and is simply used as weft in a plain weave fabric in which the warp is very fine and, being woven in a low setting, is completely covered by the tufted weft. The tufts project from the yarn in all directions as opposed to carpet chenille yarns in which the tufts are made to assume a V formation.

Chenille pile designing

The principle of designing is the same as in other pile textures in which the pattern is due to diversity of colour, the design being painted out exactly as it is required to appear when woven. On account of the means employed in producing the cloth it is of greater importance in this than in any other class of pile that the design be drafted on paper to the proper size, and for this reason a special quality of design paper is generally used.

A portion of a chenille Axminster design is illustrated in *Figure A2.25*, in which 16 different colours are represented by as many different marks. Each large square of the design paper, which represents 25 mm^2 is divided into 9 spaces vertically and 5 spaces horizontally, each vertical space corresponding to two picks of the weft which forms the chenille, and each horizontal space to one chenille thread. The design paper is thus ruled in the proportion of one-half the number of picks put in during the first weaving operation to the number of chenille threads inserted in the second weaving operation. Each small space of the design paper represents two pile tufts formed in the colour that the mark indicates.

The pitch of design paper shown in *Figure A2.25* is suitable for a texture in which the chenille threads are woven with 72 picks per 10 cm, and which contains 20 chenille threads per 10 cm, giving 14.4 tufts per cm^2. The pitch varies greatly in different cloths, ranging from 104 picks per 10 cm in the chenille and 48 chenille threads per 10 cm (giving 50 tufts per cm^2) to 32 picks per 10 cm in the chenille and 12 chenille threads per 10 cm (giving 3.8 tufts per cm^2). For the former each cm square of the design paper is divided into 13 x 12 spaces, and for the latter into 4 x 3 spaces.

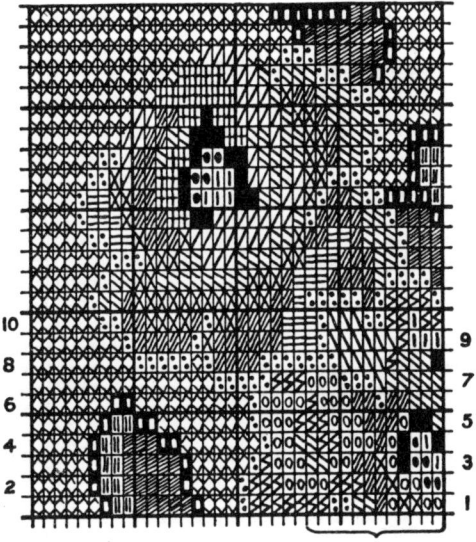

Figure A2.25

Although a design may repeat two or more times across the width, it must be extended to the full width of the texture to be woven. The horizontal spaces are numbered in consecutive order, the odd numbers on the right and the even numbers on the left, as shown in *Figure A2.25*.

Formation of the chenille

In weaving the chenille the design is turned so that the horizontal spaces are in line with the warp threads, and the cords or spaces are gone through in succession, beginning at the bottom and then at the top of succeeding cords, where the number is indicated. Two picks of the proper colour of weft are inserted to each horizontal space in a cord. This is illustrated in *Figure A2.26*, which shows the order of wefting to correspond with the bracketed portion of the first horizontal space of *Figure A2.25*, an enlarged plan of which is given on the left of *Figure A2.26*. The different colours are inserted in the order indicated in the design until the given longitudinal cord is completed, then a small space may be left without weft in order that in the setting the chenille thread will more readily turn at the sides of the cloth. Afterwards, the next longitudinal cord is gone through in the same manner, but in the opposite direction, and the process is continued until every cord in the repeat has been gone over.

The total length of chenille thread required to produce a design is equal to the length of a cord (originally a horizontal space) multiplied by the number of cords. Assuming that in the repeat of a design there are 120 chenille threads which are different from each other, and that 216 double weft picks are inserted in weaving each chenille thread the width of the cloth, there will be 120 x 216 x 2 = 51 840 picks inserted in producing the chenille for the full design. However, a large number of chenille threads may be woven alongside each other at the same time, so that one operation of chenille weaving enables very many repeats

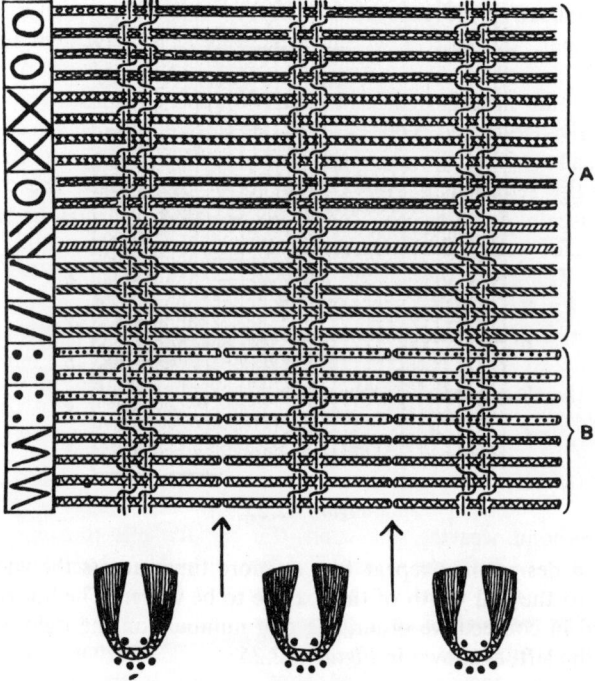

Figure A2.26

of the design to be obtained. Moreover, in the case of wheel designs and designs which are centred horizontally, it is only necessary to weave one-half of the chenille threads in the repeat in order to produce the full pattern.

The chenille is woven in a tappet loom which is fitted with a gauze mounting. The warp threads are arranged one end crossing two standard ends, and two groups of threads are reeded into consecutive splits of the reed with a space between them and the next two groups. Frequently, an ordinary form of reed is used, a number of splits being left empty between the groups of warp threads, but in some cases the reed contains splits only where the groups of threads are required to pass through. The space between the groups is varied according to the length of pile required, the pitch ranging from about 10 mm for a short pile to 25 mm and over for a very deep pile. The wefts are placed in a creel and fed through a drum-like selector with a capacity for 50 colours. Selected colour is proffered to a rigid rapier which inserts a double pick in a single insertion. At one time the colours were introduced by manual shuttle changes.

Figure A2.26 shows how the threads interlace, as viewed from the side that is underneath during the weaving of the chenille. A texture is produced across which the variously coloured picks of weft extend, being firmly bound in at intervals by the gauze interlacing, as shown in the portion lettered A in *Figure A2.26*. The next process consists of cutting the picks in the centre of the space between the groups of gauze threads, as represented at B. This is followed by a process in which the strips are subjected to heat, moisture, and pressure, which causes each to assume the form of a thread in which the severed weft picks are V shaped as illustrated below B. The threads are then indicated by a letter or number, and each is wound separately in a convenient form for subsequent use.

All the chenille threads that are woven alongside each other (with the exception of the selvedge threads which are wasted), are, of course, exactly alike, and as many threads—within the capacity of the loom—are woven at the same time as will give the required number of repeats of the design. The count of the pile weft is usually equal to 300 to 400 tex worsted, and may be two, three or four-ply, but for a very deep coarse pile a yarn ranging from 800 to 1000 tex may be used. The gauze threads are generally cotton, and 60/3 tex or 48/4 tex may be used for the crossing threads, and 48/2 tex or 48/3 tex for the standard threads. For 100 m of chenille thread about 115 m of the standard threads and from 170 to 220 m of the crossing threads are required the lengths varying according to the thickness of the weft and the number of picks per cm.

Setting

In this—the second weaving operation—the chenille pile thread, in which the differently coloured tufts are arranged in precise order according to the design, is traversed from side to side, and is bound in by means of a fine linen or cotton warp to the surface of a foundation texture. The length of each pile thread that is taken up at each horizontal traverse is equal to the width of a horizontal space of the design. The chenille thread is placed within an oblong metal case in such a manner that when it is withdrawn it is free from twists. The case is placed in a specially shaped shuttle, and the chenille is woven into the cloth in the same way as weft, except that the loom stops after the insertion of each pick of chenille while the weaver combs the thread forward and 'sets' it in the proper relative position to the preceding pick of chenille.

Structure of the fabric

The structure of the ground varies according to the purpose of the fabric—table covers, hangings, etc., being made lighter and more flexible than carpets and rugs which require to be very stiff. D in *Figure A2.27* shows the weave plan, and E a cross-section through the weft of a structure in which there are two picks to each chenille thread—one ground end to two stuffer ends, and one fine binder end or catcher to every three ground ends. F and G similarly show a weave plan and a cross-section of a structure in which there are four picks to each chenille thread, one ground end to two stuffer ends and two fine catcher ends to six ground ends. Both structures may be woven with 36 ground ends, 72 stuffer ends, and 12 catcher ends per 10 cm, while for the first example 48 picks and 24 chenille threads per 10 cm are suitable, and for the second 64 picks and 16 chenille threads per 10 cm. The catcher or stitching ends unite the chenille pile threads to the foundation, as shown in the diagrams E and G.

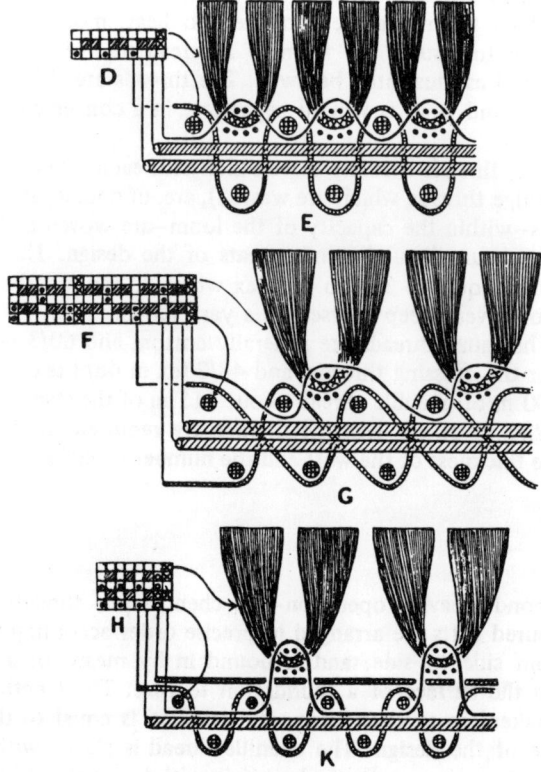

Figure A2.27

H and K in *Figure A2.27* illustrate another structure which is woven with four picks to each chenille thread. In this case the warp is arranged 1 ground end, 1 stuffer end for three times, 1 float end, and 1 fine catcher end. The float end is raised over all the picks of the foundation, but passes under the chenille thread, and the object of its insertion is to raise the chenille above the foundation and bring it more prominently to the face. In each example given in *Figure A2.27* only the fine catcher ends pass over the chenille pile threads.

WOVEN PILE FABRICS PRODUCED BY THERMAL SHRINKAGE

Useful and interesting pile constructions can be produced on normal fast running looms by utilising thermal shrinkage properties of certain synthetic materials. Some polyolefin and polyvinyl chloride/polyvinyl acetate copolymer yarns exhibit a marked tendency to shrink when subjected to the action of heat. A shrinkage of 50 to 30 per cent of the original length occurs at temperatures between 60°C to 100°C and the extent of contraction can be comparatively easily controlled either by the temperature gradient or by mechanical restraints during finishing. If in a woven structure contractible yarns are suitably interwoven with other threads which do not contract then upon the shrinkage of one set of yarns an excess is created in the other set which is thrown into a pile formation.

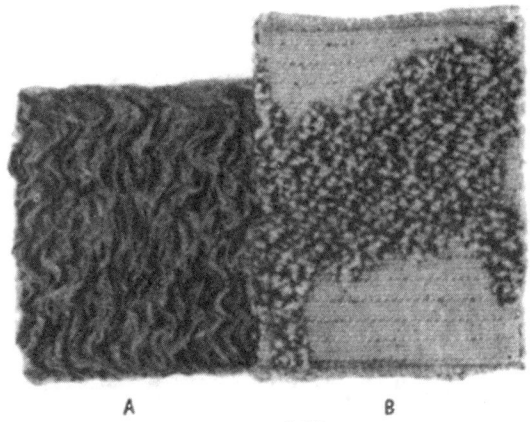

A   B

Figure A2.28

Two pile fabrics produced by the method described are shown in *Figure A2.28*. Fabric A is an all-over pile structure whilst B is a jacquard figured effect in which the pile figure is surrounded by bare ground. In the former, worsted pile yarn is used and the quality is suitable for an upholstery texture; in the latter a heavy wool and cotton blended yarn is employed for the pile in a construction of sufficient weight to serve as light rugs. The height of pile can be controlled by the degree of shrinkage achieved and by the float length, and the contractible pile threads can be used either as warp or as weft. It is preferable to use the special yarns in the warp direction as in this way the weft settings can be very low with a consequent rise in the rate of production of the cloth, the necessary consolidation being obtained on shrinking during the finishing operations. Also, when used as warp the pile effect can be magnified by tension differences between the shrinking and the shrink resistant yarn elements which are obtained by heavy weighting of the synthetic yarn beam and light weighting of the beam carrying the pile ends.

A in *Figure A2.29* shows a weave which after heat treatment results in a fabric with the appearance of a Brussels or uncut moquette structure obtained by the regimentation of the pile yarn binding marks in horizontal rows. The lifts of the contractible yarns are indicated by the dots and those of the pile ends by the crosses. It will be noted that the ratio of ends is 1:1 and that a fast pile binding order is used to secure good pile anchorage. At B a weave is shown in which the pile binding points are staggered which results in fuller surface cover. The same system of marking has been used as at A and the construction is represented by the weft sections at C and D which show the cloth appearance before and after shrinkage respectively. This structure corresponds to cloth A in *Figure A2.28*. At E in *Figure A2.29* a portion of a figured pile effect is given, one vertical row of which is represented by the weft section at F. In this cloth a shorter pile float is used and there are two contractible ends to one pile end. The 2:1 ratio of ends may be necessary in heavier cloths to overcome the considerable structural resistance to shrinkage in such fabrics.

Using the thermal shrinkage technique a rich variety of structures can be produced and at G in *Figure A2.29* a schematic diagram is given which represents

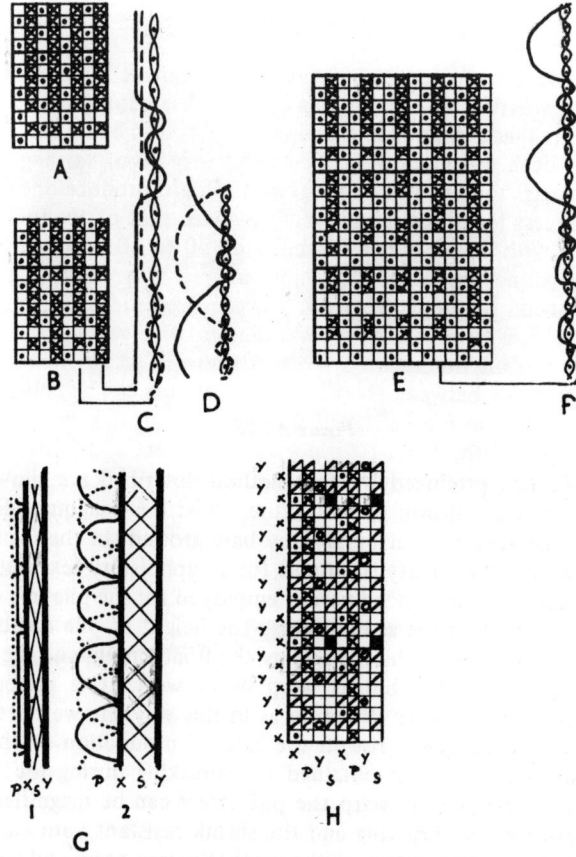

Figure A2.29

a pile fabric with a built-in resilient underlay. Diagram 1 shows the construction before, and diagram 2, after heat treatment. The ground cloth, X, and the backing cloth, Y, both contain heat contractible warp threads. The two layers are joined together by loosely bound stitching ends, and the surface of the ground cloth is covered by floating pile ends, neither of which are contractible. Upon shrinking of the ground and backing cloths the surface floats form the pile whilst the stitching ends form the resilient underlay which is structurally incorporated in the fabric as shown in diagram 2. A full weave for this construction is given at H in which the dots represent the lifts of the ground ends; the circles, of the back ends; the crosses, of the pile ends; and the solid marks, of the stitching ends. The diagonal marks represent the separating lifts.

The use of heat contractible yarns is not confined to pile constructions—they can be advantageously employed to produce seersucker and crepon effects but care must be taken to introduce them in such fabrics which do not require ironing. It will be appreciated that at temperatures of 150°C to 160°C the heat sensitive yarn elements are liable to melt.

TUCK FABRICS

Tuck fabrics are constructions in which a permanent cloth fold or plisse is created during weaving. Fabrics of this type are used for skirtings, blouses and shirtings and are made in fine yarns and settings, cotton being the most common constituent of both warp and weft. The tucks may be of varying length from 5 to 10 mm; as each tuck is, in effect, a cloth fold, to produce one it is necessary to create an excess length of fabric which is twice that of the tuck itself and in a cloth woven with 30 picks per cm, between 30 to 60 picks are inserted into each plisse portion. The folds are almost invariably produced in plain weave and a small portion of the cloth which precedes and succeeds a tuck must also be constructed in a firm weave, such as plain or fine warp rib, otherwise gaps are liable to open on the back of the cloth due to insufficient cohesion. The portions of the cloth between the firmly bound areas can be constructed in any weave and the cloth in *Figure A2.30* shows a typical tuck structure with plain tucks of varying length, 2-and-1 rib portions prior to, and following each tuck and a simple waved pique weave in the middle.

Figure A2.30

The fabrics are made with two sets of warp yarns, the ground, which does not participate in the formation of the tuck portions but is woven in the intervening areas, and the tuck ends, which weave continuously. Only one kind of weft is necessary but in some cases extra weft figuring is employed and in the fabric shown in *Figure A2.30* wadding weft is used for the pique portions. The full weave for a simple plain weave tuck fabric is given at A in *Figure A2.31* in which the ground end lifts are indicated by the dots, and the tuck end lifts by the crosses. The two sets of ends weave plain with each other outwith the tuck but in the tuck portions the plain weave is formed by the tuck ends alone whilst the ground ends float underneath. The section B represents the appearance of the cloth just prior to the formation of the tuck whilst C shows the same cloth just after the fold has been completed. For the sake of simplicity considerably fewer picks are shown in the tuck than is usually the case in an actual cloth. D in *Figure A2.31* indicates a rib ground tuck in which the warp

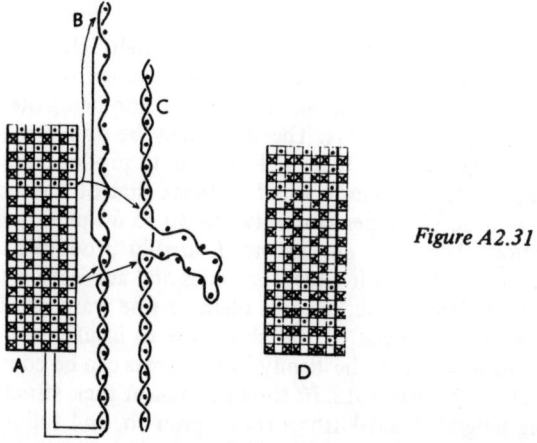

Figure A2.31

is conveniently arranged: 2 tuck ends to 1 ground end. The section C shows that a tuck is created by forcing the ground picks which follow the tuck right up to the picks which precede it along the floating ground ends. The excess cloth represented by the tuck area folds over and is permanently bound in this form into the fabric. To produce good plisse fabrics two main conditions must be observed:

(1) The ground warp must be heavily tensioned and the tuck warp comparatively slack

(2) The ground picks just prior to, and just following the tuck portion must be in the same shed.

At one time to create a tuck the loom was stopped after the insertion of the first ground pick following the tuck area with the reed fully forward, the cloth was released and the warp pulled back until the last and the first ground picks were forced together. At present possibilities exist which permit the operation to be performed without stoppages. One of the methods involves the use of an additional cloth tension bar between the take-up and the cloth rollers. When it is required to form a tuck the tension bar is released by dobby control thus providing an excess cloth length. Simultaneously, the back rest, over which the ground warp runs and which is heavily springloaded, pulls back the excess of ground warp. Thus, the first ground pick following the tuck is beaten up to the ground pick which precedes the tuck and the tuck portion representing the excess cloth length puckers up. In another method use is made of a broken-back connecting arm to the sley. At the commencement of the tuck the forward movement of the sley is progressively shortened until sufficient length is created whereupon the broken-back is straightened giving a full length beat-up which pushes up the excess cloth created into a pucker and joins the first pick which follows the tuck to the pick which precedes it.

Index

Alhambra quilts, 15
  traditional mounting for, 372
Astrakhan, imitation, 295
Axminster pile carpets, 349, 419
  chenille (*see* chenille)
  gripper, 360
  gripper, selection of pile, 362
  gripper, tuft insertion, 364
  spool, loom operation, 355
  spool presentation, 354
  spool setting, 352
  spool, tuft insertion, 358
  spool-gripper, 366

Backed cloths, figured, 55
  interchanging, 55
  interchanging treble wefted, 58
Backed cloths, imitation, 61
Backed cloths, warp, 49
  beaming and drafting, 51
  methods of backing standard weaves, 52
  methods of selecting ties for, 53
  reversible, 50
Backed cloths, weft, 44
  interchanging figured, 55
  methods of backing, 45
  reversible, 44
  with warp face weaves, 47
Backed cloths, with wadding threads, 59
Bannister or split harness, 375
Belting structures (*see* multi-ply structures)
Book muslin fabrics, figured, 73
  ground weave variation, 79
  method of designing, 76
  settings of, 79
  structure of, 74
  traditional mounting for, 372
Bottom and top douping (*see* leno fabrics)
Bottom boards, 345
Boucle or cord pile (*see* Brussels)

Brocades, compound, 89, 94, 101
  figured warp rib, 89
  figured warp rib, ornamentation, 93
  figured warp rib, traditional mounting, 372
  figured warp rib, two-colour, 91
  multi-warp, 101
  multi-weft, 94
  multi-weft, ground weaves, 96
  multi-weft, reversible, 98
  multi-weft, three-weft, 99
  multi-weft, two-weft, 94
  multi-weft, weave shading, 97
Brussels (boucle, cord) pile carpets, 296, 306
  multi-frame, 306
  self-colour, 296
  tapestry Brussels, 298

Card cutting, computerised, 9
  systems, 6
Card repeating, 9
Centre stitched double cloths, 131
  centre warp stitching, 132
  centre weft stitching, 135
Chenille Axminster pile, 419
  designing, 420
  formation of chenille yarn, 421
  setting, 423
  structure of the fabric, 423
Chintzing, 34
Cloque or crepon fabrics, 180
Comber-boards, sectional, 379
  working, 300, 311, 314, 385
Combined warp and weft tapestry (*see* tapestry structures)
Compensating rods, 324
  (*see also* easers)
Compound brocades (*see* brocades)
Conveyor belting (*see* multi-ply structures)
Cord pile fabrics (*see* Brussels)
Cordage jacquard, 314

Corded velveteen (corduroy), 268
Crepon or cloque fabrics, 180
Cut effects in double cloths, 153
  by the use of cutting threads, 156
  by yarn interchange, 154
Cutting and looping wires, 287

Damasks, 83
  bannister or split harness, 375
  card saving, 86
  diversification of effect, 88
  self-twilling jacquard, 84
Decked mail eyes, 15, 16
Designing (*see* jacquard designing)
Double cloths, classification of, 103
Double cloths, figured, 173
  by combination of fine and coarse fabrics, 176
  cloque effects, 180
  combined with float on single cloth ground, 182
  combined with warp and weft float, 177
  simple interchanging, 174
  with extra threads for wadding or figuring, 184
Double cloths, centre stitched, 131
  centre warp stitching, 132
  centre weft stitching, 135
Double cloths, interchanging, 136
  cut effects due to cutting threads, 156
  cut effects due to yarn interchange, 154
  cut effects in, 153
  double plain continuously coloured, 137
  double plain in three or four colours, 147
  double plain with changeable colour order, 142
  double twill and sateen, 152
Double cloths, self-stitched
  beaming and drafting, 117
  construction of designs, 108, 113
  relative proportions of face and back threads, 106
  reversible, 116
  selection of face and back weaves, 107
  selection of stitching positions, 120
  tying or stitching, 107
  warp wadded, 131
  weft wadded, 129
Double shuttle weaving (*see* face-to-face pile)
Double sided pile (*see* terry pile and warp pile)
Doup healds (*see* leno fabrics)

Easers or slackeners, 231, 232. 238, 245, 389

Extra thread figuring, 11
  comparison of methods, 13
  disposal of surplus, 12
  imitation, 39
  methods of introducing, 12
  principles of, 11
Extra warp figuring, 13
  Alhambra quilts (*see* Alhambra)
  binding between face floats, 17
  continuous in one extra, 14
  intermittent in one extra, 17
  planting, 23
  stitching by special picks, 23
  with two extra warps, 20
Extra warp and extra weft figuring, 36
  spot effects, 37
Extra weft figuring, 25
  chintzing, 34
  clipped spot effects, 29
  continuous in one extra, 25
  ground weave modification, 33
  intermittent in one extra, 30
  stitching by special ends, 33
  with two extra wefts, 34
Eyed doup needles, 211

Fan or paquet reed (*see* ondule fabrics)
Fancy toilet cloths (*see* figured piques)
Figured piques, 65
  classification of, 65
  fast back, 68
  five-and six-pick structures, 71
  four-pick structure, 71
  half-fast back, 67
  loose back, 67
  method of designing, 68
  traditional mounting for, 387
Figured warp rib brocades (*see* brocades)
Flat steel doups, 211, 221, 228

Gauze (*see* leno structures)
Gauze ground (*see* Madras muslin)
Ghiordes or Turkish knot, 349
Gripper Axminster (*see* Axminster)

Hairline effects in doube cloths, 137, 142, 147
Hand-knotted carpets, 349
Hand-woven tapestry, 190
Harness mountings, traditional, 371
Healds and harness mountings, 371
Healds for face-to-face pile weaving, 321

Imitation backed cloths, 61
Imitation embroidery by swivel weaving, 415
Imitation extra thread effects, 39

Imitation fur, 295
Interchanging double cloths (*see* double cloths)
Inverted hook jacquards, 242, 316, 383
Inverted key-holes in cordage jacquard, 315

Jacquard, bannister or split harness, 375
  card-cutting systems for, 6, 9
  cordage, 314
  for face-to-face pile fabrics, 345
  for leno structures, 240, 242
  for pile fabrics, 301, 311
  gripper Axminster, 363
  heald and harness mounting, 371
  inverted book, 242, 316, 383
  sectional harness, 379
  self-twilling, 84
  traditional mountings, 371
  with working comber-boards, 301, 311, 314, 385
Jacquard designing, 1
  simplified and condensed, 3

Knotted carpets, 349
  Ghiordes or Turkish knot, 350
  Sehna or Persian knot, 350

Lappet weaving, 393
  auxiliary mechanisms, 396
  common wheel construction, 401
  figuring movement, 394
  length of whip thread, 410
  multi-frame designs, 403
  presser wheel construction, 407
  presser wheel system, 406
  representation of designs, 399
  shifter bar movement, relation of, 397
  size of repeat, 406
  spacing of needles, 404
  stitching movement, 395
Leno (and gauze) structures, 207
  basic sheds of, 209
  double slotted flat steel drops, 228
  easers, 231, 232, 238, 245
  equalisation of tension in, 230
  eyed flat steel doups, 211
  methods of producing, 211
  negative easing action, 231
  point draft or counter, 214, 226
  positive easing action, 232
  principle of, 208
  Russian cords, 217
  shaker device, 234
  simple figured effects, 224
  simple net, 217
  simultaneous top and bottom douping, 219

slider frame and needle device, 255
slotted flat steel doups, 221
special lifts of standard ends, 216
with more than one assembly, 214
with two crossing ends per slot, 225
Leno structures, jacquard, 235
  easer action in, 238, 240
  marquisette styles, 237
  one-crossing-one styles, 236
  one-crossing-two styles, 237
  special mountings for, 240, 242
  two-crossing-two styles, 241
Leno structures on string doups, 388
  bottom and top douping, 388
  drafting, 389
  lifting plans for, 392
  relative position of healds, 390
  sheds formed in, 390
Loop pile (*see* terry and warp pile)
Loose-back piques, 67

Madras muslin (gauze) structures, 244
  chintzed designs in, 250
  designing, 248
  double cover structures, 251
  gauze and tug reed operation, 246
  loom mechanisms for, 245
  structure of, 244
  tape-up motion, 248
  with weft pile figure, 253
Mitcheline quilts (*see* patent satin)
Moquettes (*see* warp pile fabrics)
Multi-layer fabrics, 158
Multi-ply belting structures, 168
Multi-warp brocades (*see* brocades)
Multi-weft brocades (*see* brocades)
Muslins (*see* book and Madras muslin)

Net leno, simple, 217

Ondule fabrics, 416
  reed, 416

Pacquet or fan reed (*see* ondule reed)
Patent satin structure, 80
  method of designing, 81
  traditional mounting for, 385
Pile fabrics (*see* terry, warp and weft pile)
Pile fabrics produced by thermal shrinkage, 424
Planting, 23, 310
Plush (*see* warp and weft pile)
Presser wheel (*see* lappet)

Repp stitched weft tapestries (*see* tapestry)

Reversible double cloths, 116, 175
    multi-weft brocade, 98
    tapestry structures, 198
    treble cloths, 188
    warp-backed fabrics, 50
    warp-pile fabrics, 293
    weft-backed fabrics, 44, 55, 58
    weft-pile fabrics, 268
Rib, figured warp (*see* brocade)
Russian cords, 217

Scotch lappet wheel loom, 394
Sectional jacquard and harness, 379
    connection of hooks and needles, 380
    designing and card cutting for, 381
    draft of the warp threads, 381
    harness ties, 379
Sehna or Persian knot, 350
Self-stitched double cloths (*see* double
    cloths)
Self-twilling jacquard, 84
Selvedge shuttles, 357
Shaft monture (*see* bannister harness)
Shaker or jumper motion, 234
Slackener (*see* easer)
Slotted steel doups, 221, 228
Spool Axminster (*see* Axminster)
Spool-gripper Axminster (*see* Axminster)
Stitched double cloths (*see* double cloths)
Stitched figuring weft constructions, 73
Swivel weaving, 411
    basic operations in, 412
    elements of design, 413
    imitation embroidery, 415
    shuttle propulsion, 413

Tapestry Brussels carpets (*see* Brussels)
Tapestry structures, 190
    combined warp and weft, 202
    hand woven, 190
    heald and harness mounting for, 374
    lifting rod and heald mounting, 378
    repp stitched weft face, 200
    reversible, 198
    simple weft face, 192
    three-and four-weft, 195
    two-weft, 193
Terry pile structures, 274
    cut pile fabrics, 285
    figured fabrics, 283
    formation of pile, 274
    inverted hook jacquard for, 383
    mixed colour effects, 285
    ornamentation, 280
    special mechanisms for, 279
    stripe and check patterns, 281
    terry weaves, 276
Toilet cloths (*see* figured piques)

Top and bottom douping (*see* leno structures)
Treble cloths, 158
    beaming and drafting, 163
    figured interchanging, 186
    methods of stitching, 161
    systematic construction of, 158
    use of centre layer as wadding, 168
    with dissimilar weaves, 163
    (*see* also multi-ply belting)
Treble-wefted reversible backed cloths, 58
Tuck fabrics, 427
Turkish towelling (*see* terry pile)

Velvet (*see* warp pile)
Velveteen (*see* weft pile)

Wadding, in backed cloths, 58, 59
    in double cloths, 129, 131, 184
    in figured piques, 67
Warp-backed cloths (*see* backed cloths)
Warp pile, face-to-face, 320
    all-over (continuous) pile, 325
    carpet structures, multi-frame, 343
    carpet structures, self-coloured, 333
    continuous pile structures, 325
    fast pile structures, 330
    figured pile structures, 334
    jacquard figured constructions, 341
    jacquard, special, 345
    moquettes, 325
    special mechanisms for, 322
    use of duplicate pile threads, 335
    velvet structures, 327
    Wilton carpet structures, 343, 344, 347
Warp pile, formed over wires, 287
    all-over (continuous) pile, 289
    all the pile over each wire, 290
    alternate pile over alternate wires, 292
    Astrakhan, imitation, 295
    Brussels (boucle, cord) carpets (*see*
    Brussels)
    card cutting for, 313
    continuous pile carpets, 296
    continuous pile structures, 289
    cordage jacquard, 314
    fast pile, 291
    figured, multi-frame, 305
    figured with one series of threads, 299
    jacquard, special action, 301
    loop and cut pile figure, 303
    loop and cut pile on bare ground, 304
    method of designing, 310
    methods of weft insertion, 319
    moquettes, 308
    ornamentation, 294
    pile figure on bare ground, 300
    planting, 310

Warp pile, formed over wires (*Cont.*)
   reversible pile structures, 293
   settings, 293
   stripe and check effects, 295
   system of jacquard mounting, 311
   textural variety in, 316
   treble shed formation, 318
   uncut moquettes, 308
   velvet, 290
   Wilton carpets (*see* Wilton)
Weft-backed cloths (*see* backed cloths)
Weft pile fabrics, 257
   all-over (plain) velveteens, 258
   changing pile density, 261
   corded velveteens (corduroys), 268
   cutting of all-over velveteens, 266
   cutting of corded velveteens, 270
   density of pile, 261
   fast pile structures, 264
   figured cords, 273
   figured velveteens, 271
   length of pile, 260
   plain back velveteens, 258
   plushes, 267

Weft pile fabrics (*Cont.*)
   qualities of all-over velveteens, 266
   qualities of corded velveteens, 270
   reversible plushes, 268
   simplification of cutting, designs for, 265
   twill back velveteens, 265
Wilton pile carpets, face-to-face, 333, 342
   card cutting for, 346
   jacquard mounting for, 345
   multi-frame, 342, 344, 346
   self-colour, 296
   working bottom boards, 345
Wilton pile carpets, formed over wires, 296, 306
   card cutting for, 313
   cordage jaquard, 314
   designing for, 310
   loom mounting for, 311
   methods of weft insertion, 319
   multi-frame, 306
   planting, 310
   self-colour, 296
Wilton pile carpets, hook loom, 417
Working comber-boards, 300, 311, 314, 385